Communications in Computer and Information Science   **517**

Commenced Publication in 2007
Founding and Former Series Editors:
Alfredo Cuzzocrea, Dominik Ślęzak, and Xiaokang Yang

More information about this series at http://www.springer.com/series/7899

Lazaros Iliadis · Chrisina Jayne (Eds.)

Engineering Applications of Neural Networks

16th International Conference, EANN 2015
Rhodes, Greece, September 25–28, 2015
Proceedings

 Springer

Editors
Lazaros Iliadis
Democritus University of Thrace
Orestiada
Greece

Chrisina Jayne
Coventry University
Coventry
UK

ISSN 1865-0929   ISSN 1865-0937   (electronic)
Communications in Computer and Information Science
ISBN 978-3-319-23981-1   ISBN 978-3-319-23983-5   (eBook)
DOI 10.1007/978-3-319-23983-5

Library of Congress Control Number: 2015948718

Springer Cham Heidelberg New York Dordrecht London
© Springer International Publishing Switzerland 2015

Springer International Publishing AG Switzerland is part of Springer Science+Business Media
(www.springer.com)

Preface

It has been more than 60 years since John McCarthy introduced the term AI (Artificial Intelligence) to the scientific community. Since then, for the vast majority of the population, AI sounds more like a mythical future prediction not very close to reality. Common people use AI in their daily lives but they do not actually realize it. For example, cars make extensive use of Artificial Narrow Intelligence (ANI) to figure out when the anti-lock brakes should be used, or to control the fuel injection systems. The same use of ANI applies in our smart cell phones or in the Google search engine.

Artificial Neural Networks (ANNs) are a very important chapter of AI and more specifically a branch of Computational Intelligence and Soft Computing. Their applications are applied in diverse scientific areas, such as medical science, finance and management, environmental science, control systems, and telecommunications. For example ANNs are used in flight bookings and dynamic allocation of seats, in credit evaluation, in mortgage screening, pattern recognition, and image and video analysis. Even NASA used ANNs to develop the IFCS (Intelligent Flight Control System).

The Engineering Applications of Neural Networks (EANN) conference was established in Otaniemi, Finland in 1995, exactly 20 years ago. Since then it has become a well-established event with a very long and successful history. In the years that followed, it has achieved a continuous and dynamic presence as a major European scientific event with participants from all over the globe. An important milestone was the year 2009, when its guidance by a Steering Committee of the INNS (EANN Special Interest Group) was initiated. Thus, from that moment, the conference has been continuously supported technically by the International Neural Network Society (INNS). This year, it was scientifically and administratively supported by the Aristotle University of Thessaloniki and the Democritus University of Thrace, Greece. The event was held during September 25–28, 2015, in the "Amilia Mare" Resort and Conference Centre on Rhodes Island, Greece.

The Organizing Committee was delighted by the overwhelming response to the call for papers. In total, 84 original research papers were submitted to the EANN 2015 conference. The diverse nature of papers submitted demonstrates the vitality of neural computing and related soft computing approaches and proves the very wide range of ANN applications as well.

All papers passed through a review process by at least two independent academic referees. Where needed a third referee was consulted to resolve any conflicts. Overall 43 % of the submitted manuscripts (36 in total) were accepted for inclusion in this volume of the Communications in Computer and Information Science (CCIS) series by Springer. The accepted papers of the 16th EANN conference are related to the following thematic topics:

- Industrial Applications of ANN
- Bioinformatics
- Intelligent Medical Modelling

- Intelligent Modelling in Life-Earth Sciences
- Learning Algorithms
- Smart Telecommunications Modelling
- Fuzzy Modelling
- Robotics and Control
- Smart Cameras
- Pattern Recognition
- Emotion Recognition
- Classification
- Intelligent Financial Modelling
- Echo State Networks
- ANN Modelling

The authors of accepted papers came from 21 different countries from all over Europe (e.g., Austria, the Czech Republic, Finland, Germany, Greece, Hungary, Italy, The Netherlands, Poland, Portugal, Russia, Spain, and the UK), the Americas (e.g., Brazil, Chile, and the USA), Asia (e.g., Israel, India, and Iran), Africa (e.g., Algeria) and Oceania (New Zealand).

Three keynote speakers were invited to give lectures in timely aspects of AI and ANN.

1. Prof. Nikola Kasabov of the School of Computing and Mathematical Sciences at Auckland University of Technology, Australia, delivered a talk on: "Neuromorphic Predictive Systems Based on Deep Learning."
2. Prof. Barbara Hammer of Bielefeld University, Germany, delivered a talk on "Autonomous Model Selection for Prototype Based Architectures," and
3. Prof. Paul Verschure of the Universitat Pompeu Fabra, Barcelona, Spain, delivered a talk on "Engineering Biologically and Psychologically Grounded Living Machines: The Distributed Adaptive Control Theory of Mind, Brain and Behaviour."

In addition, two highly interesting tutorials were given within the framework of EANN 2015. The first one was delivered by Assistant Prof. Giacomo Boracchi from the Department of Electronics and Informatics, Politecnico di Milano, Italy. It was entitled "Learning under Concept Drift: Methodologies and Applications." The second tutorial was delivered by Prof. Vera Kurkova from the Institute of Computer Science, Czech Academy of Sciences, Czech Republic. The subject was "Strength and Limitations of Shallow Networks." We wish to express our sincere thanks to both distinguished scientists.

Finally, three workshops on timely AI subjects were organized successfully and collocated with EANN 2015:

1. The 4th Mining Humanistic Data Workshop (MHDW), supported by the Ionian University and the University of Patras. We wish to express our gratitude to Profs. Christos Makris and Katia Linda Kermanidis, and Dr. Ioannis Karydis for their common effort towards the organization of the 4th MHDW workshop.
2. The 5th Workshop on Artificial Intelligence Applications in Biomedicine (AIAB), supported by the Frederick University Cyprus, University of Piraeus Greece, and

the Democritus University of Thrace, Greece. We are grateful to Profs. Harris Papadopoulos, Efthyvoulos Kyriacou, Ilias Maglogiannis, and George Anastasso-poulos for their hard work in managing the 5th AIAB workshop.
3. The 2nd Innovative European Policies and Applied Measures for Developing Smart Cities (IPMSC) Workshop, supported by the Hellenic Telecommunications Organization (OTE). The IPMSC was driven by the hard work of Drs. Ioannis P. Chochliouros and Ioannis M. Stephanakis (OTE, Greece).

We hope that these proceedings will help researchers worldwide to understand and to be aware of the latest developments in ANNs. We do believe that they will be of major interest to scientists over the globe and that they will stimulate further research in the domain of Artificial Neural Networks and AI in general.

September 2015   Lazaros Iliadis
                                                                                    Chrisina Jayne

Organizing Committee

General Chair

Kostas Margaritis — University of Macedonia, Greece

Organizing Chair

Yannis Manolopoulos — Aristotle University of Thessaloniki, Greece

Program Chairs

Lazaros Iliadis — Democritus University of Thrace, Greece
Chrisina Jayne — University of Coventry, UK

Workshop Chairs

Mario Malcangi — University of Milan, Italy
Spyros Sioutas — Ionian University, Greece
Christos Makris — University of Patras, Greece

Advisory Committee

Nikola Kasabov — KEDRI Auckland University of Technology, New Zealand
Plamen Angelov — Lancaster University, UK
Marley Vellasco — Pontifical Catholic University of Rio de Janeiro, Brazil

Website and Advertising Chair

Ioannis Karydis — Ionian University, Greece

Program Committee

Shigeo Abe — Kobe University, Japan
Athanasios Alexiou — Ionian University, Greece
Georgios Anastasopoulos — Democritus University of Thrace, Greece
Andreas Andreou — University of Cyprus, Cyprus
Costin Badica — University of Craiova, Romania
Zorana Bankovic — IMDEA Software Institute, Spain
Kostas Berberidis — University of Patras, Greece
Monica Bianchini — Università di Siena, Italy

Antonis Papaleonidas	Democritus University of Thrace, Greece
Daniel Perez	University of Oviedo, Spain
Elias Pimenidis	University of the West of England, UK
Bernardete Ribeiro	University of Coimbra, Portugal
Marcello Sanguineti	Università degli Studi di Genova, Italy
Christos Schizas	University of Cyprus, Cyprus
Spyros Sioutas	Ionian University, Greece
Jefferson Souza	University of Sao Paolo, Brazil
Stefanos Spartalis	Democritus University of Thrace, Greece
Ioannis Stamelos	Aristotle University of Thessaloniki, Greece
Ioannis Stephanakis	OTE SA, Greece
Ricardo Tanscheit	Pontificia Universidade Catolica do Rio de Janeiro, Brazil
Panos Trahanias	Foundation for Research and Technology, Greece
Athanasios Tsadiras	Aristotle University of Thessaloniki, Greece
Nicolas Tsapatsoulis	Cyprus University of Technology, Cyprus
George Tsekouras	University of the Aegean, Greece
Aristeidis Tsitiridis	Swansea University, UK
Nikos Vassilas	T.E.I. of Athens, Greece
Panagiotis Vlamos	Ionian University, Greece
George Vouros	University of Piraeus, Greece
Peter Weller	City University, UK
Shigang Yue	University of Lincoln, UK
Rodolfo Zunino	Università degli Studi di Genova, Italy

External Reviewers

Gustavo Almeida	Theodoros Koutsandreas
Frederico Coelho	Mariana Lourenço
Spiros Georgakopoulos	Costas Neocleous
Georgia Kontogianni	Leonardo Silvestre

Invited Talks

Neuromorphic Predictive Systems Based on Deep Learning

Nikola Kasabov

School of Computing and Mathematical Sciences,
Auckland University of Technology, Auckland, Australia

Abstract. The current development of the third generation of artificial neural networks - the spiking neural networks (SNN) [1, 5, 9] along with the technological development of highly parallel neuromorphic hardware systems of millions of artificial spiking neurons as processing elements [2, 3], makes it possible to model big and fast data in a fast on-line manner, enabling large-scale problem solving across domain areas including building better predictive systems. The latter topic is covered in this talk. The talk first presents some principles of deep learning inspired by the human brain, such as automated feature selection, 'chain fire', polychronisation. These principles are implemented in a recent evolving SNN (eSNN) architecture called NeuCube [4] and its software development system that is made available from: www.kedri.aut.ac.nz/neucube/. These principles allow for an eSNN system to predict events and outcomes, so that once the eSNN is trained on whole spatio-temporal patterns, it can be made to spike early, when only a part of a new pattern is presented as input data. The talk presents a methodology for the design and implementation of NeuCube-based eSNN systems for deep learning and early and accurate outcome prediction from large-scale spatio-/spectro temporal data, referred here as spatio-temporal data machines (STDM). A STDM has modules for: preliminary data analysis, data encoding, pattern learning, classification, regression, prediction and knowledge discovery. This is illustrated on early event prediction tasks using benchmark large spatio/spectro-temporal data with different spatial/temporal characteristics, such as: EEG data for brain computer interfaces; personalised and climate date for stroke occurrence prediction and for the prediction of ecological and seismic events [6–8]. The talk discusses implementation on highly parallel neuromorphic hardware platforms such as the Manchester SpiNNaker [2] and the ETH Zurich chip [3, 10]. The STDM are not only significantly more accurate and faster than traditional machine learning methods and systems, but they lead to a significantly better understanding of the data and the processes that generated it.

References

1. EU Marie Curie EvoSpike Project (Kasabov, Indiveri). http://ncs.ethz.ch/projects/EvoSpike/
2. Furber, S., et al.: Overview of the SpiNNaker system architecture. IEEE Trans. Comput. **99** (2012)

3. Indiveri, G., Horiuchi, T.K.: Frontiers in neuromorphic engineering. Front. Neurosci. **5** (2011)
4. Kasabov, N.: NeuCube: a spiking neural network architecture for mapping, learning and understanding of spatio-temporal brain data. Neural Netw. **52**, 62–76 (2014)
5. Kasabov, N., Dhoble, K., Nuntalid, N., Indiveri, G.: Dynamic evolving spiking neural networks for on-line spatio- and spectro-temporal pattern recognition. Neural Netw. **41**, 188–201 (2013)
6. Kasabov, N., et al.: Evolving spiking neural networks for personalized modelling of spatio-temporal data and early prediction of events: a case study on stroke. Neurocomputing (2014)
7. Kasabov, N. (ed): The Springer Handbook of Bio- and Neuroinformatics. Springer (2014)
8. Kasabov, N.: Deep Machine Learning and Predictive Data Modelling with Spiking Neural Networks. Springer (2015)
9. Schliebs, S., Kasabov, N.: Evolving spiking neural network-a survey. Evol. Syst. **4**(2), 87–98 (2013)
10. Scott, N., Kasabov, N., Indiveri, G.: NeuCube neuromorphic framework for spatio-temporal brain data and its Python implementation. In: Lee, M., Hirose, A., Hou, Z.-G., Kil, R.M. (eds.) ICONIP 2013. LNCS, vol. 8228, pp. 78–84. Springer, Heidelberg (2013)

Bio. Professor Nikola Kasabov is Fellow of IEEE, Fellow of the Royal Society of New Zealand and DVF of the Royal Academy of Engineering, UK. He is the Director of the Knowledge Engineering and Discovery Research Institute (KEDRI), Auckland. He holds a Chair of Knowledge Engineering at the School of Computing and Mathematical Sciences at Auckland University of Technology. Kasabov is a Past President and Governors Board member of the International Neural Network Society (INNS) and also of the Asia Pacific Neural Network Assembly (APNNA). He is a member of several technical committees of IEEE Computational Intelligence Society and a Distinguished Lecturer of the IEEE CIS (2011–2013). He is a Co-Editor-in-Chief of the Springer journal Evolving Systems and has served as Associate Editor of Neural Networks, IEEE TrNN, IEEE TrFS, Information Science and other journals. Kasabov holds MSc and PhD from the TU Sofia, Bulgaria. His main research interests are in the areas of neural networks, intelligent information systems, soft computing, bioinformatics, neuroinformatics. He has published more than 560 publications that include 15 books, 180 journal papers, 90 book chapters, 30 patents and numerous conference papers. He has extensive academic experience at various academic and research organisations in Europe and Asia, including: TU Sofia, University of Essex, University of Trento, University of Otago, Guest professor at the Shanghai Jiao Tong University, Guest Professor at ETH/University of Zurich, DAA Professor TU Kaiserlautern. Prof. Kasabov has received the APNNA 'Outstanding Achievements Award', the INNS Gabor Award for 'Outstanding contributions to engineering applications of neural networks', the EU Marie Curie Fellowship, the Bayer Science Innovation Award, the APNNA Excellent Service Award, the RSNZ Science and Technology Medal, and others. He has supervised to completion 40 PhD students. More information of Prof. Kasabov can be found on the KEDRI web site: http://www.kedri.aut.ac.nz.

Autonomous Model Selection for Prototype Based Architectures

Barbara Hammer

Bielefeld University, Bielefeld, Germany

Abstract. Prototype-based learning techniques enjoy a wide popularity due to their intuitive training techniques and model interpretability. Applications include biomedical data analysis, image classification, or fault detection in technical systems. One striking property of such models consists in the fact that they represent data in terms of typical representatives; this property allows an efficient extension of the techniques to life-long learning and model adaptation for streaming data. Within the talk, we will mainly focus on modern variants of so-called learning vector quantization (LVQ) due to their strong learning theoretical background and exact mathematical derivative from explicit cost functions. We will focus on three aspects which are of particular interest if these models are used as autonomous learning models: 1) metric learning in prototype based models, 2) incremental learning with adaptive model complexity, and 3) optimum reject options. Metric learning autonomously adjusts the used metric, usually the Euclidean one, towards a richer and more problem-adjusted representation of the data. Metric learning does not only greatly enhance the model performance, but it usually also increases model interpretability, a very important property e.g. in biomedical data applications. We will discuss recent results which investigate metric learning mechanisms with a focus on their uniqueness, and we will present efficient schemes which account for a regularization of this process in particular for high dimensional data. Further, we will show that metric learning in LVQ techniques can be extended towards non-vectorial data such as sequences. Incremental learning and the possibility to reject classification are tightly interwoven aspects. These properties enable to autonomously adjust model complexity, and they enhance the system with the capability to judge its limitations in classification accuracy. We will present recent work which investigates different measurements which allow the quantification of the model insecurity, including the notion of conformal measures as one approach with very clear statistical background. Based on such measures, we will present incremental models with self-adjusted mode complexity, on the one hand, and an efficient strategy for an optimum combination of rejects in a mathematically precise sense, on the other hand.

References

1. Mokbel, B., Paassen, B., Schleif, F.-M., Hammer, B.: Metric learning for sequences in relational LVQ. Neurocomputing (2015)

2. Fischer, L., Hammer, B., Wersing, H.: Optimum reject options for prototype-based classification. CoRR abs/1503.06549 (2015)
3. Zhu, X., Schleif, F.-M., Hammer, B.: Adaptive conformal semi-supervised vector quantization for dissimilarity data. Pattern Recogn. Lett. **49,** 138–145 (2014)
4. Hammer, B., Hofmann, D., Schleif, F.-M., Zhu, X.: Learning vector quantization for (dis-) similarities. Neurocomputing **131,** 43–51 (2014)
5. Frenay, B., Hofmann, D., Schulz, A., Biehl, M., Hammer, B.: Valid interpretation of feature relevance for linear data mappings. In: *Proceedings IEEE Symposium on* Computational Intelligence and Data Mining (CIDM), pp. 149–156 (2014)
6. Biehl, M., Hammer, B., Villmann, T.: Distance measures for prototype based classification. In: Proceedings International Workshop on Brain-Inspired Computing (BrainComp), pp. 100–116 (2013)
7. Strickert, M., Hammer, B., Villmann, T., Biehl, M.: Regularization and improved interpretation of linear data mappings and adaptive distance measures. In: Proceedings IEEE Symposium on Computational Intelligence and Data Mining (CIDM), pp. 10–17 (2013)
8. Bunte, K., Schneider, P., Hammer, B., Schleif, F.-M., Villmann, T., Biehl, M.: Limited rank matrix learning, discriminative dimension reduction and visualization. Neural Netw. **26,** 159–173 (2012)
9. Schneider, P., Biehl, M., Hammer, B.: Adaptive relevance matrices in learning vector quantization. Neural Comput. **21**(12), 3532–3561 (2009)

Bio. Barbara Hammer received her Ph.D. in Computer Science in 1995 and her venia legendi in Computer Science in 2003, both from the University of Osnabrueck, Germany. From 2000–2004, she was chair of the junior research group Learning with Neural Methods on Structured Data' at University of Osnabrueck before accepting an offer as professor for Theoretical Computer Science at Clausthal University of Technology, Germany, in 2004. Since 2010, she is holding a professorship for Theoretical Computer Science for Cognitive Systems at the CITEC cluster of excellence at Bielefeld University, Germany. Several research stays have taken her to Italy, U.K., India, France, the Netherlands, and the U.S.A. Her areas of expertise include hybrid systems, self-organizing maps, clustering, and recurrent networks as well as applications in bioinformatics, industrial process monitoring, or cognitive science. She has been chairing the IEEE CIS Technical Committee on Data Mining in 2013 and 2014, and she is chair of the Fachgruppe Neural Networks of the GI and vice-chair of the GNNS. She has published more than 200 contributions to international conferences / journals, and she is coauthor/editor of four books.

Engineering Biologically and Psychologically Grounded Living Machines: The Distributed Adaptive Control Theory of Mind, Brain and Behaviour

Paul Verschure

Catalan Institute of Advanced Research (ICREA),
Technology Department, Universitat Pompeu Fabra, Barcelona, Spain

Abstract. Our society is facing a number of fundamental challenges in a range of domains that will require a new class of machines. I will call these Living Machines and will describe how their engineering will depend on extracting fundamental design principles from nature. In particular I will emphasize the emergence of a new class of machines that are based on our advancing understanding of mind and brain. The argument that this can lead to a new form of engineering is based on the Distributed Adaptive Control (DAC) that has been applied in a range of domains including robotics, the clinic and in education. DAC is based on the assumption that the brain evolved to maintain a dynamic equilibrium between an organism and its environment through action. The fundamental question that such a brain has to solve in order to deal with the how of action in a physical world is: why (motivation), what (objects), where (space), when (time) or the H4W problem. Post the Cambrian explosion of about 560M years ago, a last factor became of great importance for survival: the who of other agents. I propose that H5W defines the top-level objectives of the brain and will argue that brain and body evolved to provide specific solutions to it by relying on a layered control architecture that is captured in DAC. I will show how DAC addresses H5W through interactions across multiple layers of neuronal organization, suggesting a very specific structuring of the brain, which can be captured in robot control architectures. In explaining how the function of the brain is realized I will show how the DAC theory provides for very specific predictions that have been validated at the level of behaviour and the neuronal substrate. Subsequently I will show how the DAC theory has given rise to a qualitative new class of clinical interventions for the rehabilitation of deficits following stroke illustrating the notion of deductive medicine. These examples will show that robot based models of mind and brain do not only advance our understanding of ourselves and other animals but can also lead to novel technical solutions to complex applied problems.

References

1. Arsiwalla, X.D., Zucca, R., Betella, A., Martinez, E., Dalmazzo, D., Omedas, P., Deco, G., Verschure, P.: Network dynamics with BrainX3: a large-scale simulation of the human brain network with real-time interaction. Front. Neuroinform. **9** (2015)

2. Verschure, P., Pennartz, C.M.A., Pezzulo, G.: The why, what, where, when and how of goal-directed choice: neuronal and computational principles. In: Proceedings of the Royal Society B: Biological Sciences (2014)
3. Cameirao, M.S., i Badia, S.B., Duarte, E., Frisoli, A., Verschure, P.: The combined impact of virtual reality neurorehabilitation and its interfaces on upper extremity functional recovery in patients with chronic stroke. Stroke **43**(10), 2720–2728 (2012)
4. Verschure, P.: The distributed adaptive control architecture of the mind, brain, body nexus. Biol. Inspired Cogn. Archit. **1**(1), 55–72 (2012)
5. Mathews, Z., Verschure, P.: PASAR-DAC7: an integrated model of prediction, anticipation, sensation, attention and response for artificial sensorimotor systems. Inf. Sci. **186**(1), 1–19 (2011)
6. Verschure, P., Voegtlin, T., Douglas, R.J.: Environmentally mediated synergy between perception and behaviour in mobile robots. Nature **425**, 620–624 (2003)

Bio. Paul Verschure is an ICREA Research Professor in the Department of Information and Communication Technologies at Universitat Pompeu Fabra (UPF). He received both his Master and Ph.D. in psychology. He has pursued his research at different leading institutes: the Neurosciences Institute and the Salk Institute (both in San Diego), the University of Amsterdam, University of Zurich and the Swiss Federal Institute of Technology-ETH and currently with ICREA and UPF. Editorial Board of Acta Neurobiologiae Experimentalis (Polish Neuroscience Society). Conference Chair for the Barcelona Cognition, Brain & Technology Summer School 2008, 2009, 2010,2011, 2012, 2013 and 2014. Advisory Board Member for: Ernst Strüngmann Forum 2011 and the 2010 & 2011 CCL Linnaeus Environment (Cognition, Communication and Learning). Dr. Verschure has published over 250 peer-reviewed papers in leading scientific journals including Nature, Science, Neuron, PLoS Biology and PLoS Computational Science, Proceedings of the National Academy of Sciences USA, the Royal Society London and Public Library of Science. He holds 2 patents. Relevant technology projects include Ada: Intelligent Space at Expo'02(Switzerland). SPECS is the 1 of only 6 labs worldwide to receive the European humanoid platform iCub in the first iCub competitive call.

Tutorials

Learning Under Concept Drift:
Methodologies and Applications

Giacomo Boracchi

Department of Electronics and Informatics,
Politecnico di Milano, Milano, Italy

Abstract. Most machine learning techniques assume that the process generating the data is stationary. This guarantees that the model learned during the initial training phase remains valid during the subsequent operation. Unfortunately, stationarity is often an oversimplifying assumption because real-world processes typically change overtime. In the classification literature, changes in the data-generating process are referred to as concept drift.

Learning under concept-drift is a challenging research topic. In fact, in addition to the online-learning issues, the learner has to deal with possible changes, which would make it obsolete and unfit. Given the fact that changes are often unpredictable, as they might occur at any time and shift the data-generating process to an unforeseen state, the learner has to either undergo continuous adaptation to match the recent operating conditions (passive approach) or to steadily monitor the data stream to detect changes and, eventually, react (active approaches). In the last few years, there has been a flourishing of algorithms designed for learning under concept drift, also given the large number of applications where these techniques can be employed.

The tutorial introduces the main issues of learning under concept drift, the active and passive approaches as two extreme adaptation strategies, and few relevant applications such as those related to fraud-detection or those meant for detecting anomalies/changes in streams of signals and images

Strength and Limitations of Shallow Networks

Vera Kurkova

Institute of Computer Science,
Czech Academy of Sciences, Prague, Czech Republic

Abstract. Although originally biologically inspired neural networks were introduced as multilayer computational models, later shallow (one-hidden-layer) architectures became dominant in applications. Recently, interest in architectures with several hidden layers was renewed due to successes of deep convolutional networks. These experimental results motivated theoretical research aiming to characterize tasks which can be computed more efficiently by deep networks than by shallow ones. This tutorial will review recent developments regarding theoretical analysis of strength and limitations of shallow networks. The tutorial will focus on the following topics:

- Universality and tractability of representations of multivariable mappings by shallow networks.
- Trade-off between maximal generalization capability and model complexity.
- Limitations of computation of highly-varying functions by shallow networks.
- Probability distributions of functions which cannot be tractably represented by shallow networks.
- Examples of representations of high-dimensional classification tasks by one and two-hidden-layer networks.

Attendees will learn about consequences of these theoretical results for the methodology of choosing a neural network architecture and about open problems related to deep and shallow architectures.

Contents

Financial Intelligent Modeling

Echo State Networks

Industrial-Engineering Applications of ANN

Mixed Phenomenological and Neural Approach to Induction Motor Speed Estimation

Bartlomiej Beliczynski[✉], Lech M. Grzesiak, and Bartlomiej Ufnalski

Institute of Control and Industrial Electronics, Faculty of Electrical Engineering,
Warsaw University of Technology, 75 Koszykowa Str., 00-662 Warsaw, Poland
{bartlomiej.beliczynski,lech.grzesiak,bartlomiej.ufnalski}@ee.pw.edu.pl
http://www.ee.pw.edu.pl

Abstract. A special phenomenological model of induction motor speed estimation in the drive system is derived. The basis of approximation is calculated from the system mathematical model as a set of transformed, easily measured input variables. It is demonstrated analytically that the set suits well to speed approximation if the approximated signal is a constant or changes linearly. It is then demonstrated numerically that this set is also quite effective under non-zero jerk. Such a system could easily be implemented by widely experienced feedforward neural networks. Illustrative examples and simulation results are attached.

1 Introduction

In process of modeling of physical plant two types of models are usually considered: phenomenological models and input/output models. First of them describe natural phenomena originated in physics, chemistry, economy etc. Such models are usually continuous in time. Parameters of the models are subject of identification procedures. They have physical meaning and their values could be verified. For complex systems such models contain large number of mathematical equations not necessarily easy to handle.

The input/output models are universal models. When using them, one attempts to describe the interrelation between selected variables named inputs and others named outputs. Those variables are also called signals. The inputs are often interpreted as being cause and outputs are effects. One has to stress however that in some practical situations, labeling a signal as being cause or effect of the other could not be obvious.

Contemporary information processing technology requires that the signals are sampled. In majority of practical cases one sampling time value has to be selected for all model variables. It is the common knowledge that sampling time has to be selected within certain region: too large causes negligence of certain dynamics and too small results in a nearly linear data dependence. If selected input and output signals have very different Fourier spectra, it is difficult or impossible to find a suitable sampling time value for all of them. Thus the new input set should be considered. Quality of created model depends very much on preprocessed input and output data sets. After completing the set of signals

© Springer International Publishing Switzerland 2015
L. Iliadis and C. Jayne (Eds.): EANN 2015, CCIS 517, pp. 3–12, 2015.
DOI: 10.1007/978-3-319-23983-5_1

and sampling selection, one searches for interrelation between those sets of data. Such task could be undertaken by means of neural networks.

If originally input/output dependence has to be described via differential or difference equations, one can search for such dynamic input data preprocessing as to reduce the problem description to even complex, but algebraic only equations. Such problem could be handled by using fairly known and well applied feedforward neural networks (FFNNs). They are universal function approximates.

In this paper mixed phenomenological/neural approach will be presented and used. It means that before applying neural approximative the input and output signals will be carefully selected and transformed. The physical plant to be exercised with is the induction motor. More precisely, assuming that only easily measurable electrical signals such as voltages and currents are available we will estimate the shaft speed. For many years sensorless control of induction motor has been challenging and highly attractive task. Valuable reviews of the early achievement were presented in [8] and [9]. The majority of sensorless techniques can be classified as based on magnetic saliency (see for instance [3]) or the system mathematical models [20–22]. In implementation phase neural networks are utilized as dynamic system approximation tools. The neural model may reflect various nonlinearities and be sufficiently robust [1,10].

Neural networks are often encountered within the context of model reference adaptive systems (MRAS) as induction motor speed observers [4,5,11,12]. However, those observers do not fully (if at all) exploit the nonlinear capabilities of FFNNs. There are also numerous successful attempts to employ neural networks as inverse models of the induction motor [6,7,14,15,17]. The most common approach assumes unpreprocessed stator voltage and current signals at the neural network inputs. These signals are often raw data in the stationary reference frame or are transformed into the rotating reference frame [18].

Strong arguments in favour of phenomenological/neural approach was formulated in [1]. The information gained from phenomenological model was used to construct the input variables for neural model. Instead of using directly easily measurable signals i.e. the real and imaginary parts of the stator voltages and currents, some preprocessed variables were used as inputs to the neural network. Such approach seemed to be the key factor for successful implementation. During experiments it was demonstrated that the dynamic neural network being so called tapped delay architecture was reduced to the static one of preprocessed variables. But this was only an observation, verified by simulation [19] and the real plant experiments.

In this paper fully analytical solution is obtained. Specially selected set of transformed variables is derived from mathematical model, naturally contributing well to speed determination. It is proved that this set of the input signals suits well approximation of the speed if the approximated speed signal is a constant or changes linearly. Illustrative examples and simulation results are attached.

The neural architecture, learning algorithms and generalization are the problems for themselves. In this paper we do not pay particular attention to those standard neural elements. They are described elsewhere. They must however efficiently be solved in order to succeed in the general goal. When conducting this research the system simulation was used as a tool to examine particular steps and verify our concepts.

This paper is organized as follows. In Section 2 equivalent mathematical models of induction motor are recalled for further use. In Section 3 a method of choosing the appropriate set of preprocessed input signals is described and a condition to ensure that the problem is static formulated. In Section 4 certain practicalities are presented. Simulation results involving neural networks based speed approximation are shown in Section 5. Finally conclusions are drawn in Section 6.

One terminological comment should be added to that. In this paper we use the term "speed estimation" which will be treated as equivalent to "speed approximation". Recently quite often the term "estimation" is reserved for such situation, when stochastic environment is involved. However majority of sensorless drive system history is written with the word "estimation".

2   Induction Motor Mathematical Model

Mathematical models of a standard induction machine are commonly achieved on the basis of differential equations describing physical phenomena i.e. relationship between stator and rotor fluxes, currents, voltages and shaft angular speed. Keeping in mind simplicity of model as well as the truthful reproduction of effects taking place in the machine, it is usually aimed to make the following assumptions: machine is three-phased with stator and rotor windings symmetrical, variation of magnetic field across the air-gap is sinusoidal, losses for hysteresis and eddy currents are negligible, saturation effect of magnetic circuit is neglected, skin effect in rotor windings is also neglected, resistances and reactances of stator and rotor windings are invariable. This leads to variously formulated models.

Widely recognized and used mathematical model (for instance [13,16]) can be written as the following:

$$\underline{u}_s = R_s \underline{i}_s + \frac{d}{dt} \underline{\Psi}_s \tag{1}$$

$$0 = R_r \underline{i}_r + \frac{d}{dt} \underline{\Psi}_r - jp\Omega_m \underline{\Psi}_r \tag{2}$$

$$\frac{d}{dt}\Omega_m = \frac{1}{J}\left[\frac{-3}{2}pL_m \mathrm{Im}(\underline{i}_s^* \underline{i}_r) - T_L \right] \tag{3}$$

$$\underline{\Psi}_s = L_s \underline{i}_s + L_m \underline{i}_r \tag{4}$$

$$\underline{\Psi}_r = L_r \underline{i}_r + L_m \underline{i}_s , \tag{5}$$

where underline ($\bullet$) denotes complex variables, star (*) denotes complex conjugate and particular quantities are the following: $\underline{u}_s$ - stator voltage space vector, $\underline{i}_s$ - stator current space vector, $\underline{i}_r$ - rotor current space vector, $\underline{\Psi}_s$ -stator flux space vector, $\underline{\Psi}_r$- rotor flux space vector, L_s - stator inductance, L_r - rotor inductance, L_m - main inductance, R_s - stator resistance, R_r - rotor resistance, p - number of pole pairs, Ω_m - angular speed, T_L - load torque, J - inertia.

There are two complex variables that could easily be measured $\underline{u}_s$ and $\underline{i}_s$. Load torque T_L has to be treated as a disturbance. In next Section we will discuss possibility to estimate speed of the motor basing only on easily measured or preprocessed electrical signals.

An alternative model (for instance [13, 16]) is the following:

$$\frac{d}{dt}\begin{bmatrix} \underline{i}_s \\ \underline{i}_r \end{bmatrix} = A \begin{bmatrix} \underline{i}_s \\ \underline{i}_r \end{bmatrix} + B\underline{u}_s \tag{6}$$

$$\frac{d}{dt}\Omega_m = \frac{1}{J}\left[\frac{-3}{2}pL_m\mathrm{Im}(\underline{i}_s^*\underline{i}_r) - T_L\right] , \tag{7}$$

where

$$A = \begin{bmatrix} a_{11} & a_{12} \\ a_{21} & a_{22} \end{bmatrix} = \begin{bmatrix} \frac{-jp\Omega_m L_m^2 - R_s L_r}{L_s L_r - L_m^2} & \frac{(R_r - jp\Omega_m L_r)L_m}{L_s L_r - L_m^2} \\ \frac{(R_s + jp\Omega_m L_s)L_m}{L_s L_r - L_m^2} & \frac{-(R_r + jp\Omega_m L_r)L_s}{L_s L_r - L_m^2} \end{bmatrix} \tag{8}$$

and

$$B = \begin{bmatrix} b_1 \\ b_2 \end{bmatrix} = \begin{bmatrix} \frac{L_r}{L_s L_r - L_m^2} \\ \frac{-L_m}{L_s L_r - L_m^2} \end{bmatrix} . \tag{9}$$

The phenomenological model described by equations (1-5) or (6-9) will be used to transform the formulae and obtain a suitable model for speed estimation.

3   Preprocessed Input Variables

As mentioned before, the directly available quantities in a drive system are the stator complex voltages and currents. All together two complex or four real variables easily measured. Those variables are however poorly correlated with the speed of the shaft [1] and neural approximation is inefficient. Direct use of them should rather be avoided. One can select a set of transformed variables and use it as basis for speed estimation [1]. This process of basis selection requires engineering analysis and is partly intuitive. It ends obviously with no mathematical guarantee for anything.

In this Section we present an alternative approach. Assuming that the phenomenological models of induction motor are sufficiently good for considered application, we derive from them an equation for speed estimation. This equation contains five variables forming the basis for speed estimation. The equation is algebraic if the estimated speed derivative is a constant. Presented here the set of transformed variables has an additional advantage. The elements of the set are carrying out physically interpretable information. We will derive such model and formulate a condition that the model become static. Thus an universal function approximation scheme ensures theoretically any desired level of accuracy.

Directly from (6) one can write

$$\frac{d}{dt}\underline{i}_s = a_{11}\underline{i}_s + a_{12}\underline{i}_r + b_1\underline{u}_s \tag{10}$$

and then

$$\underline{i}_r = \frac{1}{a_{12}}\frac{d}{dt}\underline{i}_s - \frac{a_{11}}{a_{12}}\underline{i}_s - \frac{b_1}{a_{12}}\underline{u}_s \tag{11}$$

and also

$$\underline{i}_s^*\underline{i}_r = \frac{1}{a_{12}}\underline{i}_s^*\frac{d}{dt}\underline{i}_s - \frac{a_{11}}{a_{12}}|\underline{i}_s|^2 - \frac{b_1}{a_{12}}\underline{u}_s\underline{i}_s^* . \tag{12}$$

After some variables manipulation, the coefficients of the last equation might be written as follows

$$\frac{1}{a_{12}} = e_1 + je_2 , \tag{13}$$

where

$$e_1 = \frac{(L_sL_r - L_m^2)R_r}{(R_r^2 + p^2\Omega_m^2 L_r^2)L_m} \text{ and } e_2 = \frac{(L_sL_r - L_m^2)p\Omega_m L_r}{(R_r^2 + p^2\Omega_m^2 L_r^2)L_m} , \tag{14}$$

$$-\frac{a_{11}}{a_{12}} = f_1 + jf_2 , \tag{15}$$

where

$$f_1 = \frac{R_sR_rL_r - p^2\Omega_m^2 L_m^2 L_r}{(R_r^2 + p^2\Omega_m^2 L_r^2)L_m} \text{ and } f_2 = \frac{(R_sL_r^2 + R_rL_m^2)p\Omega_m}{(R_r^2 + p^2\Omega_m^2 L_r^2)L_m} , \tag{16}$$

$$-\frac{b_1}{a_{12}} = g_1 + jg_2 , \tag{17}$$

where

$$g_1 = \frac{L_rR_r}{(R_r^2 + p^2\Omega_m^2 L_r^2)L_m} \text{ and } g_2 = \frac{p\Omega_m L_r^2}{(R_r^2 + p^2\Omega_m^2 L_r^2)L_m} . \tag{18}$$

Now the imaginary part of $\underline{i}_s^*\underline{i}_r$ could be selected

$$\text{Im}(\underline{i}_s^*\underline{i}_r) = e_1(i_{s\alpha}\frac{d}{dt}i_{s\beta} - i_{s\beta}\frac{d}{dt}i_{s\alpha}) + e_2(i_{s\alpha}\frac{d}{dt}i_{s\alpha} + i_{s\beta}\frac{d}{dt}i_{s\beta}) + \tag{19}$$
$$+f_2(i_{s\alpha}^2 + i_{s\beta}^2) + e_1b_1(i_{s\alpha}u_{s\beta} - i_{s\beta}u_{s\alpha}) + e_2b_1(i_{s\alpha}u_{s\alpha} + i_{s\beta}u_{s\beta}) .$$

Looking at (3) which involves $\text{Im}(\underline{i}_s^*\underline{i}_r)$ the following new variables could be selected

$$v_1 = i_{s\alpha}\frac{d}{dt}i_{s\beta} - i_{s\beta}\frac{d}{dt}i_{s\alpha} = \text{Im}(\underline{i}_s^*\frac{d}{dt}\underline{i}_s) \tag{20}$$

$$v_2 = i_{s\alpha}\frac{d}{dt}i_{s\alpha} + i_{s\beta}\frac{d}{dt}i_{s\beta} = \text{Re}(\underline{i}_s^*\frac{d}{dt}\underline{i}_s) = \frac{1}{2}\frac{d}{dt}|\underline{i}_s|^2 \tag{21}$$

$$v_3 = i_{s\alpha}^2 + i_{s\beta}^2 = |\underline{i}_s|^2 \tag{22}$$

$$v_4 = i_{s\alpha}u_{s\beta} - i_{s\beta}u_{s\alpha} = -\text{Im}(\underline{u}_s\underline{i}_s^*) \tag{23}$$

$$v_5 = i_{s\alpha}u_{s\alpha} + i_{s\beta}u_{s\beta} = \text{Re}(\underline{u}_s\underline{i}_s^*) . \tag{24}$$

Finally taking (3), (19) and (13), (15), (17) one can write

$$\frac{d\Omega_m}{dt}(h_0 + h_1\Omega_m^2) = c_0(h_0 + h_1\Omega_m^2) + c_1v_1$$
$$+ c_2\Omega_m v_2 + c_3\Omega_m v_3 + c_4v_4 + c_5\Omega_m v_5 , \tag{25}$$

where the coefficients c_i, $i = 0, ..., 5$ are independent from Ω_m. All of them could be expressed as functions of motor parameters. Let us mention only few of them

$$h_0 = R_r^2 L_m, \quad h_1 = p^2 L_r^2 L_m \tag{26}$$

$$c_0 = -\frac{T_L}{J} . \tag{27}$$

It has to be underlined that T_L, the load torque is not known and is treated as a disturbance. The variable c_0 is assumed to be a random variable. We do not attempt to estimate it. It is the task of controller to suppress disturbances.

Solution of (25) is not known. One can learn however that because of the derivative of Ω_m on the left hand side of the equation, the variable Ω_m cannot be expressed as a function of the transformed variables $\{v_1, v_2, v_3, v_4, v_5\}$. The system is not static, but the dynamic one.

However if additionally $\frac{d\Omega_m}{dt} = \text{const} = c$, equation (25) become algebraic one and Ω_m could be expressed as an entangled function of v_1, v_2, v_3, v_4, v_5:

$$c(h_0 + h_1\Omega_m^2) = c_0(h_0 + h_1\Omega_m^2) + c_1v_1$$
$$+ c_2\Omega_m v_2 + c_3\Omega_m v_3 + c_4v_4 + c_5\Omega_m v_5 . \tag{28}$$

One can expect larger errors when Ω_m changes rapidly, specially if $\frac{d^2\Omega_m}{dt^2} \neq \text{const}$.

4   Practicalities

As one can observe from (20-24) that the base of approximation is formed via easy algebraic operation on measured voltages and currents signals, plus one more difficult one that is current differentiation. When dealing with real data, such operation must be watched carefully. Differentiator is usually associated with a lag. For such differentiator additional parameters must be selected.

Typical, realistic differentiator has the following transfer function

$$\frac{sT_d}{1 + \varepsilon sT_d} \quad \text{where} \quad \varepsilon < 1 \tag{29}$$

or discrete ZOH equivalent form

$$\frac{\frac{1}{\varepsilon}(1 - z^{-1})}{1 - \beta z^{-1}} \quad \text{where } \beta = e^{-\frac{T_s}{\varepsilon T_d}}, \text{ where } T_s \text{ is the sampling time} \tag{30}$$

or in time domain

$$y(k) = \beta y(k-1) + \frac{1}{\varepsilon}(x(k) - x(k-1)) . \tag{31}$$

So when implementing differentiation, apart from T_s and T_d one has to select ε as well.

5 Simulation Results

To justify neural network input variables selection and illustrate our approach we have run simulation of the induction motor model and collected appropriate data. The motor with the following parameters: the nominal power 1.5 kW and $p_b = 3$, $R_s = 1.54\,\Omega$, $R_r = 1.294\,\Omega$, $L_s = 0.1004\,\text{H}$, $L_r = 0.0969\,\text{H}$, $L_m = 0.0915\,\text{H}$, $J = 0.1\,\text{kg}\,\text{m}^2$ was controlled using a FOC scheme and excited by a randomly generated reference speed pattern. The load changes were simultaneously applied. We selected sampling time $T = 5\,\text{ms}$, and sampled the input signals. The simulation model responded with variables $u_{s\alpha}$, $u_{s\beta}$, $i_{s\alpha}$, $i_{s\beta}$ and Ω_m. For the realistic differentiation $T_d = 0.2$, $\varepsilon = 0.05$ were chosen, leading to $\beta = 0.61$. To determine correlation coefficients, a set of 5000 samples of 5 inputs and one output was prepared. Result of calculation for the originally easily measured four variables $u_{s\alpha}$, $u_{s\beta}$, $i_{s\alpha}$, $i_{s\beta}$ their derivatives are shown in Tab. 1. One can acknowledge that having this level of correlation between the inputs and Ω_m, despite theoretical possibility, it is really hard to obtain a good neural approximator. One needs plenty of neurons and generalization is expected to be poor. When using $\{v_1, v_2, v_3, v_4, v_5\}$ basis, the input correlation with the output is much better and is presented in Tab. 2. This result was confirmed by a simulation experiment. To estimate motor speed we used one-hidden layer neural network with 15 sigmoidal neurons and linear output neuron. Before applying neural feedforward net learning procedure, in each case the set of the inputs and the output samples were tested against being a function [2].

The network was trained on the set of 5000 samples collected with sampling time 10 ms by using standard Matlab procedure feedforwardnet. Then its generalization properties were tested by applying the set of 50000 samples acquired at 1 kHz under different reference speed and disturbance torque patterns. The result is shown in Fig. 1.

Table 1. In/out correlation coefficients

Variable	Corr. coeff. with Ω_m
$i_{s\beta}$	0.0001
$i_{s\alpha}$	0.0001
$u_{s\alpha}$	< 0.0001
$\frac{d}{dt}i_{s\beta}$	< 0.0001
$u_{s\beta}$	< 0.0001
$\frac{d}{dt}i_{s\alpha}$	< 0.0001

Table 2. The correlation coefficients of $\{v_1, v_2, v_3, v_4, v_5\}$ with Ω_m

Variable	Corr. coeff. with Ω_m
v_1	0.4680
v_2	-0.0064
v_3	0.0069
v_4	0.9665
v_5	0.1624

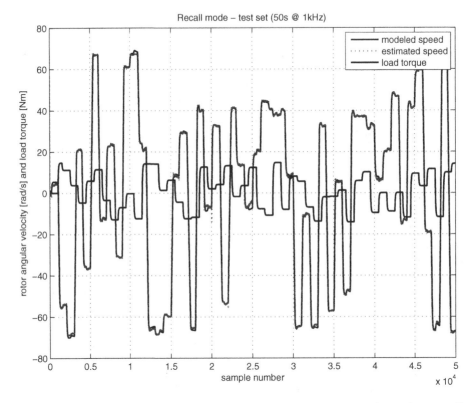

Fig. 1. Hardly distinguishable curves of the real and approximated speed in recall mode.

The selected set of preprocessed inputs could be expanded or shrunk. We do not pretend that the list presents the best possible set of preprocessed variables. One can imagine such transformation as to obtain speed calculation as a direct mathematical formula. The neural approximator will then be useless. But even now v_4 can be used alone as a rough speed estimator. Preprocessed input variables well correlated with the speed are useful for approximation.

6 Conclusions

We derived a special form of phenomenological induction motor speed estimation model. As a source of information four easily measurable electrical signals i.e. stator voltages and currents were taken. However instead of direct use of them, a particular basis for neural based approximation was derived and used. Those signals are much better correlated with the approximated signal than the original sources of information. The neural net learning process is much more efficient.

The derived phenomenological model is a model of dynamical system, however if derivative of the speed is constant, it become a static one. Our simulation experiments support our claim that this special basis ensures also good approximation capabilities of the speed signal outside of this region.

References

1. Beliczynski, B., Grzesiak, L.: Induction motor speed estimation: Neural versus phenomenological model approach. Neurocomputing **43**, 17–36 (2002)
2. Beliczynski, B., Kubowicz, K.: Appoximately static relationship between model variables. Przeglad Elektrotechniczny **53**(4), 385–389 (2004)
3. Britz, F., Degner, M., Garcia, P., Lorentz, R.: Comparison of saliency-based sensorless control techniques for ac machines. IEEE Trans. Ind. Appl. **40**(4), 1107–1115 (2004)
4. Cirrincione, M., Accetta, A., Pucci, M., Vitale, G.: MRAS speed observer for high-performance linear induction motor drives based on linear neural networks. IEEE Transactions on Power Electronics **28**(1), 123–134 (2013)
5. Dumnic, B., Popadic, B., Milicevic, D., Katic, V., Oros, D.: Speed-sensorless vector control of an wind turbine induction generator using artificial neural network. In: 16th International Power Electronics and Motion Control Conference and Exposition (PEMC), pp. 371–376, September 2014
6. Girovsky, P., Timko, J., Zilkova, J.: Shaft sensor-less foc control of an induction motor using neural estimators. Acta Polytechnica Hungarica **9**(4) (2012)
7. Goedtel, A., Nunes da Silva, I., Amaral Serni, P.J., Suetake, M., Franscisco do Nascimento, C.: Speed estimation for induction motor using neural networks method. IEEE Latin America Transactions (Revista IEEE America Latina) **11**(2), 768–778 (2013)
8. Holtz, J.: Methods for speed sensorless control of AC drives. IEEE 1, 21–29 (1998)
9. Holtz, J.: Sensorless control of induction motor drives. Proceedings of the IEEE **90**, 1359–1394 (2002)
10. Karanayil, B., Rahman, M., Grantham, C.: Online stator and rotor resistance estimation scheme using artificial neural networks for vector controlled speed sensorless induction motor drive. IEEE Trans. Ind. Electronics **54**, 167–176 (2007)
11. Maiti, S., Verma, V., Chakraborty, C., Hori, Y.: An adaptive speed sensorless induction motor drive with artificial neural network for stability enhancement. IEEE Transactions on Industrial Informatics **8**(4), 757–766 (2012)
12. Niasar, A.H., Khoei, H.R.: Sensorless direct power control of induction motor drive using artificial neural network. Advances in Artificial Neural Systems, 1–9 (2015)

13. Orlowska-Kowalska, T., Kowalski, C.T.: Neural network application for flux and speed estimation in the sensorless induction motor drive. In: Proc. Of ISIE 1997, pp. 1253–1258. IEEE (1997)
14. Ponce, P., Molina, A., Tellez, A.: Neural network and fuzzy logic in a speed close loop for DTC induction motors. In: International Caribbean Conference on Devices, Circuits and Systems (ICCDCS), pp. 1–7, April 2014
15. Santos, T.H., Goedtel, A., Silva, S.A.O., Suetake, M.: Speed estimator in closed-loop scalar control using neural networks. In: International Conference on Electrical Machines (ICEM), pp. 2570–2576, September 2014
16. Simoes, K., Bose, B.: Estimation of feedback signals for a vector controlled induction motor drive. IEEE Trans. on Industry Application, 629–639 (1995)
17. Sun, X., Chen, L., Yang, Z., Zhu, H.: Speed-sensorless vector control of a bearingless induction motor with artificial neural network inverse speed observer. IEEE/ASME Transactions on Mechatronics 18(4), 1357–1366 (2013)
18. Ufnalski, B.: Speed estimation in rotating reference frame. MATLAB Central (2014). www.mathworks.com/matlabcentral/fileexchange/48390-speed-estimation-in-rotating-reference-frame
19. Ufnalski, B.: Speed-sensorless induction motor drive. MATLAB Central (2014). www.mathworks.com/matlabcentral/fileexchange/48012-speed-sensorless-induction-motor-drive
20. Zaky, M., Khater, M.M., Shokralla, S., Yasin, H.A.: Wide-speed-range estimation with online parameter identification schemes of sensorless induction motor drives. IEEE Trans. Ind. Electronics 56, 1699–1707 (2009)
21. Zaky, M., Khater, M.M., Yasin, H., Shokralla, S.: Very low speed and zero speed estimations of sensorless induction motor drives. Electric Power Systems Research 80, 143–151 (2010)
22. Zhao, L., Huang, J., Liu, H., Li, B., Kong, W.: Second-order sliding-mode observer with online parameter identification for sensorless induction motor drives. IEEE Trans. Ind. Electronics 61, 5280–5289 (2014)

Closed Loop Identification of Nuclear Steam Generator Water Level Using ESN Network Tuned by Genetic Algorithm

Glauco Martins[1], Marley Vellasco[1(✉)], Roberto Schirru[2], and Pedro Vellasco[3]

[1] Pontifical Catholic University of Rio de Janeiro, PUC-Rio, Rio de Janeiro, Brazil
glaucopmm@gmail.com, marley@ele.puc-rio.br
[2] Federal University of Rio de Janeiro, Post-Graduation Nuclear Engineering Program,
Rio de Janeiro, Brazil
schirru@lmp.ufrj.br
[3] State University of Rio de Janeiro, UERJ, Rio de Janeiro, Brazil
vellasco@eng.uerj.br

Abstract. The behavior of Steam Generators Water Level from nuclear power plants is highly nonlinear. Its parameters change when facing different operational conditions. Simulating this system can be very useful to train people involved in the real plant operation. However, in order to simulate this system the identification process must be performed. Echo State Networks are a special type of Recurrent Neural Networks that are well suited to nonlinear dynamic systems identification, with the advantage of having a simpler and faster training algorithm than conventional Recurrent Neural Networks. Echo State Networks have an additional advantage over other conventional methods of dynamic systems identification, since it is not necessary to specify the model's structure. However, some other parameters of the Echo State Network must be tuned in order to attain its best performance. Therefore, this study proposes the use of an Echo State Network, automatically tuned by Genetic Algorithms, to a closed loop identification of a nuclear steam generator water level process. The results obtained demonstrate that the proposed Echo State Networks can correctly model dynamical nonlinear system in a large range of operation.

Keywords: Recurrent Neural Networks · Echo State Networks · Nonlinear identification systems · Closed loop identification · Genetic algorithm · Steam generator water level · Nuclear power plant

1 Introduction

Simulations are very useful to train people involved in the operation of Nuclear Power Plants. Traditional simulators used for training people in the nuclear industry were based on white box models, constructed with physics equations and mass balance [1][2]. More recently, second generation nuclear power plants have been modernized with the use of digital systems [3], resulting in more data available. So, nowadays models can be constructed based on black box system identification techniques.

© Springer International Publishing Switzerland 2015
L. Iliadis and C. Jayne (Eds.): EANN 2015, CCIS 517, pp. 13–23, 2015.
DOI: 10.1007/978-3-319-23983-5_2

In particular, the behavior of the Steam Generators Water Level from Pressurized Water Reactor (PWR) nuclear power plants changes at different operational conditions. Relevant transients exhibit its nonlinearities. Therefore, simulation of specific situations can be very useful in training people involved in the real plant operation.

Neural Networks, like Multilayer Perceptron (MLP) and Radial Basis Function (RBF), have been used in dynamical nonlinear system identification [5][6][7]. Traditional Recurrent Neural Networks (RNN), however, are more suitable to this problem, since they present a natural dynamic behavior, but their training process is more complex [8]. Echo State Network (ESN) is a kind of RNN well suited to dynamic systems identification with a simpler and faster training algorithm than conventional RNN. ESN have an additional advantage over other conventional methods of dynamic identification, since they do not require a detailed specification of the model's structure, as in Nonlinear Auto-Regressive Moving Average with eXogenous inputs NARMAX models [9][1] for instance. Some parameters of the ESN must be tuned though, in order to attain its best performance.

Therefore, this work proposes the use of ESN to closed loop identification of a steam generator using a hybrid system, named ASI-ESN (*Automatic System Identifier based on Echo State Network*) which uses a Genetic Algorithm (GA) to optimize the ESN parameters.

This paper is organized in five additional sections. Section 2 presents a brief explanation about the steam generator dynamics and reproduces a simple steam generator feed water level control based on previous linear models [10][11]. Section 3 summarises the ESN theory. Section 4 describes the proposed hybrid architecture, called ASI-ESN, applied to automatically tune ESN's parameters. Then, Section 5 discusses the results obtained in three different tests, involving data generated by a simulation model and collected from a real plant. Finally, Section 6 presents the concluding remarks.

2   Steam Generator Water Level Dynamics and Control

Steam generators are very important components to PWR nuclear power plants since they are the interface between the primary and secondary circuits of water. The heat from the primary circuit is transferred to the secondary through the steam generator that contains, in its secondary side, a mixture of water and steam. The control of water level inside the steam generator is very important. The level behavior is nonlinear, varies with generation power level and has inverse response known as shrink and swell [12].

In the early eighties, in order to verify the phenomena and study the level behavior of steam generator, a linear model, described by Equation 1, was developed for five different power levels [10], where the parameters of the linear model are altered for each different power level. In Equation 1, u is the feedwater flow rate; v is the steam flow rate; y is the water level (narrow range level); s is the Laplace variable; τ_1 and τ_2 are the damping time constants; T is the period of the mechanical oscillation; G_1 is the magnitude of the mass capacity effects; G_2 is the magnitude of the shrink and swell phenomena; and finally, G_3 is the magnitude of the mechanical oscillation [12]. The shrink and swell are a non-minimum phase phenomena caused by the thermal effects; when the steam flow-rate is increased, the water level tends to rise initially (swell), instead of falling as would be expected from the mass balance; a decrease in steam flow-rate leads then to an opposite effect, the water level tends to decrease (shrink). This model is used nowadays in the development of efficient controllers for this problem [13].

$$y(s) = \frac{G_1}{s}\left(u(s) - v(s)\right) - \frac{G_2}{1+\tau_2 s}\left(u(s) - v(s)\right) + \frac{G_3 s}{\tau_1^{-2} + 4\pi^2 T^{-2} + 2\tau_1^{-1}s + s^2}u(s) \qquad (1)$$

Figure 1 shows a steam generator and its associated equipment [14]. Valves 4 and 5, known as main and bypass valves, are the control final elements for the steam generator level control, named Feed Water Control System (FWCS). Therefore, the controller actuates at valves 4 and 5. Valve 9 is demanded by the required power level, allowing more or less steam pass towards the turbine. Valve 9 represents the governor valves that belong to other control system, called Turbine Control System.

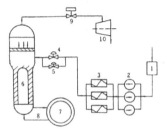

Fig. 1. Steam Generator and associated equipments: 1 – Condenser, 2- Feedwater Pumps, 3 – High pressure heater, 4 – Primary Valve, 5 – Bypass Valve, 6 – Steam Generator, 7 – Reactor, 8 – Coolant, 9 – Steam Flow Valve, 10 – Turbine

Figure 2 shows a conceptual FWCS simplified from suggested by [12]. The steam generator model refers to Equation 1. This control system has two Proportional Integral (PI) controllers.

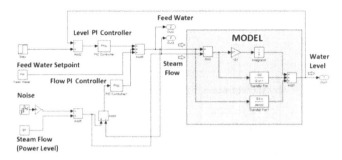

Fig. 2. Simplified FWCS with constant PI gains

The level controller corrects the level error and the flow controller is an anticipatory signal that detects the difference between mass flows. Regarding the use of the same values of proportional and integral gains in all power levels, and taking advantage of the stability produced by feedback system, it is possible to simulate the closed loop responses and verify that they are dynamically different in each power level. Real Advanced FWCS are PID adaptive. However, using the simple FWCS simulator described, it is possible to demonstrate the ESN capacity to model nonlinear dynamical systems and to represent nonlinearities in a full range of operation.

3   System Identification with Echo State Networks

Several models were proposed for identification of nonlinear dynamical systems, like NARMAX [1], Wiener and Hammerstein model, Volterra model [16] and others [4]. In general, they require specification of the model's structure, that in many cases it is not an easy task, although an unsuitable structure leads to bad results [1]. Feedforward networks, like MLP and RBF, aren't dynamic but can be used to represent the dynamic behavior of the systems by using the systems delayed input and output signals in the network's input layer [5]. Although those neural networks eliminate the need to choose nonlinear functions and to explicitly specify the relationship between regressors during the structure definition, the problem related to the definition of the order of the dynamic system remains [15].

ESN has a natural vocation for dynamic systems identification because its structure leads to the Nonlinear Auto-Regressive Moving Average (NARMA) model as can be seen in Equation 2 where G is the function, u and y are plant's inputs and outputs [9].

$$\mathbf{y}(n+1) = \mathbf{G}(..., \mathbf{u}(n), \mathbf{u}(n+1); ..., \mathbf{y}(n-1), \mathbf{y}(n)) \qquad (2)$$

ESN has a *"Dynamic Reservoir"* (DR), which is a RNN with a connectivity percentage. The DR is responsible for embodying the dynamics of the system to be identified. Sparsely connectivity provides for a relative decoupling of subnetworks, which provide the development of individual dynamics. A rich reservoir is sparsely and randomly connected [9]. So, the complete ESN is composed by a DR with N internal units, K input units and L output units, as shown in the Figure 3. The training process involves the adjustment of only the output layer weights, which is a linear and simple minimization task.

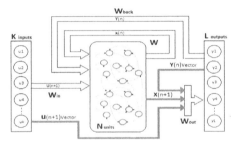

Fig. 3. ESN schematic

The weights connections are NxK matrix for the input weights $\mathbf{W}^{in} = (w_{ij}^{in})$, NxN matrix for the internal connections $\mathbf{W} = (w_{ij})$, $Lx(K + N + L)$ matrix for the connections to the output units $\mathbf{W}^{out} = (w_{ij}^{out})$, and a NxL matrix for the connections that project back from the output to the internal reservoir units $\mathbf{W}^{back} = (w_{ij}^{back})$ [9]. Regarding an input vector with K elements for a step time $n \geq 1$ $\mathbf{u}(n) = \left(u_1(n), ..., u_k(n)\right)^t$, an internal vector with N elements $\mathbf{x}(n) = \left(x_1(n), ..., x_N(n)\right)^t$ and an output vector $\mathbf{y}(n) = \left(y_1(n), ..., y_L(n)\right)^t$ with L elements, the activation of internal units is updated according to Equation 3, where $\mathbf{f}$ is the internal unit's output functions and can be typically a *sigmoid* or *tanh* function.

$$\mathbf{x}(n + 1) = \mathbf{f}(\mathbf{W}^{in}\mathbf{u}(n + 1) + \mathbf{W}\mathbf{x}(n) + \mathbf{W}^{back}\mathbf{y}(n)) \tag{3}$$

The output is computed according to Equation 4.

$$\mathbf{y}(n + 1) = \mathbf{f}^{out}(\mathbf{W}^{out}(\mathbf{u}(n + 1), \mathbf{x}(n + 1), \mathbf{y}(n))) \tag{4}$$

The ESN training is carried out in two steps: *sampling* and *weight computation*, as described in [8] and [9]. During the *sampling* period the matrix states $\mathbf{M}$ and teacher collection matrix $\mathbf{T}$ are formed. In the *weight computation*, the output weights are computed as a linear combination of the output states of the reservoir. So, the weights $\mathbf{W}^{out}$ are computed such that the mean squared training error MSE is minimized, which can be seen as a linear regression task. Computation of regression weights reduce to the computation of a pseudoinverse. The desired weights which minimize MSE are obtained by multiplying the pseudoinverse of $\mathbf{M}$ with $\mathbf{T}$ (Equation 5).

$$\mathbf{W}^{out} = \mathbf{M}^{-1}\mathbf{T} \tag{5}$$

The key concept in ESN is that the states of the reservoir depend on the prior states, that is, they have memory. So the states echoes in the new states. Echo states is a property that depends on $\mathbf{W}^{in}$, $\mathbf{W}$ and $\mathbf{W}^{back}$. The property is also relative to the type of training data. An untrained network ($\mathbf{W}^{in}$, $\mathbf{W}$, $\mathbf{W}^{back}$) with state update according to Equation 3, transfer functions *tanh*, and $\mathbf{W}$ with a spectral radius $|\lambda_{max}| > 1$, where $|\lambda_{max}|$ is the largest absolute value of an eigenvector of $\mathbf{W}$, has no echo states with respect to any input/output interval. In practice, when spectral radius of the weight matrix is smaller than unity, we do have an echo state network [9].

In order to achieve good identification performance, a fine tuning of the network's parameters and topology is usually necessary, which generally depends on the user's experience. Some rules and tricks can be applied to this task [8]. The principal parameters that can interfere in the ESN performance are: the $\mathbf{W}$ sparsely; N size of reservoir that shouldn't exceed $T/10$ to $T/2$, where T is length of the training data sampled, in order to avoid *overfitting*; the chosen of spectral radio λ; the absolute size of input weights $\mathbf{W}^{in}$; and $\mathbf{W}^{back}$ if it exists. Others actions that can improve the ESN performance are the noise injection and the use of an input bias. Avoid symmetrical inputs and scaling the inputs also are good practices when modelling difficult nonlinear problems.

The ESN are not restricted to *sigmoids* and *tanh* neurons. In this study we have used the *Leaky Integrator* neuron model [8] that has a time constant τ parameter. The control over this time constant allows the reservoir timescale to become more flexible in terms of matching input timescale [17].

4   Automatic System Identifier Based on ESN – ASI-ESN

The Automatic System Identifier using ESN (ASI-ESN) is a hybrid neuro-evolutionary model that optimizes the parameters of the ESN using Genetic Algorithms for off-line system identification. The hybrid model is considered automatic because no knowledge from the user is required to perform the identification. In other words, the user does not need to define the system's structure, nor the input sliding window or network topology. The user needs only to collect the training dataset (composed of input and target output signals) and define the range of each variable. The user can configure the GA with his preferable parameters, such as population size and steady-state gap, but some simple rules can be followed [18]. Figure 4 below presents the architecture of the proposed ASI-ESN, as well as the chromosome representation. The chromosome is composed by *Reservoir Size N*, Spectral Radius λ, Matrix $\mathbf{W}^{in}$ scale, Matrix $\mathbf{W}^{back}$ Scale, Leaky time constant τ and *Feedback scale*. The ranges can be determined by the user.

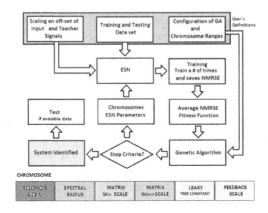

Fig. 4. ASI-ESN architecture

The *Normalized Root Mean Squared Error* (NRMSE) is used as the fitness function. *Overfitting* is prevented using a lower number of neurons in the reservoir [8]. Each individual ESN (a chromosome) is trained more than once, to obtaining an average NMRSE as the chromosome fitness function. The number of times that each ESN is trained for the same chromosome can be modified by the user, but this number will directly impact on the computational time. The evolutionary process continues until a stopping criterion is reached, which, in this case, is the number of generations.

5   Case Study - Closed Loop Nuclear Steam Generator Water Level Identification

The ASI-ESN model was evaluated to the closed loop identification of a nuclear steam generator water level. Three tests were performed in order to verify ASI-ESN performance. The first two tests were carried out using data collected from the simplified simulator presented in Section 2. Then, the proposed hybrid ASI-ESN model was evaluated using multivariable data collected from a real FWCS of a 640MW nuclear power plant. The *direct* approach for closed loop identification was used for all tests. It consists in using controller output signal as the input of the model being identified and controlled variable as the model's output [4].

The first test was performed to evaluate the generalization capacity of the best ESN produced by ASI-ESN in system identification tasks. It was considered a scenario in which the plant was operating in 5% of power level and for some reason was required more steam instantly. Using the linear model, a 10% steam step was produced in order to simulate a 10% disturbance on steam flow. 400 points of the two inputs and one output were collected. The inputs are feed water flow (the controller output) and steam flow (disturbance). Model's output is the water level. Using collected data, ASI-ESN was used to identify this situation and the best-trained ESN was found. Using the linear model, considering 5% of power level, disturbances of 8%, 5%, 12% and 15% of steam were made and collected 400 points of inputs and output data in each case. Those inputs collected were fed in the trained ESN (best ESN obtained for 10% steam step using ASI-ESN). The estimated outputs from ESN were compared with the linear model outputs and the NRMSE found are in Table 1. The trained ESN generated by ASI-ESN was able to generalize the dynamic behavior for other steam steps in 5% power level, as can be seen in Figure 5.

Table 1. NMRSE for each step response

Steam Step	10%	8%	5%	12%	15%
NRMSE	0.0034006	0.18097	0.28843	0.17503	0.11802

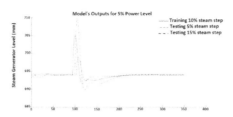

Fig. 5. Responses of ESN trained

In the second test, a nonlinear dynamical system identification task was performed in all range of operation of the simulator. The objective was to verify if only one ESN can represent all nonlinearities of the system. In order to generate data about the dynamics of the system, the linear model was excited with appropriate signal, which approximates white noise, mixed in the steam flow to simulate many random steam flow disturbances. The noise was applied, for equal period of time, in each steam generator power level of 5%, 15%, 30%, 50% and 100%. For each power level, 70% of data was collected for training and 30% for testing, forming a dataset of 25000 points for training and 12120 for testing. Collected data for training were applied in the ASI-ESN to perform system identification. For the best ESN found, the training NMRSE was 0.21595 and test NMRSE was 5.5225. Figure 6 presents the testing results for random steam steps in 5% (left) and 100% (right) power level. As can be observed, a single ESN is capable to represent the dynamics for very different regions of operation, each with a different dynamic behavior.

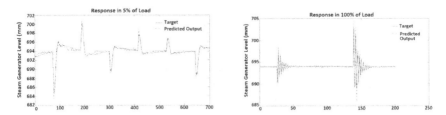

Fig. 6. Responses to steps in 5% of load (left) and 100% of load (right)

The objective of the third test was to identify an important real water level transient. It was performed using real data from the Digital Advanced FWCS of a nuclear power plant. The situation was generated by a disturbance in the steam flow. In a first moment, the steam flow increased abruptly due an unexpected problem of the condenser during operation in 100% nuclear power. In sequence, a run back of power at a big rate was implemented to decrease the power. Figure 7 (left) illustrates three inputs used for ESN training: controller output (actuates at the feed water valves); the nuclear power; and the steam flow. Water level showed is the output (controlled variable). This transient was identified using ASI-ESN and the best ESN obtained a training NMRSE of 0.16628. The resultant water level behavior can be seen in detail in Figure 7(right).

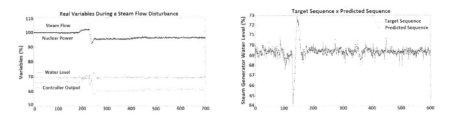

Fig. 7. Inputs and output (left); Output target x predicted (right)

The presented transient is a rare event and there is no data for testing. Considering that the ESN has a good generalization capacity with reasonable results, as seen in the first experiment, and good capacity for nonlinear dynamics representation, as seen in the second experiment, it can be inferred that the resultant model will generate acceptable results if some similar inputs were injected, what would be a good thing for simulation and training tasks. So, to check the trained ESN, some inputs with a similar behavior to the real case (but without the noise observed) were reproduced. Figure 8 (left) shows the applied generated inputs and Figure 8 (right) shows the model's response. It is possible to observe that the dynamics was very similar, which is valuable.

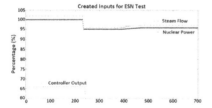 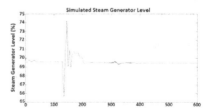

Fig. 8. Inputs simulated (left); Output simulated (right)

In all these experiments, ASI-ESN was implemented using the Simple ESN Tool Box from Jaeger [19] for the ESN and the GA tool from MATLAB [20]. The final values for the best ESN parameters and ranges used are shown in Table 2. For all cases, population of 200 individuals and elitism of 4 individuals were used, with the best result obtained around the 10th generation. The *tanh* was used in all experiments. The values for input scaling and shift, teacher scaling and shift were respectively 0.001, 1.5, 0.001 and 0.5. It is important to remember that these values can vary depending on the magnitude of the data values.

Table 2. Final parameters for ESN obtained with ASI-ESN

Tests	Training Data Lenght	Reservoir Size N	Spectral Radius	Matrix WIN Scale	Matrix Wback Scale	Leaky Time Constant	FeedBack Scale	Training NMRSE
1	400	56	0.54	19.3	39.7	0.377	0	0.0031
2	25000	1479	0.60	3.40	33.7	1.900	0	0.2015
3	699	184	0.48	35.4	30.4	1.086	0	0.1665
Range	-	40-60 1000-2000 50-200	0.1-0.99	1.0–40.0	1.0–40.0	0.000001– 5.0	0.000001- 2.0	-

6   Conclusion

Echo State Networks are Recurrent Neural Networks that are naturally appropriate to represent dynamical systems. It has been shown that they are suitable to black-box systems identification. In this work, ESN was associated to Genetic Algorithm to form a hybrid system, named ASI-ESN, that proposes to be a simpler way to do the identification of linear and nonlinear dynamical systems.

During identification tests with data from linear model of steam generator level control, ASI-ESN showed good capacity of generalization. After being trained with 10% disturbance in the 5% power level, the trained ESN was successful in generalizing the dynamic behavior for 5%, 8%, 12% and 15% steps for the same power level.

A typical strategy of system identification applying random signals as disturbance to excite the plant and use the collected data to identify the model was applied successfully using ESN. A full range steam generator operation was identified using the proposed ASI-ESN model. Only one trained ESN was able to represent all ranges and their different dynamics.

Using data from a real Feed Water Control System of a 640MW nuclear power plant, the identification of a steam flow disturbance was successful using ASI-ESN.

Therefore, the proposed ASI-ESN model is able to successfully perform system identification without having to predefine the structure or the time delay inputs.

For future work we plan to evaluate the influence of the sampling period and the noise in the real data on the performance of the proposed model.

References

1. Aguirre, L.: Introdução À Identificação de Sistemas: Técnicas Lineares e Não-Lineares Aplicadas a Sistemas Reais. 2a. UFMG (2004). ISBN 8570415842 (in Portuguese)
2. Jervis, M.: Models and Simulation in Nuclear Power Station Design and Operation. Advances in Nuclear Science and Technology **17**, 77–115 (1986)
3. Digital Instrumentation and Control Design Guide. EPRI, Product ID: 3002002989 (2014)
4. Aguirre, L.: Enciclopédia de Automática-Controle e Automação, vol. 3. 1a. Edgar Blucher (2007). ISBN 8521204108 (in Portuguese)
5. Narendra, K., Parthasarathy, K.: Identification and control of dynamical systems using neural networks. IEEE Trans. on Neural Networks **1**(1), 4–27 (1990)
6. Gang, Z., Xin, C., Weicheng, Y., Wei, P.: Identification of dynamics for nuclear steam generator water level process using RBF neural networks. In: 8th ICEMI, pp. 3–379 (2007)
7. Haykin S.: Neural Networks: a Comprehensive Foudantion, 2nd edn. Prentice Hall (1998)
8. Jaeger, H.: A tutorial on training recurrent n. networks, covering bppt, rtrl, and the echo state network approach. Tech. report, Fraunhofer Inst. for Aut. Intelligent Systems (2013)
9. Jaeger, H.: The "echo state" approach to analyzing and training recurrent neural networks. Technical report, Fraunhofer Institute for Autonomous Intelligent Systems (2010)
10. Irving, E., Miossec, C., Tassart, J.: Toward efficient full automatic operation of the PWR steam generator with water level adaptive control. In: Proc. Int. Conf. Boiler Dynamics Contr. Nuclear Power Stations, London, U.K., pp. 309–329 (1980)
11. Ablay, G.: Steam generator level control with an observer-based algebraic approach. In.: 8th Int. Conf. in Electrical and Electronics Engineering (ELECO), pp. 137–141 (2013)
12. Ahmed, Z.: Data-driven controller of nuclear steam generator by set membership function approximation. Annual of Nuclear Energy **37**, 512–521 (2010)
13. Mousa, H., Koutb, M., El-Araby, M., Ali, E.: Design of optimal fuzzy controller for water level of U-Tube steam generator in nuclear power station. J. A. Science, 629–636 (2011)
14. Zhao, F., Ou, J., Du, W.: Simulation Modeling of Nuclear Steam Generator Water Level Process-a Case Study. ISA Transactions **39**, 143–151 (2000)

15. Billings, S.: Nonlinear System Identification: NARMAX, Methods in the Time, Frequency, and Spatio-Temporal Domains. Wiley (2013)
16. Billings, S.A.: Identif. of nonlinear systems – a survey. IEE proc. Pt. D **127**(6), 272–285
17. Ferreira, A., Ludermir, T.: Comparing evolutionary methods for reservoir computing pre-training. In: 2011 Proc. Int. Joint Conf. on Neural Networks (JCNN), pp. 283–290
18. Goldberg, D.: Genetic Algorithms in Search, Optimization and Machine Learning. Addison Wesley (1989)
19. Jaeger, H.: Simple toolbox for ESNs (2009). http://reservoir-computing.org/software
20. MATLAB GA toolbox. http://www.mathworks.com/discovery/genetic-algorithm.html

On-line Surface Roughness Prediction in Grinding Using Recurrent Neural Networks

Ander Arriandiaga[1(✉)], Eva Portillo[1], Jose A. Sánchez[2], and Itziar Cabanes[1]

[1] Department of Automatic Control and System Engineering,
University of the Basque Country, C/Alameda Urquijo s/n, 48013 Bilbao, Spain
ander.arriandiaga@ehu.eus
[2] Department of Mechanical Engineering, University of the Basque Country,
C/Alameda Urquijo s/n, 48013 Bilbao, Spain

Abstract. Grinding is a key process in high-added value sectors due to its capacity for producing high surface quality and high precision parts. One of the most important parameters that indicate the grinding quality is the surface roughness (R_a). Analytical models developed to predict surface finish are not easy to apply in the industry. Therefore, many researchers have made use of Artificial Neural Networks. However, all the approaches provide a particular solution for a wheel-workpiece pair. Besides, these solutions do not give surface roughness values related to the grinding wheel status. Therefore, in this work the prediction of the surface roughness (R_a) evolution based on Recurrent Neural Networks is presented with the capability to generalize to new grinding wheels and conditions. Results show excellent prediction of the surface finish evolution. The absolute maximum error is below 0.49µm, being the average error around 0.32µm.

Keywords: Grinding · Surface roughness · Dynamic modelling · Recurrent neural networks

1   Introduction

Grinding is critical in modern manufacturing due to its capacity for producing parts with high surface quality and high precision. Thus, grinding is used in high-added value sectors such as aerospace and motor industries under very tight dimensional tolerances. Actually, grinding process parameters monitoring and control is an important research objective nowadays. One of the most important parameters that indicate the grinding quality is the surface roughness. Analytical models have been developed in order to predict the surface roughness [1-3]. However, the industrial application of these models is not easy yet since there is lack of information about the composition and performance of the grinding wheels due to the semi-hand-made production of them. Finally, the relations between process variables and process outputs are highly non-linear. Therefore, many researchers have made use of Artificial Intelligence in order to predict surface roughness through grinding parameters.

© Springer International Publishing Switzerland 2015
L. Iliadis and C. Jayne (Eds.): EANN 2015, CCIS 517, pp. 24–34, 2015.
DOI: 10.1007/978-3-319-23983-5_3

Using acoustic emission (AE) sensor, in [4] a multi-layer perceptron with 3 hidden layers was used to predict the surface roughness. A multi-layer perceptron Neural Network was also used in [5] in order to predict the surface finish in creep-feed grinding. Also for creep-feed grinding, in a more recent work [6], an Artificial Neural Network (ANN) was presented for the prediction of surface roughness. GA was used as ANN training technique for predicting the peak to valley surface roughness value in surface grinding [7]. A similar approach was presented in [8] aiming at predicting the surface finish in cylindrical grinding of steel parts. In [9] a Fuzzy-Logic controller was proposed for predicting the final surface finish in the grinding of steel.

From the literature review, it can be concluded that Artificial Intelligence and specially ANNs are suitable for modelling the surface roughness in grinding. However, all the approaches found in the literature provide a particular solution for a wheel-workpiece pair. Thus, the presented solution cannot generalize to new grinding wheels. Besides, some of these solutions do not generalize to new grinding conditions not been used during the training process. Therefore, these approaches have a very limited industrial application and are limited to the research. On the other hand, all the research works only give one surface roughness value related to the grinding wheel status. However, it is known in the industry that the performance of the wheel changes over time due to the wheel wear. In fact, the wheel loses its cutting ability while more material is removed. This is another key reason why using Artificial Intelligence techniques for predicting surface roughness have not spread in the industry. Therefore, in this work the prediction of the surface roughness (R_a) through the specific grinding energy based on ANN is presented with the capability to generalize to new grinding wheels and conditions. Unlike other signals like the acoustic emission (AE), the specific grinding energy can be easily measured without introducing sensors in the machining area, an area with very aggressive conditions and low accessibility. Besides, the presented solution predicts the surface finish to the whole life cycle of the wheel. Actually, the output of the net shows the surface finish over time while the wheel is changing its performance.

2 ANN Proposal for On-line Surface Roughness Prediction

2.1 Experimental Setup

The grinding experiments are carried out on an industrial cylindrical grinding machine (Danobat FG-600-S). In this work, the grinding of steel workpiece with non-extremely demanding surface roughness has been selected as the application field. In fact, the following aluminium oxide grinding wheels have been selected: 82AA36G6VW, 82AA36K6VW, 82AA70G6VW, 82AA100G6VW, 82AA100J6VW and 82AA701J6VW. The grit size of the wheel 82AA701J6VW is quite similar to 70. It has to be noted that the hardness of the wheels is coded with letters from A (soft) to Z (hard). For each grinding wheel, except for the wheels 82AA36G6VW and 82AA100G6VW, experiments with 9 different grinding conditions have been carried out (Table 1). In the case of the wheel 82AA100G6VW, the grinding condition with q_s=100 and Q'=4 mm^3/mm·s is not available due to its fast wear. On the other hand, the wheel 82AA36G6VW has been selected to test the generalization capabilities to new wheel characteristics, in particular, for two different grinding conditions. To sum up, 46 experiments are available to train the ANN (Table 1).

As one of the goals of the work is to predict the evolution of the surface finish, the experiments involve grinding up to 2000 mm^3/mm of specific volume (V'_w) of part material removed. In some experiments the V'_w is less than 2000 mm^3/mm because of the wheel wear. The surface roughness values of each experiment are periodically measured with a profilometer.

Table 1. The experiments carried out on the grinding machine.

Exp.	Grit size	Hardness	q_s	Q'	Exp.	Grit size	Hardness	q_s	Q'
1	36	K	60	1	24	100	G	80	4
2	36	K	60	2.5	25	100	G	100	1
3	36	K	60	4	26	100	G	100	2.5
4	36	K	80	1	27	100	J	60	1
5	36	K	80	2.5	28	100	J	60	2.5
6	36	K	80	4	29	100	J	60	4
7	36	K	100	1	30	100	J	80	1
8	36	K	100	2.5	31	100	J	80	2.5
9	36	K	100	4	32	100	J	80	4
10	70	G	60	1	33	100	J	100	1
11	70	G	60	2.5	34	100	J	100	2.5
12	70	G	60	4	35	100	J	100	4
13	70	G	80	1	36	701 (70)	J	60	1
14	70	G	80	2.5	37	701 (70)	J	60	2.5
15	70	G	80	4	38	701 (70)	J	60	4
16	70	G	100	1	39	701 (70)	J	80	1
17	70	G	100	2.5	40	701 (70)	J	80	2.5
18	70	G	100	4	41	701 (70)	J	80	4
19	100	G	60	1	42	701 (70)	J	100	1
20	100	G	60	2.5	43	701 (70)	J	100	2.5
21	100	G	60	4	44	701 (70)	J	100	4
22	100	G	80	1	45	36	G	60	2.5
23	100	G	80	2.5	46	36	G	60	4

One of the signals used in this work to predict the surface finish is the specific grinding energy (e_c). It can be easily measured on-line during the grinding process through the measurement of the power consumption of the wheel spindle:

$$e_c = \frac{P}{Q_w} \tag{1}$$

where P is the power consumption in the wheel spindle and Q_w is the material removal rate. Thus, for a reliable measuring of e_c, the power consumption in the wheel is measured using a Hall-effect based transducer (UPC-FR) by Load Controls. The NI USB-6251 A/D converter board by National Instruments is used to storage the samples during the operation at 100 Hz.

Then, using Matlab^{TM} 2012b software, both signals (surface roughness and specific grinding energy) have been interpolated with *smoothing spline* to obtain periodic signals. To avoid the oscillation of the resulting signal, the *smoothing parameter* equal to 1.91 is used. The criterion used during the interpolation process is to yield a signal value every specific volume of part material removed equal to 10 mm^3/mm.

2.2   ANN Configuration

For predicting the evolutions with ANNs time-lagged neural networks or recurrent neural networks can be used. Time-lagged neural networks use lagged inputs or a moving buffer in the input layer. These inputs can be the inputs signals and/or the past values of the output signal. Thus, every step in the future the net is fed with the lagged past values of the predicted value and/or another input signal. However, despite the lagged inputs, the network does not take into account the temporal characteristics of the time series. This solution is widely used in time-series prediction where the future values are predicted using past values [10-12]. The other solution is based on Recurrent Neural Networks (RNN). In this case, the network takes into account the temporal characteristics of the time series (it is said that the network has "memory"). Thus, it is not necessary to feed the network with the past values of the predicted value. These architectures are, also, widely used in time-series forecasting [13, 14].

In this work, the aim is to predict the evolution of the surface roughness from the beginning of the grinding operation. Thus, the option of using neural networks with lagged inputs (TDNN) is discarded because these lagged inputs are not available at the beginning of the operation. Within the different RNN architectures available, the Elman-based RNN, the Layer-Recurrent neural network [15], is selected. Unlike others RNN architectures, the Layer-Recurrent neural network does not need the initial values, and therefore, it fits to the aim of this work.

The algorithm used to train the network is Levenberg-Marquardt. Besides, back-propagation through time (BPTT) is used. The Bayesian regularization is selected to improve the generalization capabilities of the net. Finally, the tangent sigmoid activation function is used in the hidden layer because is recommended for the Bayesian regularization. The output layer activation function is linear. Thus, the input and output data are normalized within the range [-1, 1].

2.3   Training of the ANN

The aim of the training phase is to achieve the best net to predict the surface roughness. In fact, it is necessary to define the inputs to the net, the number of hidden layers, the neurons on each hidden layer, and past values (delays) to take into account in the feedback connection from the output of the hidden layer to the input.

As mentioned above, one of the net inputs is the specific grinding energy. This input is a measured value and at each step the actual e_c value is introduced in the net. Besides, grinding wheel characteristics and grinding conditions are used, also, as inputs in order to generalize to new wheel and grinding conditions once the net is trained. Although the grinding wheels cover many grinding operations, the presented work is limited to the widely used aluminium oxide abrasive, vitrified bond and wheels with structure number 6. Thus, the grinding characteristics used as network inputs are grit size and hardness. Related to the grinding conditions, speed ratio (q_s) and material removal rate (Q') have been selected as grinding process parameters due to the influence they have on part surface roughness. To sum up, the proposed net has the following structure: 5-x-1. Thus, it has 5 inputs (specific grinding energy, grit size, hardness speed ratio and material removal rate), 1 output (surface roughness) and one hidden layer with x neurons. This x is a value to determine during the training process. Besides, the delays in the feedback have also to be defined in order to predict the output properly.

For determining the number of neurons and delays the net have been trained with different structures: combinations of hidden neurons (HN) and delays (D). From previous works of the research group [16], a net with a specific structure is trained 6 times due to extensive training times. Each time, the neurons are initialized using the algorithm of Nguyen-Widrow. Besides, based on previous works, it has been seen that the range of 5-10 neurons in the hidden layer (HN) and the range of 5-10 delays (D) in the feedback connection are enough to model the relationship between the specific grinding parameter and the surface roughness. Thus, the following ANN structures have been trained: 5HN5D, 8HN8D and 10HN10D.

The experiments database is divided into the training dataset and testing dataset. The validation dataset is not necessary because Bayesian regularization is used instead of cross-validation. Thus, in the training dataset 42 experiments are included and the remaining 4 are used to generate the test dataset. In order to test the generalization to new wheel parameters and grinding conditions, the selected 4 test data are as follows:

- 2 test data with new wheel parameters not used during the training process, but with grinding conditions used during the training (Experiments data 45 and 46).
- 2 test data with wheel parameters used during the training process but under new grinding conditions (Experiments data 8 and 10).

The Table 2 summarizes the 4 data used for testing the RNN:

Table 2. The 4 experiments used for testing the net.

	Grit size	Hardness	q_s	Q'
Test data 1 (exp. 45)	36	G	60	2.5
Test data 2 (exp. 46)	36	G	60	4
Test data 3 (exp. 8)	36	K	100	2.5
Test data 4 (exp. 10)	70	G	60	1

To select the best net structure, indicators that provide measurements of the deviation between the predicted value and the real one are used:

1. The Mean Square Error (MSE): This error value within the range [-1, 1] used as the normalization range of the input and output data.
2. The Mean Absolute Maximum Error (MAME): It is calculated in units of process (μm) as:

$$MAME = \frac{1}{n}\sum_{x=1}^{n}|\max (T_x - P_x)| \qquad (2)$$

where T is the target surface roughness, P is the predicted one and n is the number of test example.

Thus, the bet net selection has two phases: First, by the MSE, the best two nets of each network structure are selected. Finally, to select the three best nets of this second networks bunch, the MAME metric is used.

3   Results

The MSE of each of the six trained nets for each net structure (HN-D) are showed in Table 3. It can be seen that the best lowest MSE results are yielded with the structure 8HN8D (0.0022 and 0.0036), coming up next 10HN10D structure (0.0037 and 0.0041). Finally, the worst results are achieved with the structure with less neurons in the hidden layer and less delays in the feedback (0.0040 and 0.0046). Moreover, analysing the results of the 8HN8D network structure, it can be observed that the results yielded are quiet low (except the net3) comparing with the other structures. In fact, the best net results are coincident with the general performance of all the nets for each structure, being the best performance with structure 8HN8D followed by the 10HN10D and been the 5HN5D structure with the overall worst performance.

Table 3. Resume of the MSE results of the six trained nets for each ANN structure.

	net1	net2	net3	net4	net5	net6
5HN5D	0.0111	0.0277	**0.0040**	**0.0046**	0.0859	0.0094
8HN8D	**0.0022**	**0.0036**	0.1457	0.0180	0.0057	0.0104
10HN10D	0.0203	0.0134	**0.0041**	0.0056	**0.0037**	0.1077

In Table 4 the MAME results for the best nets are summarized. It can be seen that the nets with the lowest MAME in units of process (μm) are the same that those with the lowest MSE. Thus, it can be concluded that the 8HN8D structure is the most suitable one to model the relationship between the specific grinding energy, wheel characteristics, operation characteristics and the surface roughness. However, in the case of the net4 with structure 5HD5D, it has the biggest MSE of the selected ones (Table 3) but the third lowest MAME. In fact, it be can concluded that MSE is not enough to select the proper net.

Table 4. Resume of the MAME errors of the best nets.

	8HN8D net1	8HN8D net2	10HN10D net5	5HN5D net3	10HN10D net3	5HN5D net4
MSE	0.0022	0.0036	0.0037	0.0040	0.0041	0.0046
MAME	**0.26**	**0.32**	0.43	0.35	0.44	**0.33**

However, as it is a prediction of an evolution, it is appropriate to see how the nets predict the surface roughness beside the numeric values. It can be seen that for test data 1 and 2 (new wheels characteristics not used during the training process), the three nets predict quite well. In the case of test data 1 (Fig. 1), it can be said that the net that more precisely predicts the surface roughness is the net 2 with structure 8HN8D. In the other two nets (8HN8D net1 and 5HN5D net4) it can be observed that the signal remains linear after 600 mm². In the three cases the maximum error is quite similar, around 0.23 μm, acceptable errors from the process point of view. For the test data 2, the results are similar. At the beginning, all of them predict well but after

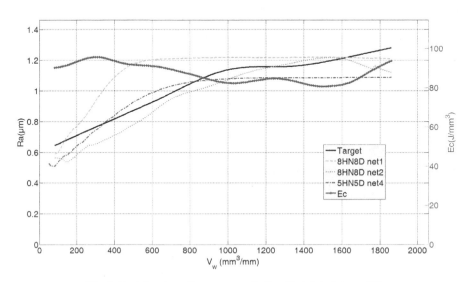

Fig. 1. Best 3 nets performance predicting Test data 1 (exp. 45).

500 mm^2, like for the test data 1, the predicted signal stays linear for the nets 8HN8D net1 and 5HN5D net4. However, for the 8HN8D net 2 the signal oscillates at a similar point. Thus, it can be said that the three nets behaviour is quite similar. In this case the maximum errors are higher than in the test data 1, and the highest one corresponds to the 5HN5D net4 (0.49µm).

In Fig. 3 and Fig. 4, the prediction for new grinding conditions but with known grinding characteristics is showed. For the test data 3 (Fig. 3) the behaviour of the three nets is quite similar. All of them start below the target value and after around 300mm^2 remain above the target signal. The net with the worst performance is the 5HN5D net4 and the net with the best behaviour is the 8HN8D net1 followed closely by the 8HN8D net2. In both nets with structure 8HN8D the maximum error is around 0.32µm. In the case of test data 4 (Fig. 4), the behaviour of 8HN8D net1 and 5HN5D net4 is similar. The surface roughness signal, in both cases, remains constant after 800mm^2 at the same point (around 0.51µm). It can be clearly seen that the worst performance corresponds to the 8HN8D net2. The maximum error for the 8HN8D net2 is 0.27µm, while the error for the other two nets is around 0.12µm. In any case, it can be said that the results in all the cases, even for the 8HN8D net2, are good from the point of view of the process. At the sight of the results, it can be said that the best net is the 8HN8D net2 because any of its prediction stays static after one point. Besides, the results in test data 1 and 2 are better, being similar, and a bit worst for test data 3 and 4, respectively.

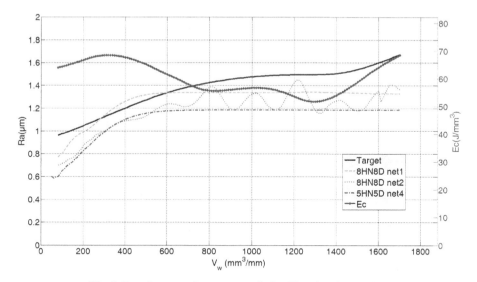

Fig. 2. Best 3 nets performance predicting Test data 2 (exp. 46).

Finally, analysing all the cases it can be concluded that for all the nets, the performance is better for test data with new wheel characteristics that for tests with new grinding conditions and known wheel characteristics. Therefore, it might be said that grinding conditions have higher influence on the surface roughness prediction than wheel characteristics. Besides, the presented solution can be an initial step to model machining parameters where the initial data are not available.

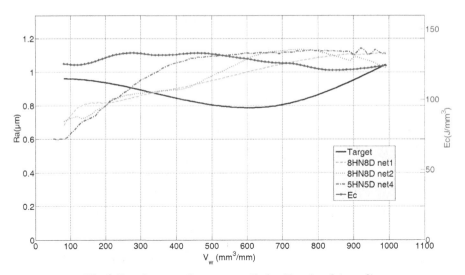

Fig. 3. Best 3 nets performance predicting Test data 3 (exp. 8).

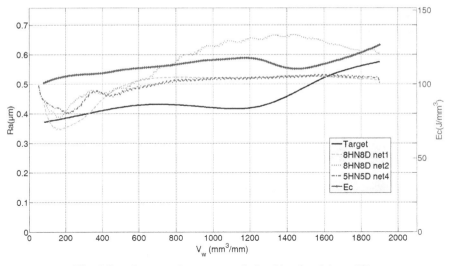

Fig. 4. Best 3 nets performance predicting Test data 3 (exp. 10).

4   Conclusions

In this work a solution based in RNN to model the evolution surface finish with the ability to generalize to new wheel characteristics and grinding conditions has been presented. From the work presented, the following conclusions can be drawn:

- Predicting the evolution of the surface roughness during the grinding process is a high interest research topic. There is not any work in the literature that gives a solution to this problem. In this work, a novel methodology based on the use of RNN has been presented.

- It is not possible to select the best net structure only by comparing MSE values. Therefore, the MAME indicator is used to help in the selection of the best configuration. In addition to the MAME, the tendency of the signal has to be analyzed so as to select the best configuration.
- The proposed RNN is capable to predict the surface finish with good results for grinding's users (errors lower than 0.48μm) without initial real values up to 2000mm^3/mm of specific volume of part material removed. .
- The selected net generalizes with good results to new wheels (not used during the grinding process), 31.96%, and new grinding conditions, 40.05%. Thus, the net generalizes slightly better to new wheels characteristics. It might be concluded that grinding characteristics have bigger influence on the prediction of the surface finish than wheels characteristics.
- The presented solution shows good generalization capabilities to new wheels and new grinding conditions for non-extremely demanding surface roughness application fields. However, further experiments have to be carried out in order to analyze the generalization capabilities to new grinding application fields (new wheels and grinding conditions) such as extremely demanding surface roughness.

References

1. Marinescu, I.D., Hitchiner, M.P., Uhlmann, E., Rowe, W.B., Inasaki, I.: Handbook of Machining with Grinding Wheels. CRC Press, Boca Raton (2006)
2. Jiang, J., Ge, P., Hong, J.: Study on micro-interacting mechanism modeling in grinding process and ground surface roughness prediction. Int J Adv Technol **67**(5–8), 1035–1052 (2008)
3. Agarwal, S., Rao, P.V.: Modeling and prediction of surface roughness in ceramic grinding. Int. J. Mach. Tools Manuf. **50**(12), 1065–1076 (2010)
4. Aguiar, P.R., Cruz, C.E.D., Paula, W.C.F.: Predicting surface roughness in grinding using neural networks. In: Advances in Robotics, Automation and Control, Vienna, pp. 33–44 (2008)
5. Vafaeesefat, A.: Optimum creep feed grinding process conditions for Rene 80 supper alloy using neural network. Int. J. Precis. Eng. Manuf. **10**, 5–11 (2009)
6. Sedighi, M., Afshari, D.: Creep feed grinding optimization by an integrated GA-NN system. J. Intell. Manuf. **21**, 657–663 (2010)
7. Yang, Q., Jin, J.: Study on machining prediction in plane grinding based on artificial neural network. In: Proceedings of International Conference on Intelligent Systems and Knowledge Engineering (ISKE), Hangzhou, China, November 15–16, 2010
8. Li, G., Liu, J.: On-line prediction of surface roughness in cylindrical traverse grinding based on BP+GA algorithm. In: Proceedings of Second International Conference on Mechanic Automation and Control Engineering (MACE), Hohhot, China, pp. 1456–1459, July 15–17, 2011
9. Nandi, A.K., Pratihar, D.K.: Design of a genetic-fuzzy system to predict surface finish and power requirement in grinding. Fuzzy Sets Syst. **148**, 487–504 (2004)
10. Ticknor, J.L.: A Bayesian regularized artificial neural network for stock market forecasting. Experts Systems with Applications **40**(14), 5501–5506 (2013)
11. Claveria, O., Torra, S.: Forecasting tourism demand in Catalonia: Neural networks vs. time series models. Economic Modelling **36**, 220–228 (2013)

12. Wu, C.L., Chau, K.W.: Prediction of rainfall time series using modular soft computing methods. Engineering Applications of Artificial Intelligence **26**(3), 997–1007 (2013)
13. Godarzi, A.A., Amiri, R.M., Talaei, A., Jamasb, T.: Predicting oil price movements: A dynamic Artificial Neural Network approach. Energy Policy **68**, 371–382 (2014)
14. Pisoni, E., Farina, M., Carnevale, C., Piroddi, L.: Forecasting peak air pollution levels using NARX models. Engineering Applications of Artificial Intelligence **22**(4–5), 593–602 (2009)
15. Beale, M.H., Hagan, M.T., Demuth, H.B.: Neural Network Toolbox™ User's Guide. The MathWorks Inc., Natick (2012)
16. Arriandiaga, A., Portillo, E., Sánchez, J.A., Cabanes, I., Pombo, I.: Virtual Sensors for On-line Wheel Wear and Part Roughness Measurement in the Grinding Process. Sensors **14**, 8756–8778 (2014)

Reliability Analysis of Post-Tensioned Bridge Using Artificial Neural Network-Based Surrogate Model

David Lehký$^{(\boxtimes)}$ and Martina Šomodíková

Brno University of Technology, Brno, Czech Republic
{lehky.d,somodikova.m}@fce.vutbr.cz

Abstract. The reliability analysis of complex structural systems requires utilization of approximation methods for calculation of reliability measures with the view of reduction of computational efforts to an acceptable level. The aim is to replace the original limit state function by an approximation, the so-called response surface, whose function values can be computed more easily. In the paper, an artificial neural network based response surface method in the combination with the small-sample simulation technique is introduced. An artificial neural network is used as a surrogate model for approximation of original limit state function. Efficiency is emphasized by utilization of the stratified simulation for the selection of neural network training set elements. The proposed method is employed for reliability assessment of post-tensioned composite bridge. Response surface obtained is independent of the type of distribution or correlations among the basic variables.

Keywords: Artificial neural network · Latin hypercube sampling · Response surface method · Reliability · Failure probability · Load-bearing capacity

1 Introduction

Structural reliability and life-time assessment is the key step when deciding about maintenance or rehabilitation of structures. Reliability assessment requires the estimation of probabilities of failure, which, in general, are of rather small magnitude:

$$p_f = P\left(Z \leq 0\right) \tag{1}$$

where $Z = g(\mathbf{X})$ is a function of basic random variables $\mathbf{X} = X_1, X_2, ..., X_N$ called safety margin. The failure probability is calculated as a probabilistic integral:

$$p_f = \int_{D_f} f_{\mathbf{X}}\left(\mathbf{X}\right) \, d\mathbf{X} \tag{2}$$

where the domain of integration of the joint probability distribution function (PDF) $f_{\mathbf{X}}(\mathbf{X})$ is limited to the failure domain D_f where $g(\mathbf{X}) \leq 0$. The function $g(\mathbf{X})$, a computational model, is a function of random vector $\mathbf{X}$ (and also of other, deterministic quantities). Random vector $\mathbf{X}$ follows a joint PDF $f_{\mathbf{X}}(\mathbf{X})$ and, in general, its marginal

© Springer International Publishing Switzerland 2015
L. Iliadis and C. Jayne (Eds.): EANN 2015, CCIS 517, pp. 35–44, 2015.
DOI: 10.1007/978-3-319-23983-5_4

variables can be statistically correlated. The explicit calculation of integral in Equation (2) is generally impossible. Therefore a large number of efficient stochastic analysis methods have been developed during the last decades.

Structural failure is most typically assessed by means of nonlinear, possibly time-variant analyses of complex structural models. In practical cases, the structural responses are being calculated by a numerical procedure such as finite element (FE) analysis. This brings another level of complexity to reliability analysis because the limit state function (LSF) $g(\mathbf{X})$ is not available as an explicit, closed-form function of the input variables. To perform reliability analysis of problems with implicit LSF Monte Carlo simulation – being the most versatile solution technique available – can be used for. Unfortunately, it is quite often not feasible in cases of complex structural models since Monte Carlo method is basically numerical experiment carried out randomly, so it requires the full analysis of the system for each generated set of loading and system parameters, respectively. This means that structural analysis based on the FE method need to be performed several hundred times resulting in large computational efforts. To defeat this problem, numerous variance reduction techniques have been proposed, for example, importance sampling [1], directional simulation [2], conditional expectation [3]. Even with these improvements, the calculation is still quite time-consuming.

Another possibility how to reduce the number of the required evaluations of the LSF is a utilization of stratified Latin Hypercube Sampling (LHS) method which is capable to cover the space of random variables using a relatively small number of samples [4]. LHS method is very efficient primarily in the estimation of statistical moments, nevertheless, a problem can occur when failure probability need to be evaluated, i.e. calculations are located close to "tail" of the PDF of the LSF. Therefore LHS method is very often combined with evaluation of Cornell's safety index β where PDF of safety margin is approximated by normal distribution, see e.g. [5].

As a suitable solution for analysis of reliability level, approximation methods can be used instead of simulation ones. In the case of first-order reliability method (FORM) or second-order reliability method (SORM), the LSF is approximated by the linear or quadratic function at the design point. Another possibility and more accurate solution is offered by a utilization of approximation methods known as response surface methods [6] where the LSF is approximated using suitable function mostly of polynomial type [7]. The failure probability calculation is then performed by utilization of classical simulation methods when the approximated function replaces the original one. The evaluation of surrogate function is significantly faster compared to original one.

In the present paper, an artificial neural network based response surface method in combination with small-sample simulation technique LHS is utilized. The artificial neural network as a powerful parallel computational system is used here as a surrogate model for approximation of the original LSF. Thanks to its ability to generalize it would be more efficient to fit the LSF even with a small number of simulations compared to the polynomial response surface. The efficiency is emphasized by utilization of stratified simulation for selection of neural network training set elements and by employment of relevant optimization algorithms for network training.

2   Response Surface Approximation

2.1   Small-Sample Simulation Method

Latin hypercube sampling belongs to the category of advanced sampling methods, which results in very good estimation of statistical moments, such as mean value and standard deviation of response using relatively small number of simulations. More accurately, LHS is considered to be one of the variance reduction techniques, because it yields lower variance of estimates of statistical moments compared to crude Monte Carlo sampling at the same sample size, see e.g. [8]. This is the reason the technique became very attractive when dealing with computationally intensive problems like e.g. complex FEM simulations.

The basic feature of LHS is that the range of each random variable is divided into N_{sim} intervals of equal probability $1/N_{sim}$, where N_{sim} is a number of simulations. The representative values (median, mean or random value) of each interval are then used in the simulation process. The corresponding samples are then calculated with respect to the PDF based on its inverse transformation. As mentioned in introduction LHS method is not suited for direct evaluation of failure probability. But thanks to its stratification principle it can be efficiently used to cover space of random variables by relatively small number of simulations and therefore we use it for stochastic preparation of training set for artificial neural network training, see section 2.3.

2.2   Polynomial Response Surface Method

The general principle of response surface method (RSM) is to replace the original LSF with suitable function mostly of polynomial type. The reliability level is then calculated by utilization of classical simulation techniques. The order of the polynomial selected for fitting to the discrete point outcomes affects the number of unknown parameters of such function, which need to be estimated, and consequently the number of the required evaluations of the original LSF. In general, a second order polynomial is most often employed:

$$\tilde{g}(\mathbf{X}) = a + \sum_{i=1}^{n} b_i x_i + \sum_{i=1}^{n} \sum_{j=1}^{n} c_{ij} x_i x_j , \qquad (3)$$

where x_i, $i = 1, \ldots, n$ are the input basic variables and parameters a, b_i, c_i are the unknown regression coefficients of the approximation function. The regression coefficients can be obtained by conducting a series of numerical "experiments", that is, a series of structural FEM analyses with input variables selected according to some "experimental design" which takes into account that the main interest is in the region of the maximum likelihood within the failure domain that is the design point [5]. Although the RSM provides sufficiently accurate results in reliability applications, there are also some difficulties. The number of experiments (calculations of actual LSF, i.e. the number of FEM analyses) significantly increases with the increase in number of variables, which is often the case in structural analysis. In order to obtain the constants in Equation (3), $1 + n + n(n + 1)/2$ experiments need to be conducted.

An iterative response surface approach was presented by Bucher & Bourgund [9]. A polynomial function is suggested in the form:

$$\tilde{g}(\mathbf{X}) = a + \sum_{i=1}^{n} b_i x_i + \sum_{i=1}^{n} c_i x_i^2 , \tag{4}$$

with the same notation as in Equation (3). It is seen that the Equation (4) doesn't contain mixed terms $x_i x_j$. Since the number of regression parameters is rather low, i.e. $2n + 1$, only few numerical experiments are required. The suggested procedure is choosing the mean values of the intervening random variables as the anchor point and calculate the actual $g(\mathbf{X})$ at the mean and, in turn, for each variable, at the mean plus three standard deviations and the mean minus three standard deviations.

2.3   Artificial Neural Network Based Response Surface Method

As an alternative to polynomial response surface method a small-sample artificial neural network based response surface method (ANN-RSM) is proposed. In this approach, the input and output parameters are related by means of relatively simple yet flexible functions, such as linear, step, or sigmoid functions which are combined by adjustable weights. The main feature of this approach lies in the possibility of adapting the input–output relations very efficiently. This is done by training the network. Since the feed-forward type network is employed a "supervised" learning is used to adjust network parameters, i.e. a set of pairs $(\mathbf{X}_i, \mathbf{g}_i)$, is introduced to the network with the aim to find a function $\tilde{g}_{ANN}(\mathbf{X})$ in the allowed class of functions that matches the examples.

A novelty of the proposed approach rests in using the small-sample simulation technique LHS to get the examples for the efficient preparation of training set elements. In case of time demanding FEM calculations, the small-sample simulation techniques based on stratified sampling of Monte Carlo type represent a rational compromise between feasibility and accuracy.

Proposed small-sample ANN-RSM is based on the general methodology of inverse analysis [10, 11]. The process of ANN-RSM application is divided into two main phases (see Figure 1): *approximation phase*, where the original LSF is approximated by suitable ANN, and, *reliability calculation phase*, where the approximated LSF is used instead of the original one for the calculation of reliability measures (reliability index, failure probability). The whole procedure is schematically illustrated using a simple flow chart in Figure 1 and can be itemized as follows:

1. Random realizations of input basic variables (possibly correlated) of analyzed stochastic problem are sampled using the stratified LHS simulation method. The number of simulations depends on the complexity and requested accuracy of analyzed problem. In general tens simulations are used. In case of poor convergence in the process of ANN training or low accuracy, additional samples can be added using Hierarchical Subset Latin Hypercube Sampling (HSLHS) strategy.

2. Evaluations of original LSF are performed repeatedly for an individual vector of realizations of input random variables and corresponding LSF values are calculated. For the calculation of original LSF the FEM analysis can be used.
3. Set of LSF values together with a corresponding set of input random variables serve as a training set for ANN training using appropriate optimization technique (back propagation methods, evolution methods, or stochastic methods).
4. Trained ANN is used as a surrogate LSF for consequent reliability analysis where reliability measures (failure probability or reliability index) are calculated by utilization of classical simulation or approximation methods. In case of simulation methods millions of simulations can be used thanks to very fast evaluation of surrogate LSF compared to the original one.

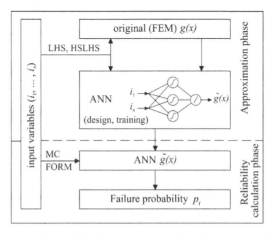

Fig. 1. Flow chart of the artificial neural network based response surface method.

3   Application to Post-Tensioned Composite Bridge

Proposed ANN-RSM was used for reliability and load-bearing capacity assessment of a single span post-tensioned composite bridge (Figure 2). Nonlinear FEM analysis was employed to provide more accurate and realistic verification. By reason of time-consuming repetitious nonlinear calculations, only serviceability limit states were investigated. In connection with a type of the bridge, the limit state of decompression, the limit state of crack initiation and the limit state of crack width were verified.

Based on the information from diagnostic survey, a numerical model (Figure 3) was created in ATENA software [12]. The following load cases were modelled: dead load of the structure, longitudinal pre-stressing, secondary dead load and traffic load according to actual loading scheme for assessment of normal load bearing capacity, V_n, defined in Czech technical standard [13]. Stochastic parameters of random input variables were defined using software FReET [14] according to recommendations of Joint Committee on Structural Safety (JCSS) [15] and Technical specifications [16] and these were updated based on the material parameters testing according to diagnostic survey. The details on the complex stochastic model can be found in [17].

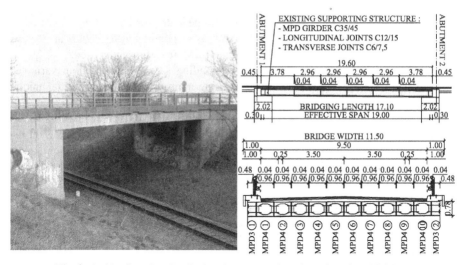

Fig. 2. A side view, longitudinal and transversal section of analyzed bridge

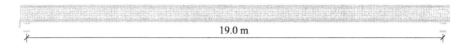

Fig. 3. A numerical model of analyzed bridge – ATENA 2D software

At first, stochastic model with a total number of 23 random input variables, corresponding to the material properties and loads, was used. Mean values and coefficients of variation (CoV) of material properties of concrete in segments and transverse joints and parameters of shear reinforcement were defined according to results obtained from diagnostic survey. Values of pre-stressing force in tendons were defined with respect to short-term as well as long-term losses of initial tension according to ČSN EN 1992-1-1 [18]. The weight of road layers were also randomized using the value of secondary dead load, g_1. Finally, traffic load for determination of normal load bearing capacity was defined using deterministic value of load according to valid loading schemes introduced in current standards. Statistical correlation between material parameters of concrete of segments and transverse joints, pre-stressing tendons and concrete reinforcement was also considered and imposed using a simulated annealing approach 19.

A thousand of simulations were generated using the LHS simulation technique and repeated numerical FEM analyses were performed to model the structural response. Afterwards, the set of structural responses was statistically analyzed. Normal load bearing capacity was assessed with respect to the target reliability level, corresponding to an investigated limit state and defined by target reliability index, β_t. These results, summarized in Table 1, served as reference values for the comparison of reliability measures obtained by various methods.

Table 1. Reference results obtained by the bridge FEM analysis

Limit state	β_t [-]	p_f [-]	V_n [t]
Decompression	0* (0.0035)	5×10^{-1}	23.4
Cracks initiation	0* (0.0037)	5×10^{-1}	27.4
Crack width	1.5** (1.51)	6.68×10^{-2}	49.1

* Reversible process according to [16]
** Irreversible process – medium consequence of damage

Sensitivity analysis of input variables was also performed to capture the most significant variables of the model. Only few of them proved to be significantly related to the structural response, and finally the number of 6 random input variables was used for subsequent calculations. Definition of random input variables is summarized in Table 2.

Table 2. Basic random variables of the analyzed bridge

Variable	Distribution	Mean	CoV
Concrete of joints – tensile strength f_t [MPa]	Weibull min.	1.913	0.35
Concrete of joints – compressive strength f_c [MPa]	Triangular	19.13	0.55
Concrete of joints – fracture energy G_f [N/m]	Weibull min.	47.82	0.25
Pre-stressing force P_1 [MN]	Normal	14.20	0.09
Pre-stressing force P_2 [MN]	Normal	10.05	0.09
Secondary dead load g_1 [kN/m]	Normal	65.55	0.05

ANN-RSM was used to approximate original implicit LFS and to calculate failure probability and corresponding reliability index. Results were compared with those obtained by polynomial RSM (POLY-RSM) according to Equation (3), including only linear terms, LHS method in combination with Cornell's reliability index and FORM. Values of reliability index β assessed using individual methods and different numbers of simulations (N_{sim} = 10, 20, 30, 50, 75 and 100) are compared with the reference value β_t according to Table 1 (exact values of β_t for the corresponding values of V_n are listed in brackets). In the case of RSM methods 1 million simulations of Monte Carlo method in combination with evaluation of approximated functions (polynomial or ANN) were performed.

Results are depicted for individual limit states in Figures 4–6. Since the limit states of decompression and cracks initiation are linear functions, linear approximations of the original function lead to good agreement with reference result (see FORM in Figs. 4–5). Also in the case of ANN-RSM, very simple network was used to approximate the limit state function. The network with only input layer with six inputs corresponding to six random variables and output layer with one output corresponding to value of $g(\mathbf{X})$ with linear transfer function was used for all numbers of simulations. Training of the ANN was carried out using gradient descent method.

The limit state of crack width was also investigated to represent the nonlinear response of the structure. From Figure 6 it is obvious that the linear response surface method leads to inaccurate results outlying the safe side for the most numbers of simulations. By contrast ANN-RSM, LHS and FORM methods reached almost identical and sufficient results in comparison with the reference value. In case of ANN-RSM the structure of the ANN was as follows: six inputs corresponding to six input random variables, an output layer having one output neuron with a linear transfer function corresponding to value of $g(\mathbf{X})$. ANN has also 1 hidden layer with 4 (for the cases with 30 and 100 simulations) or 5 neurons (for the case with 75 simulations) or 2 hidden layers with 8 + 4, 5 + 2 and 8 + 2 neurons (for 10, 20 and 50 simulations) with a nonlinear transfer function (hyperbolic tangent). Training of the ANN was carried out again using gradient descent method.

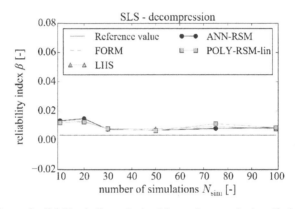

Fig. 4. Comparison of reliability indices obtained by various methods – limit state of decompression

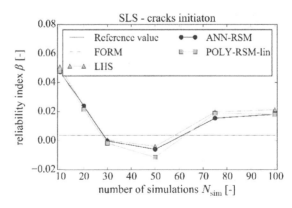

Fig. 5. Comparison of reliability indices obtained by various methods – limit state of cracks initiation

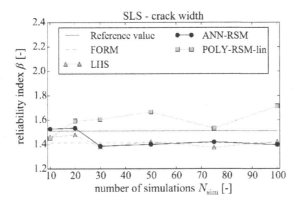

Fig. 6. Comparison of reliability indices obtained by various methods – limit state of crack width

To summarize, it can be said that for all of investigated limit states a utilization of ANN-RSM leads to sufficiently accurate results for a quite small number of simulations and can be used as a powerful tool to approximate the original limit state function also in cases of time consuming FEM analyses.

4   Conclusions

Artificial neural network based response surface method in combination with small-sample simulation technique has been employed for reliability and load-bearing capacity assessment of the single span post-tensioned composite bridge within the framework of fully probabilistic analysis. Results were compared with those obtained by other reliability methods and show that proposed method can be used efficiently for reliability analysis of complex structural systems.

Let's mention that response surface does not represent the physical model exactly, but if the structure of ANN as approximation of the physical model and numerical "experiments" used as ANN training set elements are selected carefully the results of the final reliability analysis are found to be very close to exact results.

When performing reliability and lifetime assessment of bridges and other structural systems utilization of simulation methods is often not reasonable because of extremely high computational demands. Then there is no other alternative than utilization of ANN-RSM and other approximation methods.

Acknowledgment. The authors give thanks for the support of the project FIRBO No. 15-07730S of the Grant Agency of the Czech Republic (GACR), and the project of the Specific University Research at Brno University of Technology No. FAST-J-15-2712.

References

1. Bucher, C.: Adaptive sampling-an iterative fast Monte Carlo procedure. Structural Safety **5**(2), 119–126 (1988)
2. Bjerager, P.: Probability integration by directional simulation. Journal of Engineering Mechanics ASCE **114**(8), 285–302 (1988)
3. Ayyub, B., Chia, C.: Generalised conditional expectation for structural reliability assessment. Structural Safety **11**, 131–146 (1992)
4. McKay, M.D., Conover, W.J., Beckman, R.J.: A comparison of three methods for selecting values of input variables in the analysis of output from a computer code. Technomerics **21**, 239–245 (1979)
5. Melchers, E.M.: Structural Reliability Analysis and Prediction. John Wiley & Sons Ltd., Chichester (1999)
6. Myers, R.H.: Response Surface Methodology. Allyn and Bacon, New York (1971)
7. Bucher, C.G.: Computational Analysis of Randomness in Structural Mechanics. CRC Press/Balkema, Leiden (2009)
8. Koehler, J.R., Owen, A.B.: Computer experiments. In: Ghosh, S., Rao, C.R. (eds.) Handbook of Statistics 13, pp. 261–308. Elsevier Science, New York (1996)
9. Bucher, C.G., Bourgund, U.: A fast and efficient response surface approach for structural reliability problems. Structural Safety **7**, 57–66 (1990)
10. Novák, D., Lehký, D.: ANN Inverse Analysis Based on Stochastic Small-Sample Training Set Simulation. Engineering Application of Artificial Intelligence **19**, 731–740 (2006)
11. Lehký, D., Novák, D.: Solving inverse structural reliability problem using artificial neural networks and small-sample simulation. Advances in Structural Engineering **15**, 1911–1920 (2012)
12. Červenka, V., Jendele, L., Červenka, J.: ATENA Program Documentation – Part 1: Theory. Cervenka Consulting, Prague (2007)
13. Czech technical standard – ČSN 73 6222: Load bearing capacity of road bridges. Czech Office for Standards, Metrology and Testing, Prague (in Czech) (2009)
14. Novák, D., Vořechovský, M., Teplý, B.: FReET: Software for the statistical and reliability analysis of engineering problems and FReET-D: Degradation module. Advances in Engineering Software **72**, 179–192 (2013)
15. Joint Committee on Structural Safety: Probabilistic Model Code. http://www.jcss.byg.dtu.dk/Publications/ (last updated December 6, 2013)
16. Technical specifications – TP 224: Ověřování existujících betonových mostů pozemních komunikací (Verification of existing concrete road bridges). Ministry of Transport of Czech Republic, Department of Road Infrastructure, Prague (in Czech) (2010)
17. Šomodíková, M., Doležel, J., Lehký, D.: Probabilistic load bearing capacity assessment of post-tensioned composite bridge. In: Novák, D., Vořechovský, M. (eds.) Proceedings of the 11th International Probabilistic Workshop, Brno, pp. 451–460 (2013)
18. Czech technical standard – ČSN EN 1992-1-1. Eurocode 2: Design of concrete structures – Part 1-1: General rules and rules for buildings. Czech Standardization Institute, Prague (in Czech) (2006)
19. Vořechovský, M., Novák, D.: Correlation control in small-sample Monte Carlo type simulations I: A simulated annealing approach. Probabilistic Engineering Mechanics **24**(3), 452–462 (2009)

Bioinformatics

A Grid-Enabled Modular Framework for Efficient Sequence Analysis Workflows

Olga T. Vrousgou[1], Fotis E. Psomopoulos[1,2(✉)], and Pericles A. Mitkas[1]

[1] Aristotle University of Thessaloniki, Thessaloniki, Greece
{olgav,mitkas}@auth.gr, fpsom@issel.ee.auth.gr
[2] Center for Research and Technology Hellas, Thessaloniki, Greece

Abstract. In the era of Big Data in Life Sciences, efficient processing and analysis of vast amounts of sequence data is becoming an ever daunting challenge. Among such analyses, sequence alignment is one of the most commonly used procedures, as it provides useful insights on the functionality and relationship of the involved entities. Sequence alignment is one of the most common computational bottlenecks in several bioinformatics workflows. We have designed and implemented a time-efficient distributed modular application for sequence alignment, phylogenetic profiling and clustering of protein sequences, by utilizing the European Grid Infrastructure. The optimal utilization of the Grid with regards to the respective modules, allowed us to achieve significant speedups to the order of 1400%.

Keywords: Bioinformatics · Grid computing · Comparative genomics · Sequence alignment · Protein clustering · Phylogenetic profiles · Parallel processing · Modular software engineering

1 Introduction

When it comes to tools for analyzing and interpreting bio-data, the research community has always been one step behind the actual acquisition and production methods. Starting from the first amino acid sequences and moving on to whole genome, epigenome, transcriptome analyses and genome wide association studies (GWASs) on the gene level, and to proteome, reactome and metabolome on the enzymatic and protein level, the same pattern holds for the next generation of biodata. Although the amount of data currently available is considered vast, the existing methods and widely used techniques can only hint at the knowledge that can be potentially extracted and consequently applied for addressing a plethora of key issues, ranging from personalized healthcare and drug design to sustainable agriculture, food production and nutrition, and environmental protection.

Researchers in genomics, medicine and other life sciences are using big data to tackle big issues, but big data requires more networking and computing power. "Big data" is one of today's hottest concepts, but it can be misleading. The name itself suggests mountains of data, but that's just the start. Overall, big data consists of three v's: volume of data, velocity of processing the data, and variability of data sources.

© Springer International Publishing Switzerland 2015
L. Iliadis and C. Jayne (Eds.): EANN 2015, CCIS 517, pp. 47–56, 2015.
DOI: 10.1007/978-3-319-23983-5_5

These are the key features of information that require big-data tools. In order to address these features, current approaches in Life Science research favor the use of established workflows which have been proven to facilitate the first steps in data analysis.

One of the major computational bottlenecks in the vast majority of these approaches is the comparative phase of the involved workflows, which includes the production of similarity scores and therefore the construction of gene (or protein) families. There have been several attempts to address this issue in recent literature, ranging from highly specialized algorithms and tools [1][2][3], to Cloud-enabled frameworks [4][5] and platforms [6] (such as MapReduce). However, although such efforts clearly provide a computational edge against their vanilla counterparts, they often require the expertise to setup and fully employ a sophisticated software system, an expertise that most Life Science researchers lack. Moreover, specialized systems tend to be updated at a much lower pace, if at all, as compared to the more widely used vanilla approaches. In order to address these two issues, we have developed a user-friendly Grid-enabled solution for comparative genomics that employs the vanilla version of the necessary tools, while harnessing and fully utilizing the potential of the computational Grid. As such, we achieve significant speedup in the process while maintaining full compatibility with future updates of the involved tools.

2   Background

The proposed framework spans across two distinct research areas; Grid Computing and Bioinformatics. In this Section, we will describe briefly the different platform and tools employed, as well as their impact in the overall structure of the framework.

2.1   Grid Computing

Grid Computing is an established method of high performance computing that is mostly utilized by embarrassingly parallel processes. As an innovative method to perform distributed computational and storage tasks over geographically distributed resources, Grid computing as an infrastructure exists for over a decade now. The European Grid Infrastructure (EGI) is the result of pioneering work that has, over the last decade, built a collaborative production infrastructure of uniform services through the federation of national resource providers that support multi-disciplinary science across Europe and around the world. An ecosystem of national and European funding agencies, research communities and technology providers, over 350 resource centers and other functions have now emerged to serve over 21,000 researchers in their intensive data analysis across over 15 research disciplines, carried out by over 1.4 million computing jobs a day. In order to showcase our proposed framework, we employed the HellasGrid part of the EGI infrastructure, which is the biggest infrastructure for Grid computing in the South-Eastern European area, providing High Performance Computing and High Throughput Computing services to Greek educational and research centers.

2.2 Bioinformatics

Although Bioinformatics is a very general term, we have focused particularly on the common steps which are computationally expensive and further required in the vast majority of bioinformatics research approaches. These steps comprise the alignment of sequences, the identification of families as well as the construction of phylogenetic profiles.

Sequence Alignment (BLAST)

Sequence alignment of different types of sequences (DNA, RNA and protein) is traditionally performed using the BLAST (Basic Local Alignment Search Tool) algorithm. It is provided by the NCBI Toolkit and is the predominant algorithm for sequence alignment [7] where each alignment is characterized by a number of parameters. In our framework, but without any loss of generality as the parameter selection is user-dependent, we utilized two of them, i.e. the identity and the e-value. Identity refers to the extent to which two (nucleotide or amino acid) sequences have the same residues at the same positions in an alignment, and is often expressed as a percentage. E-value (or expectation value or expect value) represents the number of different alignments with scores equivalent to or better than S that is expected to occur in a database search by chance. The lower the E-value, the more significant the score and the alignment. Finally, despite the popularity of the BLAST algorithm, running this process is still extremely computationally demanding. For example, a simple sequence alignment between ~0.5 million protein sequences, can take up to a week on a single high-end PC. Even when employing high-performance infrastructures BLAST requires significant time as well as the expertise to both run and maintain an HPC BLAST variant.

Gene/Protein Families

Based on sequence alignment data, a common practice is to construct the corresponding families (either at the gene or the protein level). A gene family is a set of several similar genes, formed by duplication of a single original gene, and generally with similar biochemical functions. A protein family is a group of proteins that share a common evolutionary origin, reflected by their related functions and similarities in sequence or structure. The use of the family construct has several advantages; from the functional annotation of novel sequences, to insights on the evolutionary histories of gene groups. As such, algorithms that allow for a fast and efficient identification of families are widely used. MCL (Markov Cluster Algorithm) is one of the most well known, and is a fast and scalable unsupervised clustering algorithm for graphs based on simulation of stochastic flow graphs [8]. However, although MCL is a fast and efficient algorithm, it also requires a significant amount of resources and especially RAM resources. Therefore, an approach that allows for the better scaling of the application with larger datasets, is an essential step towards the Big Data analytics in Life Sciences.

Phylogenetic Profiles

Phylogenetic profiling is a bioinformatics technique in which the joint presence or joint absence of two traits across large numbers of species is used to infer a meaningful biological connection, such as involvement of two different proteins in the same biological pathway. By definition, a phylogenetic profile of a single sequence of a gene/protein family is a vector that contains the presence or absence of the particular entity across a number of known genomes that participate in the study. The biological premise underlying the phylogenetic profile construct is that proteins that exhibit the same or similar profile strongly tend to be functionally linked [9]. As such, profiles are an essential step in large scale analyses, as they can provide in an elegant manner, insights on the organization and interconnection of novel entities based solely on their sequence. Moreover, beyond the traditional binary phylogenetic profiles (denoting presence or absence), there are several different variants of in literature [10], such as extended profiles (i.e. number of homologues) and fuzzy (level of presence) among others. However, it is important to note that the construction of phylogenetic profiles is a computationally expensive process. Based on the sequence alignment data, each profile requires the comparison and identification of all homologues across the different number of genomes in the study. Therefore, a scalable approach that caters to this specific need will be a measurable asset for time efficiency in the respective bioinformatics workflows.

3   Materials and Methods

We have designed and implemented a time-efficient distributed modular application for sequence alignment, phylogenetic profiling and clustering of protein sequences, by utilizing the European Grid Infrastructure. The guiding requirements for our approach are (a) maximizing the efficiency of a given workflow using the computational resources provided by EGI, (b) providing an automated approach and therefore a more user-friendly interface for researchers with no technical experience and (c) using the established (vanilla) applications and tools in order to maintain backwards compatibility and maintenance, which is a usual issue in most of the custom approaches.

A second design aspect of the framework is the modular paradigm. In the era of Big Data, there is a significant trend towards analysis pipelines; an arbitrary combination of tools and applications connected through common interfaces towards a user-specific goal [11]. We have adopted the same approach, by implementing the overall system as a set of different components that communicate at the data level.

The application is comprised of three main components; (a) BLAST alignment, (b) construction of phylogenetic profiles based on the produced alignment scores and (c) clustering of entities using the MCL algorithm. These modules have been selected as they represent a common aspect of a vast majority of bioinformatics workflows. The modules can be combined independently, and ultimately provide 4 different modes of operation. These modes loosely correspond to different goals by the end user, that range from the identification of gene families and the construction of phylogenetic profiles, to a pangenome analysis of the participating genomes [12].

Based on the selected mode of operation, our proposed framework proceeds with the distribution of both processes and data across the provided resources. The distribution is performed automatically, based on the selected mode as well as the data under study. Moreover, the framework continuously monitors and evaluates the execution of the spawned processes at all steps of the workflow. It is important to note that a Grid infrastructure has an inherently high number of jobs that fail to execute properly and therefore require resubmission, due to a number of unexpected issues such as extremely long queue times. The proposed framework through constant monitoring of all processes, can evaluate which jobs require resubmission thus achieving higher efficiency in the overall process. Finally, after successful completion of all intermediate processes, the framework gathers the respective output for presentation to the end user. A brief outline of this approach can be seen in Fig. 1.

3.1   Modes of Operation

The 4 modes of operation are presented below:

1. MCL clusters of the protein query and database sequences where the clustering criteria is the BLAST output (identity or e-value), based on the preference of the user. This mode is mostly utilized when trying to evaluate the potential function of novel sequences (query) by assigning them to protein/gene families produced from a given set (database).
2. Phylogenetic profiles of each query sequence, where the genomes into consideration are the ones whose proteins form the database.
3. MCL clusters of the protein query sequences and database genomes, and phylogenetic profiles.
4. This mode is essentially a combination of the output produced in modes 1 and 3. In practice, this is the most common approach, as it combines the functional gene/protein families with evolutionary insights provided by the produced phylogenetic profiles.

It is important to note that, based on the mode of operation, data distribution across the computational resources is handled in a different way. In the first mode, the query file is distributed across the participating nodes, whereas the database file is copied multiple times. In modes 2 through 4, the situation is reversed; the data that is distributed is that of the database file, whereas the query file is copied multiple times. Although this seems an arbitrary choice in data distribution, it has been validated through rigorous experimentation. However, an automated optimization of the internal parameters based on the characteristics of the input files, is still an open issue and will be addressed in future work.

Beyond the aforementioned mode, there exists also a fifth mode that provides the same output as the fourth one, with the only difference being that the same file is used both as a database and a query. This is the case of an all-vs-all sequence comparison, widely used when performing a pangenome analysis. A key requirement in such case is to identify the number of protein families evident in the dataset, as well as their distribution across the different participating genomes. This sort of analysis can provide significant insights into the organization of the pangenome, as well as the functional relationships of entities across the different members of the pangenome.

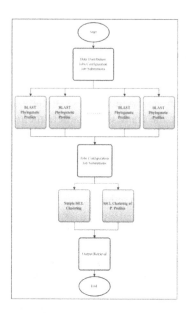

Fig. 1. General flow chart showcasing the connection of the three components.

The proposed framework in this mode, distributed the involved data by splitting the query file instead of the database file, as opposed to the process followed in the similar forth mode. This is due to the fact that in the case of all-vs-all, comparisons are executed faster when splitting the query file. The database file can only be distributed in so many pieces as the genomes participating in its creation, in order for the phylogenetic profiles to be created faster. In the case of all-vs-all where the database file is also the query file, the number of splits that can be done to the query file are more than that of the database file, since the creation of phylogenetic profiles can be done easily, regardless of how the query file is split. The speedup achieved when distributing the data in this fashion is bigger.

3.2   Input

The input comprises of the following files; (1) two files containing the query protein sequences and the database protein sequences to be aligned, in FASTA format, which is a text-based format for representing nucleotide or peptide sequences, (2) a text file with the genomes whose protein sequences form the database file, and (3) a configuration file for the application to run on a specific mode. All the input files are identical with the files required in most bioinformatics workflows and therefore we comply with the established requirements. The only exception is the custom configuration file, however, it is designed to be intuitive aiming for the non-expert end user.

3.3   Program Flow

The application is automated, and needs only the configuration file to operate. Since BLAST execution and the creation of phylogenetic profiles are the most time-consuming modules, and as they both are embarrassingly parallel in nature, we identify this as the critical execution path that we need to parallelize. By having parallel jobs executing in the participating Grid Infrastructure Nodes on a fraction of the input data, we achieve parallelism without the need of implementing common parallelization techniques such as MPI or threads.

In case of the BLAST algorithm and the construction of a phylogenetic profile for each query sequence, the application submits a number of jobs to the Grid. Each job is assigned a fraction of the input data that is assigned according to the mode of operation the application is running. Sequence alignment is then executed, and the construction of phylogenetic profiles follows. Therefore, each job processes a piece of the input data, producing a piece of the output data. On each step, the data used and produced is uploaded to the Grid's storage elements, making it accessible to all executed jobs and to the end user. Jobs are tracked to assure their successful execution at all times, preventing any information loss.

3.4   Output

The output may consist of (a) MCL clusters that were shaped using BLAST output as clustering criteria, (b) MCL clusters that were shaped using phylogenetic profiles as clustering criteria, and (c) phylogenetic profiles for each query protein sequence.

The output is formatted in such a way that makes further processing easier. As mentioned earlier, the proposed framework addresses the initial, computationally expensive steps in the majority of bioinformatics workflows. Therefore it is expected that the produced results will be further analyzed using a wide variety of additional tools. To better facilitate interaction and further analysis, a series of parsing scripts are provided aiming to filter the output with specific areas of interest. In any case, the complete output of a given workflow is also available for the user to download in raw format.

4   Results

We have implemented the proposed framework[1] using a number of scripts suitable for a Unix environment of a Grid Infrastructure. We have evaluated the application through several different scenarios, ranging from targeted investigations of enzymes participating in selected pathways against a custom database to produce functional groups, to large scale comparisons. As a database, we used all genomes available for the bacillus genus from the Ensembl database (data from bacteria.ensembl.org, R16). Specifically, the produced database comprises 78 different bacillus genomes, with a total of 418,590 protein sequences. As a query file we used three different inputs, and as such accounted three different test case scenarios.

---

[1] Current version of source code: https://github.com/BioDAG/BPM

Test Case 1 - Leucine: 24 protein sequences of the Escherichia coli K-12 organism that participate in the leucine biosynthesis pathway.

Test Case 2 - Bacillus5: 32,747 protein sequences from the top five largest genomes used in the bacillus database.

Test Case 3 - Bacillus: In this test case, the FASTA file with all available sequences was used for query and for a database files.

The results of these test cases are shown in Table 1. For each module, we present the average execution time, along with the average waiting time (time in-queue) for each job in the Grid's job queues. It is important to note that we aim to explore the impact of the proposed framework with regard to the efficiency in execution times. Other aspects, such as clustering accuracy or performance of the alignment, although very significant in the overall process, are beyond the scope of the current work.

Table 1. Results of the three test cases for each mode of operation (Q: Query, D: Database). The in-queue time is computed as the time from job submission until initiation of execution.

Test Case	Mode Of Operation		BLAST (min)	Phylogenetic Profiles (min)	MCL Clustering (min)	Time in Queue (min)	No. of Jobs	File Distributed
1	1	mean	1.17	-	1	15.99	24	Q
		σ	0.086	-	-	25.19		
	2,3,4	mean	2.75	1	0.03	19.22	78	D
		σ	0.046	0	-	50.53		
2	1	mean	4.58	-	19	79.75	500	Q
		σ	1.13	-	-	145.67		
	2,3,4	mean	65.77	20.22	20.11	67.54	78	D
		σ	9.92	3.59	-	88.95		
3	1,5	mean	44.45	16	107.8	214.17	500	Q
		σ	9.73	1.88	-	163.25		
	2,3,4	mean	756.6	2972.4	188	364	78	D
		σ	118.64	589.88	-	566.21		

The fifth mode of operation, all-vs-all, was further tested to observe the trade-off between the modules execution time versus the time in-queue that the Grid inserts to the total run time. The same data files where submitted, but with a different number of jobs running each time.The net run time of each job significantly decreases when the number of submitted jobs increases, but then in-queue time also increases (see Fig. 2.). When more jobs are submitted, each job has to process a smaller piece of input data. However, this leads to a larger number of jobs to be handled by the respective Grid infrastructure, thus leading to longer waiting times. We can see that it is essential to find the optimal ratio of these two parameters, i.e. number of jobs and estimated in-queue time.

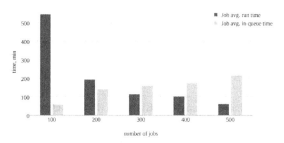

Fig. 2. Average run and in-queue time per job, as a function of the number of jobs submitted

The overall speedup that can be achieved when distributing the query file, increases with the size of the respective query. The speedup of each module is significant, as expected. It is important to highlight the high difference between the net speedup (speedup of average job run time, with zero in-queue time, when comparing it with the sequential alternative) and the actual speedup (including the in-queue time). The importance of the infrastructure used is clear (see Fig. 3.).

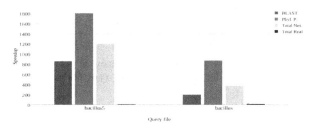

Fig. 3. Speedup achieved for test cases 2 and 3, compared to their sequential run time, when the query file is distributed

5   Conclusions

As the volume of bioinformatic data to be processed increases, utilization of Big Data Techniques and the Grid Infrastructure is a necessity, in order to reap the benefits of parallelization. The proposed framework uses well-known protein sequence comparison tools and combines them in a way that maximizes the benefits of parallelization, where it is possible, and that connects the output of one tool with the input of the other, offering an automated bioinformatic workflow.

The application achieves significant speedup that may be increased when the performance of the underlying infrastructure improves. Even with the delay time that today's Grid introduces, when the right mode of data distribution is used the results are more than satisfactory, speeding up the comparison of half a million sequences by a factor of 14. Protein sequence comparison of large data sets is possible at a reasonable time frame, and with no intervention from the user side. Furthermore, the modular composition of the application provides the means for updating any components that may need updating, and for adding further functionality easily.

The current implementation provides a proof-of-concept approach to the proposed framework. However, there are several outstanding issues that will be addressed in future work, such as a rigorous complexity analysis of each step, the automated optimization of the data distribution process, as well as integration with existing web-interfaces, such as the Galaxy [13] platform.

Acknowledgments. We thank Athanassios Kinstakis (PhD Student at AUTH) for valuable discussions and help with the experimentation process. This work used the European Grid Infrastructure (EGI) through the National Grid Infrastructure NGI_GRNET – HellasGRID.

References

1. Hach, F., et al.: SCALCE: boosting sequence compression algorithms using locally consistent encoding. Bioinformatics. **28**(23), 3051–3057 (2012)
2. Jourdren, L., et al.: Eoulsan: a cloud computing-based framework facilitating high throughput sequencing analyses. Bioinformatics. **28**(11), 1542–1543 (2012)
3. Vouzis, P., et al.: GPU-BLAST: using graphics processors to accelerate protein sequence alignment. Bioinformatics **27**(2), 182–188 (2011)
4. Chung, W.C., et al.: CloudDOE: a user-friendly tool for deploying Hadoop clouds and analyzing high-throughput sequencing data with MapReduce. PLoS One **9**(6), e98146 (2014)
5. Jun, G., et al.: An efficient and scalable analysis framework for variant extraction and refinement from population scale DNA sequence data. Genome Res. 16. pii: gr.176552.114 (2015)
6. Decap, D., et al.: Halvade: scalable sequence analysis with MapReduce. Bioinformatics. 26. pii: btv179 (2015)
7. Lobo, I.: Basic Local Alignment Search Tool (BLAST). Nature Education **1**(1), 215 (2008)
8. Enright, A.J., Van Dongen, S.: C. A. Ouzounis.: An efficient algorithm for large-scale detection of protein families. Nucleic Acids Res. **30**(7), 1575–1584 (2002)
9. Pellegrini, M., et al.: Assigning protein functions by comparative genome analysis: Protein phylogenetic profiles. Proc. Natl. Acad. Sci. USA **96**, 4285–4288 (1999)
10. Psomopoulos, F.E., Mitkas, P.A., Ouzounis, C.A.: Detection of Genomic Idiosyncrasies Using Fuzzy Phylogenetic Profiles. PLoS ONE **8**(1), e52854 (2013)
11. Gómez, J., et al.: BioJS: an open source JavaScript framework for biological data visualization. Bioinformatics **29**(8), 1103–1104 (2013)
12. Psomopoulos, F.E, et al.: The Chlamydiales Pangenome Revisited: Structural Stability and Functional Coherence. Genes **3**(2), 291–319
13. Goecks, J., et al.: Galaxy: a comprehensive approach for supporting accessible, reproducible, and transparent computational research in the life sciences. Genome Biol. **11**(8), R86 (2010)

Application of Elastic Nets Using Phase Transition for Detection of Gene Expression Patterns with Different Carbon Sources

Marcos Lévano$^{(\boxtimes)}$

Escuela Ingeniería Informática, Universidad Católica de Temuco,
Av. Manuel Montt 56, Casilla 15-D, Temuco, Chile
mlevano@inf.uct.cl

Abstract. This work develops data interpretation by elastic nets based on statistical mechanics, for the detection of clusters of data structures in an n-dimensional space. The problem is to find patterns of activity in genes of E. coli, which are stressed by different carbon sources that allow them to activate various expressiveness. The results of the study show that for a string node using distinct temperatures distinct phase transitions are shown. As result of phase transitions, changed centroids of respective groupings are observed. At each node, different behaviors revealed by carbon sources applied to genes are observed.

Keywords: Phase transitions · Genes · Elastic net

1 Introduction

In general there exist many problems in different areas of science, where classification of groups of similar objects are a key issue. As an example we mention the classification of patterns of genes expressions in genetics and vegetation in landscapes in satellite imagery. Clustering methods are one major tool for the analysis of these objects in multivariate space without knowledge of an *a priori* distribution in this method. The objects are mapped on data points in multidimensional space. The general way to solve the problem of clustering is to find a partition of the data points into subgroups which should be as homogeneous as possible. This can be done by defining a cost or energy function, where the data points are distributed over an arbitrary number of clusters, defined by a probability distribution and its centroid. The number of data points in every cluster and the positions of the clusters are found by minimizing the cost function to its lowest minimum, the global minimum. Generally the cost functions have several local minima and a deterministic algorithm tends to get trapped in a local minimum.

A way to solve this difficulty is the so called *deterministic annealing* [1]. In analogy with statistical mechanics, Rose et.al [2] used as the cost function the energy of the data points assigned to the clusters by a square distance function

© Springer International Publishing Switzerland 2015
L. Iliadis and C. Jayne (Eds.): EANN 2015, CCIS 517, pp. 57–66, 2015.
DOI: 10.1007/978-3-319-23983-5_6

with respect to the centroid of the clusters. The system is deterministically optimized sequentially at each temperature, starting from high temperatures and going down. The distribution of the data points in clusters must than maximize the entropy of the system. This gives the optimal probability of the distribution of the data points on the clusters. This is done for a variable number of clusters in such a way that for high temperatures the global minimum corresponds to one cluster which is the entire data set. Decreasing the temperature there will be a splitting of this cluster into smaller clusters and this will happen at well defined temperatures. The process may be viewed as a sequence of second order phase transitions were the new clusters with their centroid's correspond to the order parameters. Rose et.al. [2] derived the critical values of the temperatures where the transitions occur.

In comparison with Rose et.al.[2], where one finds a sequence of phase transitions with a growing number of clusters and their centroid's, in the case of the presence of nodes one associates a cluster to every node, given by its nearest data points. The centroid's of these clusters are in general not in the node positions. But at every phase transition one finds for a growing number of clusters the same position of centroid's and nodes. One would than expect that the order parameter is proportional to the number of equal nodes and centroid's. At high temperatures the chain of nodes contracts to nearly a point, the centroid of the cluster of all data points. Decreasing the temperature the system, nodes and data points, will pass a sequence of second order phase transitions which will be analyzed by discussing the internal energy and its derivative. In most of the problems one has a large number of data points and only a few numbers of nodes. So the chain of nodes may be viewed as a small probe which in interacting with the data points shows directly and more clearly the effect of the phase transitions and gives information on the formation of clusters.

In the next section we derive the elastic net algorithm and discuss the annealing process with its phase transitions [3]. In the third section we apply the method to the 7312 genes of the Escherichia Coli genome where exist data on the level of gene expressions obtained by micro array technique for their activity in different carbon atmosphere [4]. The idea is to find clusters of genes which show especially high growing levels of gene expressions (up-regulated genes) [5]. We discuss the results for different temperatures and as a function of the node distribution. Finally we discuss possible improvements and formulate the conclusions.

2   The Method and Phase Transition

In our approach we use as the cost function the interaction energy of the data points x_i with a number of nodes y_j and a next neighbor interaction between the nodes. Data points and nodes are vectors in a multidimensional space. The next neighbor interaction between the nodes ensure topological neighborhoods. In our case we use initially a chain of nodes which will be deformed during the deterministic annealing, but will not change their topology. In analogy with the work of K.Rose et.al.[2] we use the maximum entropy clustering of statistical mechanics. For the energy of the data point x_i with the node y_j a square distance will be used:

$$E_{ij} = \frac{1}{2}|x_i - y_j|^2 \tag{1}$$

The data x_i and y_j are without dimensions. The factor $\frac{1}{2}$ is introduced for convenience. The average total energy of the system for a given number of nodes is than defined as:

$$<E> = \sum_i \sum_j P_{ij} E_{ij} \tag{2}$$

where P_{ij} is the probability to find the distance $|x_i - y_j|$. The maximum entropy principle gives under the constraint (2) for the probability to find the distance $|x_i - y_j|$

$$P_{ij} = \frac{e^{-\beta E_{ij}}}{Z_i} \tag{3}$$

with the partition function for the data point Z_i

$$Z_i = \sum_j e^{-\beta E_{ij}}, \tag{4}$$

where the parameter β is inversely proportional to the temperature T and is without dimension. One notes that for $\beta = 0$ each data point is equally associated with all nodes, while for $\beta \to \infty$ each node will be in exactly one data point with probability 1, which will be the centroid of a corresponding cluster.

For a given number of nodes one assumes that the probabilities which associate different data points to the nodes are independent, so that the total partition function is

$$Z = \prod_i Z_i \tag{5}$$

From the partition function one derives the Helmholtz free energy as

$$F = -\frac{1}{\beta}\ln Z + E_{nod} = -\frac{1}{\beta}\sum_i \ln \sum_j e^{-\beta E_{ij}} + E_{nod}, \tag{6}$$

where we have used relations (3),(4) and (5). We have added an interaction energy between the nodes, E_{nod} with a square distance interaction between next neighbor nodes:

$$E_{nod} = \frac{1}{2}\lambda \sum_j |y_j - y_{j-1}|^2 \tag{7}$$

λ is the spring constant. The final result for the Helmholtz free energy is the same as in the work of Ball et.al. [6],[7], where they used a slightly different approach.

The positions of the nodes which optimize the cost function is the minimum of the Helmholtz free energy with respect to y_j:

$$\frac{\partial}{\partial y_j} F = 0, \ \forall j. \tag{8}$$

which gives the implicit none linear equation

$$\sum_i P_{ij}(x_i - y_j) + \lambda(y_{j+1} - 2y_j + y_{j-1}) = 0, \ \forall j. \tag{9}$$

to be resolved for the node positions y_j. If we neglect the interaction energy between the nodes, relation (10) is an implicit equation for the optimized nodes as centroid's

$$y_j = \frac{\sum_i x_i P_{ij}}{\sum_i P_{ij}} \ \text{ for } \ \lambda = 0. \tag{10}$$

Equation (11) is the same as in the work of Rose et.al [2], where it gives for a fixed value of β a number of optimal centroid's for the corresponding clusters. One notes that the nodes are optimal centroid's of *fuzzy clusters*, where each data point is associated *in probability* with each cluster. For $\beta = 0$ it is one cluster with its centroid, at a certain value of β the cluster splits into smaller clusters and for a higher value of β there will be a new splitting and so on. The formation of the sub clusters and there centroid's are described as second order phase transitions. The presence of a fixed number of nodes does not change the formation of sub clusters at certain values of β, but may split them additionally and shift the phase transitions.

Let us assume that we use a fixed number of nodes and there is no interaction between the nodes. Equation (11) gives the optimal node positions for the corresponding fuzzy clusters in a given phase and if the number of nodes is the same as the number of existing clusters, every node position would be the same as the optimal centroid of these clusters. If we have more nodes than clusters, some of the nodes would not coincide with the optimal centroid's but take intermediate positions or accumulate at the node for the optimal centroid's. For more clusters than nodes these will not be detected anymore. The number of nodes gives a limit of the cluster size to be found.

One can estimate the node for the optimal clusters in the following way: The clusters which exist in a phase should be *well defined* in the sense that their probability distribution has large values for the data points of the cluster and goes fast to zero for other data points. For the node y_j associated as the optimal centroid of this well defined cluster, we define a *hard cluster*,determined by the nearest data points of the node. All data points are so subdivided in hard clusters for every node. The centroid $\hat{y}_j$ of the hard cluster, the *hard centroid*, is given by

$$\hat{y}_j = \frac{1}{n_j} \sum_{i \in C_j} x_i \ \text{ for } \ \lambda = 0, \tag{11}$$

where C_j forms the cluster of nearest neighbors of node y_j with n_j members. This corresponds to a Voronoi tesselation, where the whole cluster is subdivided in non overlapping Voronoi cells. As a consequence one expects that for a well defined cluster the hard centroid should be in good accordance with the node position. The criterion for the well defined clusters and the Gaussian form of the probability distribution, given by equations (4) and (1), can be used to estimate the size d_c of the forming clusters by

$$\frac{1}{2}\beta d_c^2 \approx 1 \tag{12}$$

In many of our problems we use for a large number of data points only very few nodes, so that with a not to large spring constant λ the node interaction is a small perturbation or may be neglected at all. For any of the two cases, equations (10) or (11) can be resolved iteratively. We use the steepest descent and ascent algorithm [6]

$$\triangle y_j = -\triangle \tau \frac{\partial F}{\partial y_j} = \triangle \tau \sum_i P_{ij}(x_i - y_j) + \triangle \tau \lambda (y_{j+1} - 2y_j + y_{j-1}) \ \forall j. \tag{13}$$

with the time step $\triangle \tau$ chosen adequately.

The internal energy of the system,

$$U = \frac{\partial \beta F}{\partial \beta} = <E> + E_{nod} \tag{14}$$

shows these phase transitions as a function of β. At every transition we expect a new deformation of the chain of nodes. As the energy of the chain E_{nod} is generally much smaller than the interaction energy of the nodes with the data points, the change in the deformation should be of the order of the energy of the chain and E_{nod} shows the phase transitions much better than U.

3   Application

The data we use as an example correspond to the level of gene expression of 7,312 genes obtained by the micro array technique of E.Coli [5]. Each gene is described by 15 different experiments (which correspond to the dimensions for the representation of each gene) whose gene expression response is measured [5] on glucose sources. Specifically there are 5 sources of glucose, 2 sources of glycerol, 2 sources of succinate, 2 sources of alanine, 2 sources of acetate and 2 sources of proline. These data are found in the GEO database [8] (Gene Expression Omnibus) of the National Center for Biotechnology Information [9].The original data have expression level values between zero and hundreds of thousands. Such extensive scale does not offer an adequate resolution to compare expression levels; therefore a logarithmic normalization is carried out.

The work of Liu et. al.[5] provides 345 up-regulated genes that were tested experimentally. The definition of up-regulated genes according to [5] is given in relation to their response the series of sources of glucose considering two factors: that its level of expression is greater than 8.5 on a log_2 scale (or 5.9 on a ln

scale), and that its level of expression increases at least 3 times from the first to the last experiment on the same scale. In our evaluation we analyze those clusters which contain a large number of the up-regulated genes. We use a less restrictive definition for the up-regulation that includes the genes that have only an increasing activity of the level of expression with the experiments, since the definition given in [5] for up-regulated genes contain very elaborate biological information which requires a precise identification of the kind of gene to be detected.

We apply to these data the Elastic Net Algorithm using $M = 20$ initially equally spaced nodes on a chain which covers the whole cluster and passes through its center. The algorithm was formulated as a *Fortran* program and run on a laptop with an Intel Core(TM)2 Duo CPU T7500 with 2.20GHz. The calculation from $\beta = 0.0005$ to $\beta = 4.0$ with 8000 iterations took about 20 minutes. We tested the results of the cluster formation with different values for the spring constant $\lambda = 0$, 0.1, 2.0 and found practically no changes in the phase transitions and the cluster properties. This can be understood because of the large number of data points ($N = 7312$) and the small number of nodes ($M = 20$) which give a much smaller node energy E_{nod} with respect to the internal energy of the system and the chain may be seen as a perturbation. In order to preserve the topology of the chain, we use in our evaluation the value of $\lambda = 0.1$. To see a series of possible phase transitions, it is sufficient to apply relation (13) for values of β between 0 and $\beta_{max} = 4.0$, using an iteration step of $\triangle\beta = 0.0005$ and a time step $\triangle\tau = 0.001$.

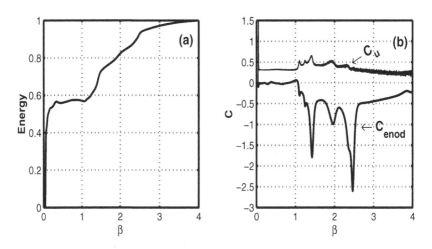

Fig. 1. (a) Normalized node interaction energy $\tilde{E}_{nod}$ as a function of β, (b) Derivatives of normalized internal and node interaction energy $C_u = -\beta^2 \frac{d\tilde{U}}{d\beta}$ and $C_{enod} = -\beta^2 \frac{d\tilde{E}_{nod}}{d\beta}$.

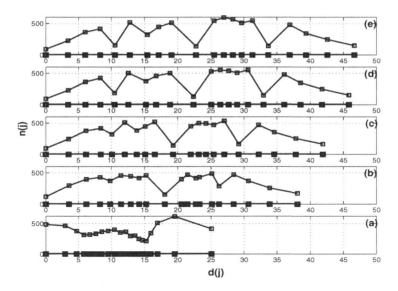

Fig. 2. Number of data points $n(j)$ in every hard cluster j as a function of the node distance $d(j)$ with respect to the first node for the phases $\beta = 0.1$ (a), $\beta = 1.55$ (b), $\beta = 2.1$ (c), $\beta = 3.0$ (d) and $\beta = 4.0$ (e).

One notes a first phase transition at $\beta = 0.1$ where the whole cluster splits into a number of smaller clusters. A next triple transitions occur between $\beta = 1.0 - 1.5$ and is not clearly seen. Between $\beta = 1.7 - 2.5$ there is another anomaly in U which should correspond to a phase transition. One notes additionally inside every phase that U decreases with increasing β because of the better formation of the clusters.

In figure 1a we show the normalized node energy $\tilde{E}_{nod} = E_{nod}(\beta)/E_{nod}(\beta_{max})$, where one notes much more clearly the phase changes. Another way to extract the transitions is to take the derivative of the internal energy with respect to β. Figure 1b shows the functions $C_u(\beta) = -\beta^2 \frac{d\tilde{U}}{d\beta}$ and $C_{enod} = -\beta^2 \frac{d\tilde{E}_{nod}}{d\beta}$ which correspond to the specific heat of the corresponding energies.

We calculated the cluster properties for 6 values of β, $(0.015, 0.1, 1.55, 2.10, 3.00, 4.00)$. The first of these values gives only one cluster for the whole system, the next value, $\beta = 0.1$, is taken shortly after the the first transition. The other values are picked such that they show a new phase. For every node y_j, which is the centroid of the corresponding *fuzzy cluster*, we determine the nearest data points and its centroid, which define the *hard cluster* and the *hard centroid* $\hat{y}_j$. The differences between the positions of the nodes and the hard centroid's are analyzed.

Firstly we calculated the node distribution for $\beta = 0.015$, which is before the first transition. The chain of nodes showed a length of 0.011. Compared to the cluster extension of approximately 11 for every dimension shows that the nodes are practically reduced to one point and form the centroid of the whole cluster.

In figure 2 we show for 5 different values of β in subsequent phases, $(0.1, 1.55, 2.10, 3.00, 4.00)$, the number of the nearest neighbor data points $n(j)$, for every hard cluster as a function of the distance $d(j)$ between the corresponding node and the first node. The length of the chain of nodes increase in subsequent phases. After the first transition at $\beta = 0.05$ the system splits into a number of smaller clusters. The second phase exist for values of β between $0.05 - 1.0$ and cluster formation can be seen by the decreasing internal energy. Fig.2a shows the node distribution for $\beta = 0.1$, which is shortly after the first transition. The transition changes the length of the chain of nodes from 0.011 to 25. The distribution of the hits over the nodes shows a not too large variation.

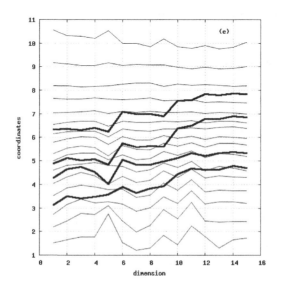

Fig. 3. (coordinates of cluster centroids for phase (e), bold lines mark centroids with increasing activity.

In figure 3 we show for phase (e) the hard centroids for all 20 clusters as a function of the dimensions. One notes clearly that the four clusters, 5,7,10,16 have maximal increasing activity.

Fig.2b shows the distribution of the number of nearest neighbors of the nodes $n(j)$ as a function of the node distances with respect to the first node position after the triple transition for $\beta = 1.55$. This phase exist for β between 1.5 and 1.7 and is much smaller than the second phase. The length of the chain has increased to 38 and the nodes separate more and show more variation in the distribution of the nearest neighbors $n(j)$.

From the phase (c) on, one finds hard centroid's where the value of the coordinates increase with the dimension number. This corresponds to increasing activity of the level of genes expression and these hard clusters must contain the up-regulated genes of the work of Liu [5].

The results of this analysis can be resumed by the following points:

(i) The elastic net algorithm applied to the E.Coli data gives well defined clusters for the 20 nodes, when the centroid's of the fuzzy clusters are nearly the same as the centroid's of hard clusters.

(ii) Fig. 2 shows for this optimal case that 5 clusters, 1, 5, 10, 16 and 20, have a much smaller number of data points then the other clusters. For clusters 1 and 20 this may be a result of the begin and end of the chain of nodes, but clusters 5, 10 and 16 show additionally to the low number of data points an increasing activity. They show some anomaly in their behavior.

(iii) Comparing the data points with the list of up-regulated genes, one finds more than half of them in the two clusters 10 and 16. The other two clusters with a strongly increasing activity have too low genes expressions, nearly all below 5.0 and do not contribute to the up-regulated genes.

It seems that the elastic net algorithm sorts out for the optimal conditions some clusters with data points which show different properties then the rest. This seems to be a good result for this simple and fast algorithm presented here.

4   Conclusions

For the problem of finding clusters with the help of an elastic chain of nodes, we use the Durbin-Willshaw elastic net algorithm [10] and together with a mechanical statistics formulation show that during the deterministic annealing a series of phase transitions go together with the formation of clusters.

We find that the number of nodes of the elastic chain determines the number of well defined clusters and the final temperature of the annealing process. Well defined clusters means that the resulting fuzzy clusters are the same as the clusters given by the nearest neighbors of every node, the hard clusters. This equivalence, which is formulated as the equivalence of the node positions y_j with the position of the centroid of the hard clusters $\hat{y}_j$. This happens from a certain temperature or parameter β on and can be used as a stopping criterion for the annealing process. The minimal cluster size for this β can be estimated with the relation (12). Shorter cluster sizes or correlations can be found by increasing the number of nodes.

Additionally we find that the node interaction energy E_{nod}, which can be treated as a perturbation on the data cloud, shows much more clearly the different phases and phase transitions. As E_{nod} appears directly in the sum for the free energy, one may interpret the phase transitions on the elastic chain as first order phase transitions.

Finally we note that the elastic net algorithm for large numbers of data points but no too large numbers of nodes is very fast, the algorithm can be parallelized and the stopping criterion may help additionally to shorten the process.

References

1. Duda, R.O, et al.: Patteren classification. A Wiley-Interscience Publication (2001)
2. Rose, K., et al.: Phys. Rev. Let. **65**(8) (1990)
3. Lévano, M., Nowak, H.: New aspects of the elastic net algorithm for cluster analysis. In: Palmer-Brown, D., Draganova, C., Pimenidis, E., Mouratidis, H. (eds.) EANN 2009. CCIS, vol. 43, pp. 281–290. Springer, Heidelberg (2009)
4. Tamayo, P., Slonim, D., Mesirov, J., Zhu, Q., Kitareewan, S., Dmitrovsky, E., Lander, E., Golub, T.: Interpreting patterns of expression with self-organizing maps: Methods and application to hematopoietic differentiation. Genetics **96**, 2907–12 (1999)
5. Liu, M., Durfee, T., Cabrera, T., Zhao, K., Jin, D., Blattner, F.: Global transcriptional programs reveal a carbon source foraging strategy by E. Coli. J. Biol. Chem. **280**, 15921–15927 (2005)
6. Ball, K.D., et al.: Journal of Comp. Chemistry **23**(1) (2001)
7. Yuille, A.L.: Neural Computation **2**(1) (1990)
8. National Center for Biotechnology Information. http://www.ncbi.nlm.nih.gov
9. Edgar, R., Domrachev, M., Lash, A.: Gene expression omnibus: NCBI gene expression and hybridization array data repository. Nucleic Acids Research **30**(1), 207–210 (2002)
10. Durbin, R., Willshaw, D.: Nature **326**, 689–691 (1987)

Intelligent Medical Modeling

Automatic Detection of Microaneurysms for Diabetic Retinopathy Screening Using Fuzzy Image Processing

Sarni Suhaila Rahim[1,2]([✉]), Vasile Palade[1], James Shuttleworth[1],
Chrisina Jayne[1], and Raja Norliza Raja Omar[3]

[1] Faculty of Engineering and Computing, Coventry University, Priory Street,
Coventry, CV1 5FB, UK
rahims3@uni.coventry.ac.uk, sarni@utem.edu.my,
{vasile.palade,csx239,ab1527}@coventry.ac.uk
[2] Faculty of Information and Communication Technology,
Universiti Teknikal Malaysia Melaka, Hang Tuah Jaya, 76100,
Durian Tunggal, Melaka, Malaysia
[3] Department of Ophthalmology, Hospital Melaka,
Jalan Mufti Haji Khalil 75400, Melaka, Malaysia
norliza_eye@yahoo.com

Abstract. Fuzzy image processing was proven to help improve the image quality for both medical and non-medical images. This paper presents a fuzzy techniques-based eye screening system for the detection of one of the most important visible signs of diabetic retinopathy; microaneurysms, small red spot on the retina with sharp margins. The proposed ophthalmic decision support system consists of an automatic acquisition, screening and classification of eye fundus images, which can assist in the diagnosis of the diabetic retinopathy. The developed system contains four main parts, namely the image acquisition, the image preprocessing with fuzzy techniques, the microaneurysms localisation and detection, and finally the image classification. The fuzzy image processing approach provides better results in the detection of microaneurysms.

Keywords: Diabetic Retinopathy · Eye screening · Colour fundus images · Fuzzy image processing · Microaneurysms

1   Introduction

Diabetic Retinopathy (DR) is one of the chronic complications of Diabetes Mellitus (DM). Diabetes Mellitus is defined as the chronic condition due to an excess of glucose circulating on the bloodstream [1]. Diabetic Retinopathy refers to the capillaries, arterioles and venules in the retina and the effects from the small vessels leakage.

The critical lesions that occur in diabetic retinopathy, which created the diabetic retinopathy severity scale described by The Early Treatment Diabetic Retinopathy Study (ETDRS) are microaneurysms, small retinal haemorrhages, larger retinal haemorrhages, hard exudates, cotton wool spots, intraretinal microvascular abnormality, venous

© Springer International Publishing Switzerland 2015
L. Iliadis and C. Jayne (Eds.): EANN 2015, CCIS 517, pp. 69–79, 2015.
DOI: 10.1007/978-3-319-23983-5_7

abnormalities, arteriolar abnormalities, fibrous proliferation at the disc, fibrous proliferation elsewhere, new vessels on disc, new vessels elsewhere, vitreous haemorrhage,
preretinal haemorrhage and also post-laser-treatment-laser scars [2]. In addition to the
early classification by ETDRS, Wilkinson and others [3] proposed an international
clinical diabetic retinopathy severity scales in relation to the lesions.

Microaneurysm, one of the lesions in diabetic retinopathy is defined as a red spot
with the size less than 125 microns with sharp margins [1]. It is formed by ballooning
out of a weak part of the capillary wall that appears as a dot to the observer. Microaneurysm is one of the earliest signs of diabetic retinopathy; the detection of microaneurysms at an early stage is vital and it is the first step in preventing the diabetic
retinopathy. Therefore, the accurate detection of the microaneurysms through a reliable eye screening is essential. Regular screening can help for early detection and early
treatment in order to assist the management of diabetic retinopathy as the affected
eyes can lead to vision loss if it is not treated.

The paper is organised as follows. Section 2 presents the related work on the microaneurysms detection, while Section 3 explains the proposed microaneurysms detection system. Section 4 describes the image preprocessing stage followed by Section
5, which explains the localisation and detection of microaneurysms. Section 6 describes the classification, while Section 7 presents the results of the system and, finally, Section 8 details the conclusions of the work and a future plan.

2   Related Work on Microaneursyms Detection

Automatic detection of microaneurysms for the diabetic retinopathy screening is a
challenging task and there has been extensive research on the localisation and detection of the first signs of diabetic retinopathy. An accurate detection of microaneurysms is required for early detection of diabetic retinopathy. The automatic detection
of such an important sign of diabetic retinopathy, which is microaneurysms, has been
proposed by some of researchers. Rahim et al. [4] proposed an automatic diabetic
retinopathy screening system, followed by a microaneurysms detection system [5].
One of the effective techniques for the localisation and detection of the microaneurysms used by researchers on microaneurysms detection is the Circular Hough Transform (CHT). The Hough Transform can be used to detect lines, circles or other parametric curves. Since the microaneurysms are circular in shape, the Circular Hough
Transform is suitable to be used to locate and detect objects of circular shapes. In
addition, CHT is also useful to detect the optic disc in a diabetic retinopathy screening
system. Amiri et al. [6] and Abdelazeem [7] proposed the use of CHT for the detection of microaneurysms in retinal angiography images.

The Retinopathy Online Challenge (ROC) presents an online competition for numerous methods in microaneurysms detection to compare with each other on the same
data [8]. Other recent works on the detection of microaneurysms are [9-12].

Fuzzy image processing employs different fuzzy approaches to process the images. Fuzzy edge detection, fuzzy histogram equalisation and fuzzy filtering are among the fuzzy processing techniques that can be performed on images. Sheet et al. [13] proposed a modified technique of the brightness preserving dynamic histogram equalisation. The representation and processing of images in the fuzzy domain allows the technique to handle the inexactness of grey level values in a better way and improve the performance. The technique proposed by Sheet et al. was used for contrast enhancements in digital pathology images by Garud et al. [14]. Meanwhile, fuzzy filters have been used for image filtering by Patil and Chaudhari [15], Toh [16] and also Kwan [17]. Edge detection works by comparing the intensity of neighbouring pixels. Fuzzy edge detection allows the use of membership functions to define the degree to which a pixel belongs to an edge or a uniform region.

3   Proposed System

In this paper, a novel automatic detection method of microaneurysms in colour fundus images for diabetic retinopathy screening is presented. The proposed system is different from other microaneurysms detection systems as it implements fuzzy techniques in the image preprocessing part. In addition to the new proposed techniques, the system is evaluated with the combination of normal and Diabetic Retinopathy (DR) fundus images from a novel dataset which was developed as part of this project.

The fundus images are collected from the Eye Clinic, Department of Ophthalmology, Hospital Melaka, Malaysia. The dataset consists of colour fundus images from 2401 patient's folders collected. Each of the patient's folders has two images minimum, at least one for the right side and one for the left side, where two different angles were captured; the optic disc centre and the macula centre. The original images, which are of size 3872 x 2592 in JPEG format, provide a high quality and details of the images, and were captured with a KOWA VX-10 digital fundus camera. Three experts from the Department of Ophthalmology, Hospital Melaka, Malaysia were involved to diagnose the 600 fundus images from 300 patients into ten retinopathy stages: No Diabetic Retinopathy, Mild DR without maculopathy, Mild DR with maculopathy, Moderate DR without maculopathy, Moderate DR with maculopathy, Severe DR without maculopathy, Severe DR with maculopathy, Proliferative DR without maculopathy, Proliferative DR with maculopathy and Advanced Diabetic Eye Disease (ADED). An Excel file containing the fundus images link and retinopathy stages drop down list was provided to each of the experts in three separate devices to avoid bias. As a result from the expert diagnosis analysis performed by using the SPSS software, the total numbers of No DR and DR classes are mostly balanced. However, the total numbers of severe and proliferative cases are hugely imbalanced. This is due to the fact that, for severe cases, the patient will be referred directly for laser treatment rather than capturing their fundus images and the other reason is that the diabetic retinopathy takes a long time to progress on the severity of the diabetes mellitus. Nonetheless, the dataset offers several advantages as it represents South East Asian population, particularly Malaysian. Besides, it balances the total number of No DR/Normal and DR/Abnormal, where in other medical images diagnosis, it is difficult to find a large number of normal cases. Moreover, the categorisation of the expert diagnosis is following the standard practice based on the International Clinical Retinopathy and Diabetic Macula Oedema Disease Severity Scale. The classification of data involves the maculopathy, which is the yellow lesion near the macula. This is a very detailed categorisation compared to other datasets. The detection of maculopathy is very important as the macula is responsible for central vision and it is a sensitive part and also vital to detect the urgency of the referral. We plan to make the dataset available for public use in the near future.

The proposed screening system has been developed using the Matlab R2014a software. The system starts with the image acquisition process, where the system selects images for further processing. The function *imgetfile* is used to display the Open Image dialog box, followed by the *imread* function to read a JPEG colour fundus images from the file specified. The selected images undergo preprocessing in order to improve the image contrast as well as perform other enhancements. After that, the preprocessed images are used to locate and detect the retinopathy signs, the microaneurysms. The number of microaneurysms detected are displayed. Finally, in the classification phase, the images are classified into two main groups: microaneurysms detected or microaneurysms not detected, based on the number of microaneurysms detected. The system could be used to help the clinician.

Figure 1 presents the block diagram of the proposed automatic detection of microaneurysms system for diabetic retinopathy screening. Individual stages are discussed in more detail in the following sections.

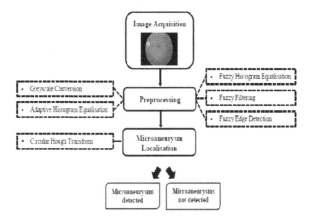

Fig. 1. Block diagram of the proposed automatic detection of microaneurysms for diabetic retinopathy screening

4   Image Preprocessing

Image preprocessing is the operation performed for further processing, which leads to an improved quality of the image. The image preprocessing techniques involved in the present work include Greyscale Conversion, Adaptive Histogram Equalisation, Fuzzy Histogram Equalisation, Fuzzy Filtering and Fuzzy Edge Detection.

4.1   Greyscale Conversion

The first preprocessing technique used is converting the colour fundus image into a greyscale image, as greyscale is usually the ideal format for image processing. The intensity is calculated by using a common formula combination of 30% of red, 59% of green and 11% of blue.

4.2   Adaptive Histogram Equalisation

Histogram equalisation is the re-distribution of grey-level values uniformly. Adaptive histogram equalisation, a more advanced version of histogram equalisation, divides the image into smaller tiles, applies histogram equalisation to each tile, and then interpolates the results. Adaptive histogram equalisation includes limits on how much the contrast is allowed to be changed, called the Contrast-Limited Adaptive Histogram Equalisation (CLAHE).

4.3   Fuzzy Histogram Equalisation

Histogram equalisation is a computer image processing technique for improving the image's contrast. The proposed preprocessing techniques are used to help increase the contrast of the colour fundus images. The uses of the colour fundus images are more challenging compared to the other modes of fundus photography examination which are angiography and red-free. Therefore, appropriate techniques need to be implemented in order to improve the contrast of the fundus images for better visualisation and detection. Due to the good performance of the Brightness Preserving Dynamic Fuzzy Histogram Equalisation (BPDFHE) technique proposed by Sheet et al. [13] which was proven to work well in the medical images such as pathology images presented by Garud et al. [14], the technique has been chosen as a preprocessing technique for the proposed microaneurysms detection in diabetic retinopathy screening.

4.4   Fuzzy Filtering

Image filtering is used to improve the image quality or restore the digital image which has been corrupted by some noise. The proposed system implements the median filter by employing fuzzy techniques. The Fuzzy Switching Median (FSM) filter by Toh et al. [16] was working well in removing salt-and-pepper noise while preserving image details and textures very well, by incorporating fuzzy reasoning in correcting the detected noisy pixel. The technique composed of two modules which are the salt-and-pepper noise detection and the fuzzy noise cancellation module. The first module works by searching for the two intensities of positive and negative noise pulses in the noisy image, while the second module (the detection) would only cancel noise if the central pixel is noisy.

4.5   Fuzzy Edge Detection

An edge is a boundary between two uniform regions. The edge can be detected by comparing the intensity of neighbouring pixels. However, since uniform regions are not crisply defined, small intensity differences between two neighbouring pixels do not always represent an edge or it might represent a shading effect. Therefore, the use of membership functions would overcome the problems by defining the degree with which the pixel belongs to an edge or a uniform region. The image gradients are created as the inputs of a Fuzzy Inference System (FIS) using a Mamdani system. For each input, a zero-mean Gaussian membership function is specified, where if the gradient value for a pixel is 0 (region), then it belongs to the zero membership function with a degree of 1. Another membership function is added, which specifies the standard deviation for the zero membership function for the gradient inputs. As an output of the FIS, which is the intensity of the edge-detected image, two triangular membership functions, white and black, are specified. The values of the standard deviation and also the values of the triangular membership functions for the output can be changed in order to adjust the edge detector performance. Next, the FIS rules are specified. The first rule indicates that if both inputs are zero, then the output is white. On the other hand, the second rule generates the output as black if either one of the inputs is not zero. Figure 2 shows the membership functions of the inputs and outputs for the edge detection, while Figure 3 shows the output after each of the fuzzy preprocessing operations on an image selected, as explained previously.

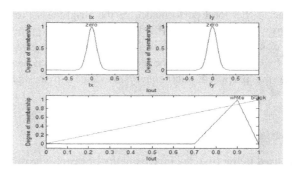

Fig. 2. Membership functions of the inputs and outputs

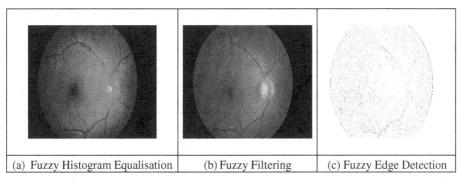

| (a) Fuzzy Histogram Equalisation | (b) Fuzzy Filtering | (c) Fuzzy Edge Detection |

Fig. 3. Preprocessing the output image with Fuzzy approaches

5   Localisation and Detection of Microaneurysms

After performing the preprocessing techniques, localisation and microaneurysms detection take place. In order to detect well the microaneurysms, which are circular in shape, Circular Hough Transform (CHT) technique is implemented in the proposed system. The function *imdistline* in Matlab is used to calculate the radius of the optic disc as well as for the microaneurysms in the fundus images. Based on the specified radius range, the system will find the circles in the fundus images using the CHT with the use of the *imfindcircles* function offered by Matlab. After finding the circles in the image based on the radius range, the function *viscircles* in Matlab is used to create the circle. The function draws circles in the fundus image with a specified centre and radius onto the current axes.

6   Classification

After performing the microaneurysms localisation and detection, the counting of microaneurysms takes place. The presence of any number of microaneurysms is represented as Diabetic Retinopathy, DR (microaneurysms detected), and if there is no microaneurysm detected, it is considered as normal or no retinopathy (microaneurysms not detected). The categorical types of the output whether Normal/No DR or DR are used for the analysis of the system performance later. The categorical types from the expert diagnosis and the system generated are compared. The nominal inputs from both the expert and the system are represented as 1 for Normal/No DR, and 2 for DR. The average from the three experts is used for the overall expert diagnosis. Later, the comparison between the overall expert diagnosis and the proposed system are generated and an analysis of both performances is produced.

7   Results and Discussion

Figure 4 shows one user interface snapshot of the proposed developed system. The performance analysis summary of the proposed system, which utilises different techniques (System Variants I-IV), is presented in Table 1. System Variant I implemented the greyscale conversion, histogram equalisation and Circular Hough Transform (CHT), while System Variant II proposed the implementation of the Fuzzy Histogram Equalisation in addition to the greyscale conversion and CHT. System Variant III and System Variant IV proposed the implementation of other fuzzy image processing techniques, such as Fuzzy Filtering and Fuzzy Edge Detection, respectively. Two types of statistical tests were performed to test the system performance compared to the expert diagnosis. Six hundreds fundus images represent the number of fundus images diagnosed by the expert and the systems involved for both tests. T-test is used to test the differences in mean between the annotated images and the system output. On the other hand, Chi-Square test is used to compare the two groups: the expert diagnosis and the system diagnosis by using the same six hundreds fundus images. The results generated show, with the 95% confidence interval, that the means of the annotated images (1.54) and the system with the fuzzy techniques, i.e., the Fuzzy Histogram Equalisation (1.49) and the Fuzzy Edge Detection output (1.59), are not significantly different. Thus, it can be concluded that the system findings are more likely to be not different from the expert findings. However, the annotated images and the system with the fuzzy median filter (with mean 1.43) show the opposite. The inferential statistical analysis (i.e., T-test) result indicates a p-value of 0.00 (System Variant I), 0.92 (System Variant II), 0.00 (System Variant III) and 0.73 (System Variant IV), respectively. The results show that the Chi-square test indicates a p-value of 0.00, 0.94, 0.00 and 0.81 for the four systems. Table 1 shows the results generated to descriptively compare the methods of assessment between the two groups: expert and system diagnosis for both normal and diabetic retinopathy categories. As a conclusion for the analysis based on the presented results, the implementation of the fuzzy preprocessing techniques provides better contrast enhancement and other improvements for fundus images, and, hence, it greatly assists to detect the microaneurysms and a more efficient and reliable performance of the diagnosis system can be produced.

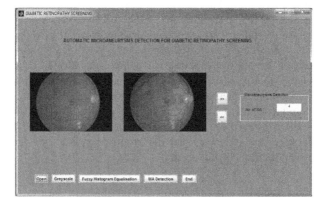

Fig. 4. Snapshot of the proposed system user interface

Table 1. Results of the proposed systems

		System I	System II	System III	System IV
Techniques		Greyscale, Histogram Equalisation, CHT	Greyscale, Fuzzy Histogram Equalisation, CHT	Greyscale, Fuzzy Median Filter, CHT	Greyscale, Fuzzy Edge Detection, CHT
T-Test Mean:					
	Expert	1.54	1.54	1.54	1.54
	System	1.67	1.49	1.43	1.59
T-Test *p*-value		0.00	0.92	0.00	0.73
Chi-Square *p*-value		0.00	0.94	0.00	0.81
Expert Count:					
	No DR	276	276	276	276
	DR	324	324	324	324
System Count:					
	No DR	201	305	343	246
	DR	399	295	257	354

8   Conclusions and Future Work

An automatic microneurysms detection system for diabetic retinopathy screening using colour fundus images has been developed using fuzzy image processing. It can be concluded that the use of fuzzy image processing techniques plays an important role in producing better image quality and improved performance analysis. The system will further be enhanced by implementing different preprocessing techniques. These include optic disc localisation, vessel segmentation, other fuzzy techniques for image preprocessing and the use of Fuzzy Circular Hough Transform for the microaneurysms detection. In addition, the system will also be extended by combining together more fuzzy preprocessing techniques for the preprocessing part, namely, the Fuzzy Histogram Equalisation, Fuzzy Filtering and Fuzzy Edge Detection, in one microaneurysms detection system. Such a complete and accurate system could be used to help the diabetic retinopathy screening team to perform the screening in a better and more efficient way. Moreover, it can help achieve the overall aim of the screening process, which is to detect earlier the sight threatening diseases and to ensure timely treatment in order to prevent vision loss.

Acknowledgement. This project is a part of PhD research currently being carried out at the Faculty of Engineering and Computing, Coventry University, United Kingdom. The deepest gratitude and thanks go to the Universiti Teknikal Malaysia Melaka (UTeM) and Ministry of Education Malaysia for sponsoring this PhD research. The authors are thankful to the Ministry of Health Malaysia and Hospital Melaka, Malaysia, for providing the database of retinal images and also for the manual grading by the three experts.

References

1. Scanlon, P.H., Wilkinson, C.P., Aldington, S.J., Matthews, D.R.: A Practical Manual of Diabetic Retinopathy Management. Wiley-Blackwell, Chicester (2009)
2. Early Treatment Diabetic Retinopathy Study Research Group: Grading diabetic retinopathy from stereoscopic color fundus photographs- an extension of the modified Airlie House classification. ETDRS report number 10. Ophthalmology **98**(5 Suppl), 823–833 (1991)
3. Wilkinson, C.P., Ferris, F.L., Klein, R.E., Lee, P.P., Agardh, C.D., Davis, M., Dills, D., Kampik, A., Pararajasegaram, R., Verdaguer, J.T.: Proposed International Clinical Diabetic Retinopathy and Diabetic Macula Edema Disease Severity Scales. American Academy of Ophthalmology **110**(9), 1677–1682 (2003)
4. Jayne, C., Rahim, S.S., Palade, V., Shuttleworth, J.: Automatic screening and classification of diabetic retinopathy fundus images. In: Mladenov, V., Jayne, C., Iliadis, L. (eds.) EANN 2014. CCIS, vol. 459, pp. 113–122. Springer, Heidelberg (2014)
5. Rahim, S.S., Jayne, C., Palade, V., Shuttleworth, J.: Automatic detection of microaneurysms in colour fundus images for diabetic retinopathy screening. Journal of Neural Computing and Applications (2015) (in press). doi:10.1007/s00521-051-1929-5
6. Amiri, S.A., Hassanpour, H., Shahiri, M., Ghaderi, R.: Detection of microaneurysms in retinal angiography images using the Circular Hough Transform. Journal of Advances in Computer Research **3**(1), 1–12 (2008)
7. Abdelazeem, S.: Microaneurysm detection using vessels removal and circular hough transform. In: Nineteenth National Radio Science Conference (NRSC 2002), pp. 421–426. IEEE Press, New York (2002)
8. Niemeijer, M., van Ginnerken, B., Cree, M.J., Mizutani, A., Quellec, G., Sanchez, C.I., Zhang, B., Hornero, R., Lamard, M., Muramatsu, C., Wu, X., Cazuquel, G., You, J., Mayo, A., Li, Q., Hatanaka, Y., Cochener, B., Roux, C., Karray, F., Garcia, M., Fujita, H., Abramoff, M.D.: Retinopathy Online Challenge: Automatic detection of microaneurysms in digital color fundus photographs. IEEE Transactions on Medical Imaging **29**(1), 185–195 (2010)
9. Akram, M.U., Khalid, S., Tariq, A., Khan, S.A., Azam, F.: Detection and classification of retinal lesions for grading of diabetic retinopathy. Computers in Biology and Medicine **45**, 161–171 (2014)
10. Sundhar, C., Archana, D.: Automatic screening of fundus images for detection of diabetic retinopathy. International Journal of Communication and Computer Technologies **2**(1), 100–105 (2014)
11. Lim, G., Lee, M.L., Hsu, W., Wong, T.Y.: Transformed representations for convolutional neural networks in diabetic retinopathy screening. Modern Artificial Intelligent for Health Analytics, 21–25 (2014)
12. Adal, K.M., Sidibe, D., Ali, S., Chaum, E., Karnowski, T.P., Meriaudeau, F.: Automated detection of microaneurysms using scale-adapted blob analysis and semi-supervised learning. Computer Methods and Programs in Biomedicine **114**, 1–10 (2014)
13. Sheet, D., Garud, H., Suveer, A., Mahadevappa, M., Chatterjee, J.: Brightness preserving dynamic Fuzzy Histogram Equalization. IEEE Transactions on Consumer Electronics **56**(4), 2475–2480 (2010)

14. Garud, H., Sheet, D., Suveer, A., Karri, P.K., Ray, A.K., Mahadevappa, M., Chatterjee, J.: Brightness preserving contrast enhancement in digital pathology. In: 2011 International Conference on Image Information Processing (ICIIP 2011), pp. 1–5. IEEE, New York (2011)
15. Patil, J., Chaudhari, A.L.: Development of digital image processing using Fuzzy Gaussian filter tool for diagnosis of eye infection. International Journal of Computer Applications **51**(19), 10–12 (2012)
16. Toh, K.K.V., Mat Isa, N.A.: Noise adaptive Fuzzy switching median filter for salt-and-pepper noise reduction. IEEE Signal Processing Letters **17**(3), 281–284 (2010)
17. Kwan, H.K.: Fuzzy filters for noisy image filtering. In: IEEE International Symposium on Circuits and Systems 2003 (ISCAS 2003), vol. 4, pp. 161–164. IEEE, New York (2003)

Endotracheal Tube Position Confirmation System Using Neural Networks

Dror Lederman[✉]

Faculty of Engineering, Holon Institute of Technology, 5810201, Holon, Israel
drorl@hit.ac.il

Abstract. Endotracheal intubation is a complex medical procedure in which a ventilating tube is inserted into the human trachea. Improper positioning carries potentially fatal consequences and therefore confirmation of correct positioning is mandatory. In this paper we report the results of using a neural network-based image classification system for endotracheal tube position confirmation. The proposed system comprises a miniature complementary metal oxide silicon sensor (CMOS) attached to the tip of a semi rigid stylet and connected to a digital signal processor (DSP) with an integrated video acquisition component. Video signals are acquired and processed by a confirmation algorithm implemented on the processor. The performance of the proposed algorithm was evaluated using two datasets: a dataset of 250 images of the upper airways. The results, obtained using a leave-one-case-out method, show that the system correctly classified 240 out of 250 (96.0%).

Keywords: Endotracheal intubation confirmation · Esophageal intubation detection · Medical image classification · One-lung intubation detection

1   Introduction

Endotracheal intubation is a relatively common procedure (~25 million intubations per year in US) performed by trained health care providers in emergency and scheduled medical conditions in which the patient ceases or cannot breathe on his own. During the procedure, a tube is positioned in the patient's trachea, thus securing access to the lungs and enabling artificial ventilation. On rare occasions, the anatomy of the patient does not allow easy insertion of the endotracheal tube (ETT) and consequently the ETT might be incorrectly positioned, either in the esophagus or too deep in the trachea (i.e. right main bronchus) [1, 2]. Unrecognized esophageal intubation carries fatal risk as the stomach is ventilated rather than the lungs resulting in severe, incompatible with life, oxygen deficiency in the vital organs. In cases of right bronchus intubation (also termed one-lung intubation (OLI)), only one lung is ventilated. One lung ventilation might cause serious complications such as collapse of the non-ventilated lung, yielding reduction in oxygen tension in the blood or lung infection, hyperinflation or rupture of the ventilated lung resulting in shock. Both esophageal intubation and OLI might also occur after the ETT was properly positioned ("dislodgement") from many reasons such as neck flexion [3, 4].

© Springer International Publishing Switzerland 2015
L. Iliadis and C. Jayne (Eds.): EANN 2015, CCIS 517, pp. 80–85, 2015.
DOI: 10.1007/978-3-319-23983-5_8

Several methods have been proposed to verify correct tube positioning. For many years, auscultation of breath sounds over the lungs using a stethoscope was the most widely used technique as it utilizes a very accessible tool. This technique requires high attention, lack of environmental noise and it suffers significant false positive and false negative rates significantly affecting its reliability [4-6]. Recently measurement of carbon dioxide (CO_2) in the expired air has become the standard practice. Since one of the main functions of the lung is to eliminate CO_2 by exhalation then measuring a substantial amount of CO_2 in the exhaled air (end-tidal CO_2 ($ETCO_2$)) proves that the endotracheal tube is located in the airways. De-facto it has become the gold standard for confirming correct ETT positioning. This method suffers however from significant shortcomings. It is not capable of differentiating proper ETT positioning from a too deep one because in both cases CO_2 is exhaled in the same concentration [6, 7]. Additionally, $ETCO_2$ is unreliable for endotracheal intubation confirmation in many medical emergencies such as cardiac arrest [5-9]. Other techniques have been proposed [10-14], but none of them has been proven better. Auscultation and $ETCO_2$ measurement are therefore the common practice in the absence of a better method.

Based on our previous preliminary work (e.g. [15]), we further developed and tested our novel approach for automatic endotracheal intubation confirmation. The approach is based on direct visual cues, i.e., identification of specific anatomical landmarks as indicators of correct or incorrect tube positioning. In this study, the system is further developed and evaluated using animal and human tissue model.

2   Material and Methods

The correct position of an ETT tip is 2-5 cm above the carina (the bifurcation of the trachea into the two main bronchi). The image of the carina is therefore the definitive anatomical landmark for confirming endotracheal intubation. The proposed approach is based on identifying the carina in the acquired video images, and discriminating between the carina and other anatomical structures. The method combines an artificial neural network scheme which is employed in a textural-based feature space. In the following subsections, an overview of the proposed system is given.

2.1   The Video-Stylet

We designed and assembled a system which resembles an intubating stylet. The tip comprises a miniature CMOS sensor. The inner part of the stylet contains wires to transfer the image and a narrow lumen to spray water or air in order to clear blood and secretions away from the camera sensor (Fig. 1). The image sensor is connected to a processor with an integrated image acquisition component. During intubation, this rigid stylet is inserted into a commercial endotracheal tube with its camera at the tip. Video signals are continuously acquired and processed by the confirmation algorithm implemented on the processor.

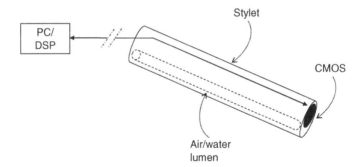

Fig. 1. A schematic drawing of the video-stylet which includes the stylet and complementary metal oxide silicon (CMOS) sensor connected to a digital signal processor (DSP) or a personal computer (PC).

2.2   Pre-Processing and Features Extraction

The confirmation algorithm is based on classification of specific anatomical land-marks, including the carina, tracheal rings (upper trachea) and esophagus. The pre-processing and feature extraction methods were described elsewhere [15]. Briefly, we use textural features [16] that contain important information about the structural arrangement of surfaces and their relationship to the surrounding environment. In particular, features based on gray level co-occurrence matrices (GLCM) are utilized. These features are based on the assumption that texture information on an image is contained in the overall or "average" spatial relationship, which the gray tones in the image have to one another. More specifically, it is assumed that this texture information is adequately specified by a set of gray tone spatial dependence matrices which are computed for various angular relationships and distances between neighboring resolution cell pairs on the image. One of the advantages of these features is that they are robust to imaging angles and scaling. This property is of great importance to the task in hand, as during intubation the tube may be inserted in different angles and directions, depending on the technique employed by the person performing the procedure. It was therefore hypothesized that textural features will allow reliable classification of the images, independently of the angle at which the tube was inserted. Typical examples of images and the corresponding textural features appear in Fig. 2.

2.3   The Neural Network

In order to classify the video frames, we utilized in this work a feed-forward artificial neural network classifier (ANN) which consists of three layers. The first (input) layer includes neurons that connect to selected features, the second layer includes hidden neurons, and the third (decision) layer includes one neuron that generates a likelihood score of a test case belonging to one of the three categories. To minimize over-fitting and maintain robustness of the ANN performance, a limited number of training iterations (1000), and a large ratio between the momentum (0.9) and learning rate (0.01), is used. The likelihood scores obtained by the ANN classifier in leave-one-out tests are used to make the classification decision.

3   Results

3.1   Classification of Cow Intubation Video Images

The proposed system was evaluated off-line using an available annotated database that includes 250 images which were randomly-selected from a larger database of 1600 images of the upper airways. The images are divided into three categories: upper-trachea (82 images), carina (101 images) and esophagus (67 images). Evaluation of the proposed approach was performed using a leave-one-case-out validation method: in each iteration, the images extracted from 9 intubation-videos were used to train the models, i.e. estimate the ANN and the rest of the images were used to test system performance. This process was repeated 10 times, such that each video participated once in the testing phase.

The classification results are summarized in Table 1 where the rows represent the predicted (recognized) classes and the columns represent the actual classes. The system achieved an overall classification rate of 96.0% (240 out of 250 images).

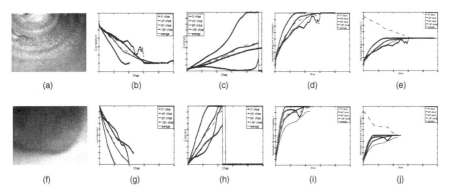

Fig. 2. Two examples of non-carina images: Upper-tracheal intubation (a) and esophageal intubation (f), and the calculated textural features: correlation (b, g) and contrast (c, h).

Table 1. Summary of classification results

	Actual		
Recognized	Upper-trachea	Carina	Esophagus
Upper-trachea	77	0	0
Carina	5	98	2
Esophagus	0	3	65
Total	82	101	67

4   Discussion

The ANN-based classification system achieved a high precision of 96.0%. Despite the encouraging results, we recognize that this is a very preliminary study. The available database consists of images taken from only 10 intubation videos. A much larger database is required in order to reliably validate system performance. Various factors might challenge the system performance, especially fog and secretions, which could result in poor image quality. In addition, the effect of possible anatomical variability between patients on system performance is yet to be evaluated. Clearly, more work is needed in order to evaluate system performance in real-time detection of misplaced and dislodged intubations. With these challenges in mind, successful implementation of the proposed method into a real-time confirmation system can serve as a major contribution to patient safety.

References

1. Silvestri, S., Ralls, G.A., Krauss, B.: The effectiveness of out-of-hospital use of continuous end-tidal carbon dioxide monitoring on the rate of unrecognized misplaced intubation within a regional emergency medical services system. Ann. Emerg. Med. **45**, 497–503 (2005)
2. Timmermann, A., Russo, S.G., Eich, C., Roessler, M., Braun, U., Rosenblatt, W.H., et al.: The out-of-hospital esophageal and endobronchial intubations performed by emergency physicians. Crit. Care and Trauma. **104**, 619–623 (2007)
3. Yap, S.J., Morris, R.W., Pybus, D.A.: Alterations in endotracheal tube position during general anesthesia. Anaesth Crit. Care, 586–588 (1994)
4. Vergese, S.T., Hannallah, R.S., Slack, M.C., Cross, R.R., Patel, K.M.: Auscultation of bilateral breath sounds does not rule out endobronchial intubation in children. Anesth Analg, 56–58 (2004)
5. Wang, H.E., Lave, J.R., Sirion, C.A., Yealy, M.: Paramedic intubation errors: isolated events or symptoms of larger problems? Health Affairs **25**, 501–509 (2006)
6. Webb, R.K., Walt, J.H.V.D., Runciman, W.B., Williamson, J.A., Cockings, J., Russel, W.J., et al.: Which monitor? an analysis of 2000 indicent reports. Anaesth Intens. Care **21**, 529–542 (1993)
7. Gravenstein, J.S., Jaffe, M.B., Paulus, D.A.: Capnograhpy clinical aspects. Cambridge University Press (2004)
8. Bhende, S.M., Thompson, A.E.: Evaluation of an end-tidal CO2 detector during pediatric cardiopulmonary resuscitation. Pediat. **95**, 395–399 (1995)
9. Li, J.: Capnography alone is imperfect for endotracheal tube placement confirmation during emergency intubation. J. Emerg. Med. **20**, 223–229 (2001)
10. O'connor, C.J., Mansy, H., Balk, R.A., Tuman, K.J., Sandler, R.H.: Identification of endotracheal tube malpopsitions using computerized analysis of breath sounds via electronic stethoscopes. Anesth Analg. **101**, 735–739 (2005)
11. Tejman-Yarden, S., Zlotnik, A., Weizman, L., Tabrikian, J., Cohen, A., Weksler, N., et al.: Acoustic monitoring of lung sounds for the detection of one-lung intubation. Anesth Analg. **105**, 397–404 (2007)

12. Tejman-Yarden, S., Lederman, D., Weksler, N., Gurman, G.: Acoustic monitoring of double lumen ventilated lungs for the detection of selective unilateral lung ventilation. Anesth Analg. **103**, 1489–1492 (2006)
13. Lederman, D.: An energy ratio test for one lung intubation detection. In: 18th Biennial International EURASIP Conference, Brno, Czech Republic (2006)
14. Weizman, L., Tabrikian, J., Cohen, A.: Detection of one-lung intubation incidents. Annals of Biomed. Eng. **36**, 1844–1855 (2008)
15. Lederman, D., Lampotang, S., Shamir, M.: Automatic endotracheal tube position confirmation system based on image classification- a preliminary assessment. Med. Eng. & Phys., October 2011
16. Haralick, R.M., Shanmugam, M., Dinstein, I.: Textural features for image classification. IEEE Trans on Systems, Man, and Cybernatics **SMC-3**, 610–621 (1973)

Life-Earth Sciences Intelligent Modeling

Intelligent Bio-Inspired Detection of Food Borne Pathogen by DNA Barcodes: The Case of Invasive Fish Species Lagocephalus Sceleratus

Konstantinos Demertzis$^{(\boxtimes)}$ and Lazaros Iliadis$^{(\boxtimes)}$

Department of Forestry & Management of the Environment & Natural Resources,
Democritus University of Thrace, 193 Pandazidou st. 68200, Orestiada, Greece
{kdemertz,liliadis}@fmenr.duth.gr

Abstract. Climate change combined with the increase of extreme weather phenomena, has significantly influenced marine ecosystems, resulting in water overheating, increase of sea level and rising of the acidity of surface waters. The potential impacts in the biodiversity of sensitive ecosystems (such as Mediterranean sea) are obvious. Many organisms are under extinction, whereas other dangerous invasive species are multiplied and thus they are destroying the ecological equilibrium. This research paper presents the development of a sophisticated, fast and accurate Food Pathogen Detection (FPD) system, which uses the biologically inspired Artificial Intelligence algorithm of Extreme Learning Machines. The aim is the automated identification and control of the extremely dangerous for human health invasive fish species "Lagocephalus Sceleratus". The matching is achieved through extensive comparisons of protein and DNA sequences, known also as DNA barcodes following an ensemble learning approach.

Keywords: Extreme learning machines · Ensemble learning · Food pathogen detection · DNA barcoding · Lagocephalus sceleratus · Invasive species · Climate change

1 Introduction

1.1 Climate Change and Invasive Species

The effect of climate change in marine ecosystems is obvious in various levels of biological organization and especially in the disturbances of biodiversity and in the extinction of organisms due to the appearance of invasive species (IS). The IS are intruders in new strange for them habitats where they can disturb the natural flora and fauna, harming the environment. The social and financial consequences are considered very crucial. For example they can affect human health, agriculture, fishing and food production. Especially, regarding marine ecosystems, various species like fish are traveling searching for colder water, either due to the fact that their natural environment does not satisfy the range of temperatures required for their survival, or because they follow various plant-organism species moving to colder waters [1].

The Lagocephalus Sceleratus is a characteristic case of invasive species whose presence in the Mediterranean Sea, causes serious problems [2]. It is probably the

© Springer International Publishing Switzerland 2015
L. Iliadis and C. Jayne (Eds.): EANN 2015, CCIS 517, pp. 89–99, 2015.
DOI: 10.1007/978-3-319-23983-5_9

most dangerous from the 29 "lessepsian" immigrants that passed in the Mediterranean Sea through the Suez Canal. Its uncontrolled invasion and its reproduction, threatens the marine environment with an irreparable imbalance. Its presence causes an intense competition with the native fish regarding the available food. Moreover its consumption by humans is extremely dangerous as it contains the toxic substance "tetradotoxine" which causes stomach pain, then diarrhea, vomiting, breathlessness, paralysis and even death. Pattern recognition of the Lagocephalus Sceleratus exclusively with phenotypic markers is an extremely difficult and dangerous process. This is due to the fact that this species is unknown to the public and also neither the big morphological differences nor the significant similarities reflect the affinity of the organisms. The need for a comprehensive and absolutely valid identification of the specific species is vital in cases of food production or food canning as it is obviously a serious threat for public health.

1.2   Literature Review

Akova et al. [3] proposed a novel method using an advanced statistical model that boosts the ability of computers to detect the presence of bacterial contamination in tested samples. These formulae drive machine-learning, making possible the identification of known and unknown classes of food pathogens. In [4] an optical sensor system is proposed, for rapid and noninvasive classification of FBP. This system was used for image acquisition and focused in comparing its potential with multivariate calibrations in classifying three categories of popular bacteria. Both linear (LDA, KNN and PLSDA) and nonlinear (BPANN, SVM and OSELM) pattern recognition methods were employed comparatively for modeling. Also [5] describes an application of the BARDOT (Bacteria Rapid Detection using Optical scattering Technology), which is a technique for pathogen belongs to the broad class of optical sensors and relies on forward-scatter phenotyping to classify bacteria belonging to the Salmonella class in a non-exhaustive framework. They use a Bayesian approach in learning with a non-exhaustive training dataset to allow for the automated detection of unknown bacterial classes. Similarly [6] demonstrates a pattern recognition technique to classify particles on the basis of their discrete scatter patterns collected at just five different angles, and accompanied by the measurement of axial light loss. The proposed approach can be potentially used with existing instruments because it requires only the addition of a compact enhanced scatter detector. It has been shown that information provided just by five angles of scatter and axial light loss can be sufficient to recognize various bacteria with 68-99% success rate.

In [7] Pan Yi discusses the use of machine learning methods with various advanced encoding schemes and classifiers in order to improve the accuracy of protein structure prediction. Also [8] proposes a machine learning method for classifying DNA-binding proteins from non-binding proteins based on sequence information. Finally paper [9] introduces three ensemble machine learning methods for analysis of biological DNA binding by transcription factors (TFs). The goal is to identify both TF target genes and their binding motifs. Subspace-valued weak learners (formed from an ensemble of different motif finding algorithms) combine candidate motifs as probability weight matrices (PWM), which are then translated into subspaces of a DNA k-mer (string) feature space. Assessing and then integrating highly informative subspaces by machine methods gives more reliable target classification and motif prediction.

2 Conventional Methods for Food Pathogen Detection

Conventional methods for testing food-borne pathogens are based on the cultivation of pathogens, a process that is complicated and time consuming. Three methods for Food Pathogen Detection can be used:

- ✓ Polymerase Chain Reaction (PCR) sequencing,
- ✓ Restriction Fragment Length Polymorphisms (RFLP) and
- ✓ DNA barcoding.

The most common method, PCR, involves extracting DNA from the food followed by amplification of specific pieces of DNA through an enzymatic process. The amplified DNA fragments are separated by size using a technique called agarose gel electrophoresis and are compared with DNA fragments of known size to enable their identification.

In RFLP analysis, the DNA sample is broken into pieces by restriction enzymes i.e. enzymes that can recognize specific base sequences in DNA and cut the DNA at that site (the restriction site). The resulting restriction fragments are separated according to their size using gel electrophoresis. RFLP analysis was the first DNA profiling technique inexpensive enough for widespread application.

DNA barcoding is a molecular based system, which is based on the analysis of a short genetic marker called the "DNA barcode" in an organism's DNA. The most commonly used barcode region, for animals, at least, is a segment of approximately 600 base pairs of the mitochondrial gene cytochrome oxidase I (COI). By comparing the DNA barcode to a compiled database of barcodes it can be identified as belonging to a particular species. It differs from molecular phylogeny in that the main goal is not to determine patterns of relationship but to identify an unknown sample in terms of a preexisting classification. So there is demand for alternative methods to test for food-borne pathogens that are simpler, quick and applicable to a wide range of potential applications [10].

3 Novelty of the Proposed Intelligent System

The main contribution of this research is the development of the sophisticated ISDNAEF approach (Intelligent System for DNA Extraction in Food testing) embedded in Droplet Digital System [11], for the automated recognition of the Lagocephalus Sceleratus fish. This process includes extended comparisons of proteins and DNA sequences (barcodes). Initially a mapping of the existing DNA barcodes to a set of spatial points is done, in order to determine the complex relations that characterize these datasets. Pattern recognition in this research is performed by Employing Ensemble Learning (ENL) an approach that incorporates the biologically inspired Extreme Learning Machine (ELM) Artificial Intelligence (AI) algorithm. ELMs are a modern, fast and reliable Machine Learning (ML) approach used successfully for classification problems. A comparative analysis of the convergence of the ISDNAEF bio-inspired algorithm with the performance of other relative ML methodologies has shown its reliability for cases of high complexity datasets.

The actual innovation of this research is the AI incorporation in the digital machines that perform the accurate identification, offering safety in food production and protection of consumers' health. Not only the DNA extraction is done extremely accurately but also the required time is minimized significantly, due to the exploitation of the ELMs.

Another strong advantage of this research is the gathering and selection of data which is done not only by employing heuristic and bibliographic approaches but also with the use of the FASTA algorithm [12] that creates scenarios of high complexity and rational data sets that can generalize to new ones.

Feature selection was performed by using Particle Swarm Optimization (PSO) [13] in order to have the optimal convergence and the minimum requirement of computational resources. To the best of our knowledge the ELMs are used for the first time in the case of a Food borne Pathogen Detection System, performing DNA barcode analysis.

4   Machine Learning Methodologies

4.1   Ensemble Learning

Ensemble methods [14] use multiple learning algorithms to obtain better predictive performance than could be obtained from any of the constituent learning algorithms. Usually they refer only to a concrete finite set of alternative models, but typically they allow for much more flexible structures to exist between those alternatives. Also, they are primarily used to improve the performance of a model, or to reduce the likelihood of an unfortunate selection of a poor one. Other applications of ensemble learning include assigning a confidence to the decision made by the model, selecting optimal (or near optimal) features, data fusion, incremental learning, non-stationary learning and error-correcting. The novel concept of combining learning algorithms is proposed as a new direction of ensemble methods for the improvement of the performance of individual algorithms. These algorithms could be based on a variety of learning methodologies and could achieve different ratios of individual results. The goal of the ensembles of algorithms is to generate more certain, precise and accurate system results. Numerous methods have been suggested for the creation of ensembles of learning algorithms:

- ✓ Using different subsets of training data with a single learning method.
- ✓ Using different training parameters with a single training method (e.g. using different initial weights or learning methods for each neural network in an ensemble).
- ✓ Using different learning methods.

Herein the 1st approach was applied in order to develop the ELM ensembles.

4.2   Extreme Learning Machines (ELM)

The ELMs are applying the Single hidden Layer Feed Forward Networks' approach (SLFNs) and the conventional (both single hidden layer and multi hidden layer) feed forward ANN one, with the particularity that the hidden layer (known as feature mapping) of the SLFNs does not necessarily work in concert. In other words the hidden nodes/neurons of the generalized feed forward networks might have been developed randomly. Also all the hidden node parameters are independent from the target functions or from the training datasets. The output weights of the ELMs might be determined in various ways (e.g. with or without iterations, with or without incremental implementations). Finally the ELMs might create the hidden node/neuron parameters randomly before seeing the training data vectors and they can handle non-differentiable activation functions without any problems, like stopping criterion, learning rate and learning epochs [15].

5   Description of the Proposed Algorithm

Before applying ensembles of algorithms using different subsets of training data with the ELM learning method, resampling of the datasets was performed by employing the Bootstrap with replacement methodology. Bootstrapping is a statistical method for estimating the sampling distribution of an estimator by sampling with replacement from the original sample, with the purpose of deriving robust estimates of standard errors and confidence intervals of a population parameter, like mean, median, proportion, odds ratio and correlation coefficient. In the methodology proposed herein, three successive data resampling's were performed in order to obtain the average performance of the algorithm and determine its accuracy. The overall algorithmic approach that was proposed herein is described clearly and in details in the figure 1.

Subsequently, the ELMs were used in order to classify the developed datasets. This was done by employing the Gaussian Radial Basis Function kernel according to equation 1 [16].

$$K(u,v)=\exp(-\gamma\|u-v\|^2) \tag{1}$$

The hidden neurons $k=20$, w_i are the assigned random input weights and b_i the biases, where i=1,...,N and H is the hidden layer output matrix.

$$H=\begin{bmatrix} h(x_1) \\ \vdots \\ h(x_N) \end{bmatrix}=\begin{bmatrix} h_1(x_1) & \cdots & h_L(x_1) \\ \vdots & & \vdots \\ h_1(x_N) & \cdots & h_L(x_N) \end{bmatrix} \tag{2}$$

$h(x) = [h_1(x), \ldots , h_L(x)]$ is the output (row) vector of the hidden layer with respect to the input x.

Function $h(x)$ actually maps the data from the d-dimensional input space to the L-dimensional hidden-layer feature space (ELM feature space) H and thus, $h(x)$ is indeed a feature mapping. ELM aims to minimize the training error as well as the norm of the output weights as shown in equation 3 [16]:

$$\text{Minimize} : \|H\beta - T\|^2 \text{ and } \|\beta\| \qquad (3)$$

To minimize the norm of the output weights $\|\beta\|$ is actually to maximize the distance of the separating margins of the two different classes in the ELM feature space $2/\|\beta\|$. The calculation of the output weights β is done according to equation 4 [16]:

$$\beta(\frac{I}{C} + H^T H)^{-1} H^T T = \qquad (4)$$

where c is a positive constant and T is obtained from the *Function Approximation of SLFNs* with additive neurons

$$T = \begin{bmatrix} t_1^T \\ \vdots \\ t_N^T \end{bmatrix} \qquad (5)$$

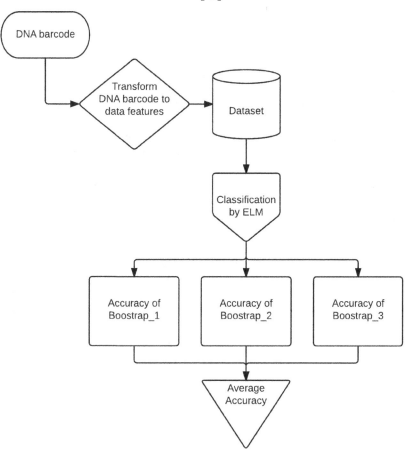

Fig. 1. The structure of the proposed algorithm

6   Data and Testing

Three datasets were developed to be used in the training process of the algorithm. These datasets emerged from the numerical conversion of the genetic information encoded as A, T, C and G (abbreviations of adenine, thymine, cytosine, and guanine) included in the DNA barcodes of the species under examination [17] [18].

The first *lago_family_dataset* (Lagocephalus Family Dataset) comprised of 582 independent parameters and 11 classes, coming from the DNA barcodes 88 fish samples of the Lagocephalus family as shown in the following graph 1. It should be mentioned that three out of the eleven classes include only one case.

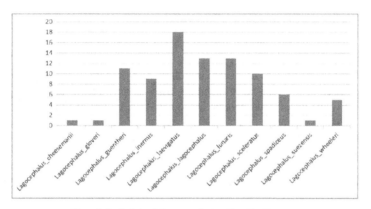

Graph 1. Original Lagocephalus Family Dataset

Actually an imbalanced dataset was developed. Imbalanced datasets are a special case of classification where the distribution is not uniform among the classes. The problem in such cases is that standard learners are often biased towards the majority class. That is because these classifiers attempt to reduce global quantities such as the error rate, without taking the data distribution into consideration. As a result examples from the overwhelming class are well-classified whereas examples from the minority class tend to be misclassified.

The Synthetic Minority Oversampling technique (SMOTE) was employed to resolve the above problem by performing re sampling of the minority class [19]. Re sampling the minority class provides a simple way of biasing the generalization process. It can do so by generating synthetic samples accordingly biased and controlling the amount and placement of the new samples. SMOTE is a technique which combines informed oversampling of the minority class with random undersampling of the majority class. Also it produces the best results as far as re-sampling and modifying the probabilistic estimate techniques. For each minority sample, SMOTE works as follows:

- ✓ Find its *k*-nearest minority neighbors.
- ✓ Randomly select *j* of these neighbors.
- ✓ Randomly generate synthetic samples along the lines joining the minority sample and its *j* selected neighbors (*j* depends on the amount of oversampling desired).

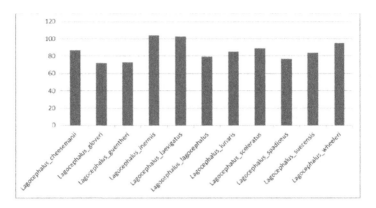

Graph 2. Lagocephalus Family Dataset after application of SMOTE

The *lago_family_dataset*, was re-created by following the SMOTE technique and its new version comprised of 948 cases, distributed in classes as shown in the following graph 2.

Additionally the *usual_fish_dataset* was created, including 1127 fish samples using 519 independent parameters and 14 classes representing widely used Mediterranean fish that can be confused as Lagocephalus Sceleratus.

Finally extended protein and DNA sequence comparisons related to Lagocephalus Sceleratus, were performed with fish species that have high genetic similarities, in order to create a new dataset of high complexity by following the Protein Similarity Search (PSS) method. PSS method provides sequence similarity searching against protein databases using the FASTA algorithm. The FASTA algorithm searches protein and deoxyribonucleic acid (DNA) databases for sequences with statistically significant similarity and compares proteins, DNA, short peptides and oligonucleotides [20].

The *lago_fasta_dataset* (Lagocephalus FASTA Dataset) that have emerged includes the DNA barcode of 772 fish, related to 558 independent parameters and 16 classes, corresponding to fish of high genetic similarity with Lagocephalus Sceleratus.

6.1   Results

The basic hypothesis that the use of datasets with high genetic similarity produces lower efficiency machine learning models was confirmed. We have found higher correlation between the independent parameters and the classes. More specifically for the *usual_fish_dataset*, whose data are related to usual Mediterranean fish with high level of phenotype similarity and low similarity of genetic indices to the Lagocephalus Sceleratus Machine Learning algorithms managed to classify the data with an accuracy as high as 99%. Respectively in the case of the *lago_family_dataset* which includes only data of the *Lagocephalus* species the classification performance was quite lower which is something due to the high genetic correlation between the fish species under examination. This led to high complexity between the independent parameters and the dataset classes. This problem was re confirmed more intensively in the *lago_fasta_dataset* where the data had the highest similarity index in terms of protein and DNA sequences. These similarities were determined after using the

Protein Similarity Search method and the FASTA algorithm. However the fact that the classification efficiency results still remain very high (in the worst case 94%) is very encouraging. This fact enhances and confirms that our proposal to embed AI algorithms in automatic devices for DNA extraction for food testing is correct.

6.1.1 Comparative Analysis

An analysis was performed for the datasets under consideration, between the ELMs method that we have used, versus Multilayer Perceptron (MLP), Support Vector Machines (SVM) and Random Forest (RF). The results are presented in the following table 1. In this case the hold out approach was used (70% Training, 15% Validation and 15% Testing).

Table 1. Accuracy (ACC) and Performance Matrices (PM) comparison between ELM, MLP, SVM and RF algorithms in *usual_fish_dataset*

| Classifier | usual_fish_dataset | | | lago_family_dataset | | | lago_fasta_dataset | | |
| | ACC and PM | | | ACC and PM | | | ACC and PM | | |
	ACC	RMSE	ROC	ACC	RMSE	ROC	ACC	RMSE	ROC
ELM	**98.6%**	**0.0416**	**0.998**	**96.7%**	**0.0879**	**0.987**	**95.9%**	**0.0810**	**0.983**
MLP	98.2%	0.0473	0.994	95.6%	0.1091	0.986	94.4%	0.0899	0.979
SVM	97.1%	0.0635	0.985	94.8%	0.1139	0.973	94.1%	0.0817	0.980
RF	97.9%	0.0609	0.994	95.5%	0.1055	0.974	94.8%	0.0826	0.979

Taking into consideration the superiority of the ELMs for every dataset we have used the ensembles of ELMs approach with different subsets of training data. The Bootstrap with replacement was used for the re-sampling three times for each dataset and the final average classification accuracy was considered. The results are displayed in table 2.

Table 2. Accuracy (ACC) and Performance Matrices (PM) of ELM after resampling by Bootstrap with replacement method

| | usual_fish_dataset | | | lago_family_dataset | | | lago_fasta_dataset | | |
| | ACC and PM | | | ACC and PM | | | ACC and PM | | |
	ACC	RMSE	ROC	ACC	RMSE	ROC	ACC	RMSE	ROC
Boostrap_1	99.1%	0.0314	0.999	96.9%	0.0647	0.990	95.8%	0.0768	0.987
Boostrap_2	98.6%	0.0418	0.998	96.8%	0.0662	0.994	96.1%	0.0715	0.994
Boostrap_3	99.0%	0.0394	0.998	97.0%	0.0658	0.995	96.2%	0.0709	0.995
Average	**98.9%**	**0.0375**	**0.998**	**96.9%**	**0.0656**	**0.993**	**96.0%**	**0.0731**	**0.992**

Due to the fact that the number of the used features was too high for all three datasets (lago_family_dataset 582 independent parameters, usual_fish_dataset 519 and lago_fasta_dataset 558) several feature selection attempts were done for the reduction of the training time and for the enhancement of the generalization in order to avoid overfitting. The Particle Swarm Optimization was used to search for the optimal feature subset. The assessment of each subset was done by considering the value of each subset which is based on the contribution and the degree of redundancy of each characteristic. The parameters considered for the final decision are related to the classification accuracy and to the correlation of the classification errors in comparison to the accuracy of the initial parameters set. After the feature selection the lago_family_dataset has 231 independent parameters (reduction by 60%), the usual_fish_dataset 197 ones (reduction by 62%) and the lago_fasta_dataset 235 (reduction by 57%). The final results are presented in the following table 3.

Table 3. Accuracy (ACC) and Performance Matrices (PM) of ELM after feature selection by PSO search method

	usual_fish_dataset			lago_family_dataset			lago_fasta_dataset		
	ACC and PM			ACC and PM			ACC and PM		
	ACC	RMSE	ROC	ACC	RMSE	ROC	ACC	RMSE	ROC
Boostrap_1	99.3%	0.0309	0.999	97.0%	0.0642	0.991	96.2%	0.0688	0.995
Boostrap_2	98.9%	0.0406	0.999	97.1%	0.0639	0.996	96.3%	0.0685	0.994
Boostrap_3	99.2%	0.0382	0.999	97.3%	0.0631	0.996	96.4%	0.0679	0.996
Average	**99.1%**	**0.0366**	**0.999**	**97.1%**	**0.0637**	**0.994**	**96.3%**	**0.0684**	**0.995**

The final conclusion is that the proposed method has proven to be reliable and efficient and has outperformed at least for these datasets the other approaches.

7   Discussion–Conclusions

This research paper presents an innovative ensemble classifier method using ELMs which was embedded successfully in an advanced food pathogen detection system, aiming in the pattern recognition of the dangerous invasive species Lagocephalus Sceleratus by digital systems. Machine Learning was employed successfully that performs the pattern recognition based on extended comparisons of protein and DNA sequences known as bar codes. It is really important to identify this species in food production and canning. The proposed algorithm uses a sophisticated technique of combined learning, improving the potential efficiency of the classification and the generalization ability. The whole process is automated by running fast bio-inspired machine learning approaches instead of complicated and costly biochemical processes. The most significant innovation of this methodology is that it uses embedded AI algorithms in digital control machines of food quality control that ensure public health. The system was tested thoroughly under various scenarios showing very high accuracy and reliable performance. Future research could involve its extension under a hybrid scheme, which will combine semi supervised methods and online learning for the trace and exploitation of hidden knowledge between the inhomogeneous data that might emerge. Also the use of the unsupervised learning for the re sampling of the datasets will create subsets of higher homogeneity, that will potentially train a model with higher accuracy. Finally, it would be very interesting to perform optimization of the proposed algorithm, by using heuristic approaches like genetic or particle swarm optimization based selective ensemble.

References

1. Frank, J.R., Olden, J.D.: Assessing the Effects of Climate Change on Aquatic Invasive Species. Conservation Biology **22**(3), 521–533 (2008). doi:10.1111/j.1523-1739.2008.00950. x. Society for Conservation Biology
2. Kheifets, J., Rozhavsky, B., Solomonovich, Z.G., Rodman, M., Soroksky, A.: Severe Tetrodotoxin Poisoning after Consumption of Lagocephalus sceleratus (Pufferfish, Fugu) Fished in Mediterranean Sea, Treated with Cholinesterase Inhibitor. Case Reports in Critical Care **2012**, Article ID 782507, 3 p. (2012). doi:10.1155/2012/782507
3. Akova, F., Dundar, M., Davisson, V.J., Hirleman, D.E., Bhunia, A.K., Robinson, J.P., Rajwa, B.: A Machine-Learning Approach to Detecting Unknown Bacterial Serovars. Statistical Analysis and Data Mining (2011). doi:10.1002/sam.10085
4. Pan, W., Zhao, J., Chen, Q.: Classification of foodborne pathogens using near infrared laser scatter imaging system with multivariate calibration (2015). doi:10.1038/srep09524
5. Rajwa, B., Dundar, M.M., Akova, F., Bettasso, A., Patsekin, V., Hirleman, E.D., Bhunia, A.K., Robinson, J.P.: Discovering the Unknown: Detection of Emerging Pathogens Using a Label-Free Light-Scattering System. Cytometry Part A **77A**, 1103–1112 (2010)
6. Rajwa, B., Venkatapathi, M., Ragheb, K., Banada, P.P., Hirleman, E.D., Lary, T., Robinson, J.P.: Automated classification and recognition of bacterial particles in flow by multiangle scatter measurement and a support-vector machine classifier. Cytometry A **73**(4), 369–379 (2008). doi:10.1002/cyto.a.20515
7. Pan, Y.: Protein structure prediction and understanding using machine learning methods. In: 2005 IEEE Granular Computing, vol. 1 (2005). doi:10.1109/GRC.2005.1547225
8. Ma, X., Hu, L.: Extracting sequence features to predict DNA-binding proteins using support vector machine. In: 2013 Fifth International Conference on Computational and Information Sciences (ICCIS) (2013). doi:10.1109/ICCIS.2013.48
9. Yu, D.-J., Hu, J., Li, Q.M., Tang, Z.M., Yang, J.Y., Shen, H.B.: Constructing Query-Driven Dynamic Machine Learning Model With Application to Protein-Ligand Binding Sites Prediction. NanoBioscience, IEEE, **14**(1) (2015)
10. Leigh, D., Thredgold, E.A.V., Lenehan, C.E.: Direct detection of histamine in fish flesh using microchip electrophoresis with capacitively coupled contactless conductivity detection. Anal. Methods, 1802–1808 (2015). doi:10.1039/C4AY02866J
11. http://www.bio-rad.com/
12. Lipman, D.J., Pearson, W.R.: Rapid and sensitive protein similarity searches. Science **227**(4693), 1435–1441 (1985). doi:10.1126/science.2983426. PMID 2983426
13. Moraglio, A., Di Chio, C., Poli, R.: Geometric Particle Swarm Optimization **2008**, Article ID 143624, 14 p. (2008). doi:10.1155/2008/143624
14. Rokach, Lior: Ensemble-based classifiers. Artificial Intelligence Review **33**(1–2), 1–39 (2010). doi:10.1007/s10462-009-9124-7
15. Cambria, E., Huang, G.-B.: Extreme Learning Machines. IEEE Intelligent Systems (2013)
16. Huang, G.-B.: An Insight into Extreme Learning Machines: Random Neurons, Random Features and Kernels (2014). doi:10.1007/s12559-014-9255-2, Springer
17. http://www.cabi.org/isc/
18. http://www.boldsystems.org/
19. Nitesh, V., Chawla, B.K.W., Hall, L.O., Kegelmeyer, W.P.: SMOTE: Synthetic Minority Over-sampling Technique. Journal of Artificial Intelligence Research **16** (2002)
20. http://www.ebi.ac.uk/Tools/sss/fasta/

Modelling of NO$_x$ Emissions in Natural Gas Fired Hot Water Boilers

Pekka Kumpulainen[1(✉)], Timo Korpela[1], Yrjö Majanne[1], and Anna Häyrinen[2]

[1] Department of Automation Science and Engineering, Tampere University of Technology,
P.O. BOX 692 33101, Tampere, Finland
{pekka.kumpulainen,timo.korpela,yrjo.majanne}@tut.fi
[2] Helen PLC, 00090 Helsinki, Finland
anna.hayrinen@helen.fi

Abstract. Nitrogen oxides (NO$_x$) are one of the main pollutants produced by combustion processes. New European emission regulations (IED) extent emission monitoring requirements to smaller boilers. Heating grid operators may have a notable number of such boilers and therefore appreciate affordable monitoring solutions. This paper studies several types of regression models for estimating NO$_x$ emissions in natural gas fired boilers. The objective is to predict the emissions utilising the existing process measurements for monitoring, without an external NO$_x$ analyser. The performance of linear regression is compared with three nonlinear methods: multilayer perceptron, support vector regression and fuzzy inference system. The focus is on generalisation ability. The results on the two boilers in the study suggest that linear regression and multilayer perceptron network outperform the others in predicting with new, unseen data.

Keywords: Nox emission · Nonlinear regression · Multilayer perceptron · Support vector regression · Anfis

1 Introduction

Nitrogen oxides (NO$_x$) are related to nitrogen monoxide (NO) and nitrogen dioxide (NO$_2$) that are present in flue gases of combustion processes. NO$_x$ emissions are responsible for health and environmental issues. Therefore, the NO$_x$ emissions are restricted by further tightening emission regulations that include e.g. emission limits and monitoring requirements.

In Europe, Industrial Emission Directive [1] brings new smaller (15–50 MW) boilers under the directive, so inexpensive NO$_x$ emission monitoring tools are appreciated. One attractive monitoring approach is to estimate the emissions indirectly by existing process measurements and NO$_x$ emission model. In literature, there are numerous data based NO$_x$ emission models for various fuels and boiler types, most of which utilize nonlinear models. Radial basis function (RBF) neural network outperformed multilayer perceptron (MLP) network when applied to a 160 MW gas boiler [2]. Ikonen et al. [3] used neuro-fuzzy modelling for fluidised-bed combustion process. Eng-genes and MLP neural networks were applied to a power plant fed with oil and methane [4]. An ensemble model consisting of fuzzy clustering, least squares

© Springer International Publishing Switzerland 2015
L. Iliadis and C. Jayne (Eds.): EANN 2015, CCIS 517, pp. 100–108, 2015.
DOI: 10.1007/978-3-319-23983-5_10

support vector machine (LSSVM) and partial least squares (PLS) was applied to a coal boiler [5]. Liukkonen & Hiltunen presented an emission warning system based on self-organizing map and fussy clustering [6]. As a conclusion, the literature review proposes to use nonlinear models and multiple input variables for NO$_x$ monitoring, which provide fairly accurate NO$_x$ predictions especially in research environment. The objective is to investigate the performance of several monitoring methods to find practical solution to a real case, i.e. to estimate NO$_x$ emission in two fairly similar natural gas fired municipal hot water boilers.

The paper is organised as follows. Chapter 2 presents the process and case description. Chapter 3 focuses on modelling methods and Chapter 4 on results. Chapter 5 includes discussion and Chapter 6 conclusions.

2   Target Process and Test Setup

The process in this study is a municipal back-up heating plant in Helsinki, Finland. The plant consists of six hot water boilers which are connected to a single chimney. Three boilers have heavy fuel oil burners and three are equipped with both, heavy fuel oil, and natural gas burners. Each boiler produces up to 43 MW power, yielding total power of 258 MW. The data for this study were acquired from two gas boilers, referred here as Boilers 2 and 3. They were identical, except for some modifications made for the burners.

The data to be used as estimators were collected from the process automation system. The data include values from the following measurements: fuel flow, moist flue gas O$_2$ content (w.b.), and temperatures of flue gas, feed water, return water, and boiler room; and pressures of combustion chamber and combustion air. The emissions were measured with a portable flue gas analyser that consists of a Horiba PG-350 SRM portable flue gas analyser with NO$_x$, SO$_2$, CO, CO$_2$ and dry O$_2$ (d.b.) measurements, gas conditioning and sampling system PSS-5, and a portable gas sample probe PSP4000-H. The data from automation system and external NO$_x$ analyser were aligned and resampled to a combined set with 15 seconds sampling rate.

2.1   Test Setup

One trial run was conducted with each boiler, and another independent test run with Boiler 3. Only the boiler under study was in operation at each run, all others were shut down. The beginning of the trial with Boiler 2 was an exception. The gas flow (i.e. fuel power) and flue gas oxygen content dominate the NO$_x$ emissions. The preliminary tests verified that they had the highest correlation with the NO$_x$ level. This is also confirmed by Ferretti & Piroddi [7], which states that the NO$_x$ emissions can be expressed in general form as a function of pressure, residence time, temperature and excess air ratio. The first three are related to boiler power level and therefore directly proportional to the gas flow. The last one is proportional to the flue gas oxygen content. Including additional variables was tested by linear stepwise regression and adding any variables decreased the estimation accuracy in test data.

The trial plan consisted of constant power levels within the normal operational range. Air feed steps to vary oxygen levels were performed at each power level exceeding the typical operation regions. The sections of constant power were used for training the models and the transition sections in normal automation mode were used for testing the model performance.

The trial for Boiler 2 lasted almost eight hours, total of 1895 observations. 1016 observations were used as training and 879 as test data. The power and O_2 levels are presented in Fig. 1. During the first 40 minutes there were also other boilers in operation. This was due to external requirements to provide more power to the heating grid. This was not planned but provides a very interesting testing period for the emission estimation. The constant fuel power levels in Boiler 2 were approximately 37.4, 30.4, 24.6, and 17.2 MW. The dry O_2 steps, measured with the gas analyser, range from 1 to 3.8% (d.b.) at each power level.

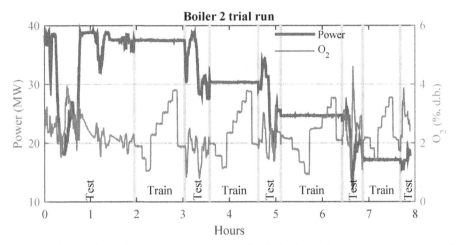

Fig. 1. Power and O_2 levels in the trial setup of Boiler 2. Sections of constant power were used for training of the models and the rest of the run for testing.

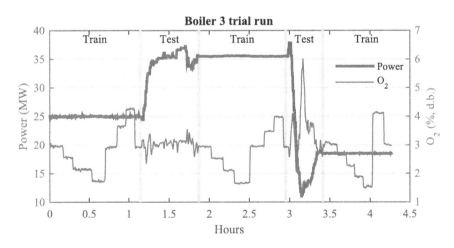

Fig. 2. Power and O_2 levels in the trial setup of Boiler 3. Sections of constant power were used for training of the models and the rest of the run for testing.

The usable duration of the trial run for Boiler 3 was four hours and 16 minutes, total of 1026 observations. 743 observations were used as training and 283 as test data. The constant power levels were 25.0, 35.5, and 18.5 MW as depicted in Fig. 2. The O$_2$ was varied between 1.5 and 4.2 % (d.b.).

The additional test run with Boiler 3 is presented in Fig. 3. The duration of the run was five hours and 36 minutes, total of 1347 observations. This run was conducted in automation mode, and the changes to process states were conducted by power and O$_2$ controller set point changes. The fuel power levels were approximately 19.7, 26.9, 31.8 and 36.1 MW. The O$_2$ set points were between two and four % (d.b.).

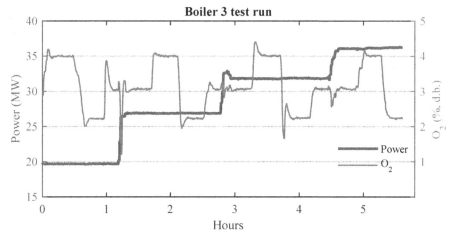

Fig. 3. Power and O$_2$ levels of separate test run on Boiler 3.

2.2   Data Selection and Preprocessing

In this paper two process variables were selected to estimate the NO$_x$ emission: fuel flow and moist flue gas O$_2$ content (w.b.). These have the most effect on the production of NO$_x$. The variables were scaled before applying the modelling methods. All variables x_i were scaled to $x^s{}_i$ by transforming their normal operational range [OR_{min} OR_{max}] to the range [-1 1]: $x^s{}_i = 2 * (x_i - OR_{min})/(OR_{max} - OR_{min}) - 1$. The operational ranges were [0 1.2] Nm³/s for fuel flow and [0 6] for O$_2$. The target variable, NO$_x$ was scaled similarly from operational range [70 180].

3   Methods

The objective in this study was to find a functional application for NO$_x$ emission monitoring which introduces additional requirements to the methods. In addition to prediction accuracy, the model needs to be reasonably straightforward to identify and maintain in varying process conditions. The selected methods are widely used and easily available.

Linear multivariate regression is the most basic method [8]. It is straightforward to identify without any parameters to optimise. Therefore it is also easy to update and maintain.

Multilayer perceptron (MLP) networks [9] with one hidden layer containing sufficient number of neurons act as a universal approximator. Therefore they are very commonly used for nonlinear regression model. We trained the MLP networks using Levenberg-Marquardt method [10].

Sugeno type fuzzy inference system (FIS) is intended for nonlinear regression [11]. The membership functions of the output are either constant or linear functions. The final output is a weighted average of the output functions, weighted by the membership functions of the inputs. The membership functions can be identified from the data by ANFIS (adaptive-network-based fuzzy inference system), a hybrid learning algorithm, which combines the least-squares method, and the backpropagation gradient descent method [12]. In this study we used an ANFIS model with linear output functions and bell shape membership functions for inputs.

The basic form Support Vector Machine (SVM) is a classifier that uses hypothesis space of linear functions in a high-dimensional kernel-induced feature space [13]. Support Vector Regression, such as, v-SVR [14] can be used for nonlinear modelling of continuous data. Parameter v controls the number of support vectors in the model. We used Radial Basis Function (RBF) kernel $K(\mathbf{x}_i,\mathbf{x}_j) = \exp(-\gamma\|\mathbf{x}_i - \mathbf{x}_j\|^2)$, $\gamma > 0$, which is versatile and works well in most applications [15]. In this study we used a software package LIBSVM [16].

4   Results

The training sections with constant power levels were used for training the estimation models. The test sets and the separate test run of Boiler 3 were used for model validation only.

4.1   Cross Validation

In order to avoid over fitting, the optimal parameters for all nonlinear models were selected by cross validation with the training data sets.

The optimal number of neurons in the hidden layer of the MLP network was determined by 10-fold cross validation, including ten Monte Carlo repetitions for each partitioning. The minimum cross validation errors were achieved with four neurons for Boiler 2 and two neurons for Boiler 3.

The numbers of bell shape membership functions for inputs in the FIS were also selected by ten-fold cross validation, including ten Monte Carlo repetitions for each partitioning. The minimum cross validation errors we achieved with three functions for Boiler 2 and two functions for Boiler 3.

The optimal values for the parameters v and γ for v-SVR were selected by ten-fold cross validation. The cross validation errors are presented in Fig. 4. The darker grayscale refers to higher error. Absolute values are not relevant here. The minimum error for Boiler 2 was provided with parameters $v = 0.65$ and $\gamma = 2$ ($\log_2(\gamma) = 1$). For Boiler 3 the optimal parameters were $v = 0.7$ and $\gamma = 2$.

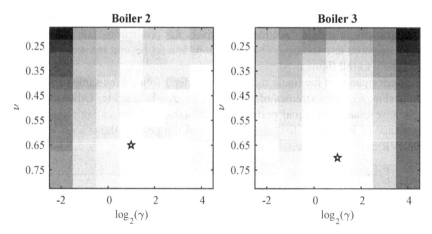

Fig. 4. Ten-fold cross validation errors of ν-SVR regression.

4.2 NO_x Estimates

The parameters detected by cross validation were used to train all models separately for both boilers. The final models were trained with the whole training data sections.

The results for the trial on Boiler 2 are presented in Fig. 5. The training and testing sections are separated by vertical lines and the training sections marked with text 'Tr'. All nonlinear models perform well in training sections. The linear model has difficulties in the first training section when both, the power and O_2, have high or the lowest levels (see Fig.1). However, the performance during test sections is very similar to those of MLP and SVR. The first test section, which is the most challenging due to

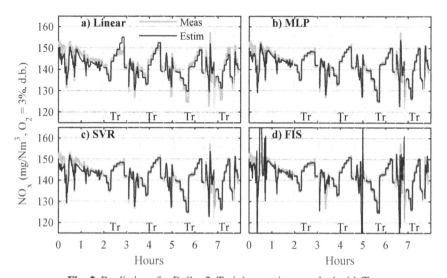

Fig. 5. Predictions for Boiler 2. Training sections marked with Tr.

other boilers also in operation, is best predicted by MLP network. The performance of FIS is excellent in training sections. However, the predictions fail totally at some points in all three test sections. The predictions extend beyond the axis range, which is set equally for all figures.

The performance of the linear model with Boiler 3 follows the same pattern as with Boiler 2 (Fig. 6). The extremes of the O_2 levels are challenging. Other than that, all models have essentially equal performance. Notable differences are slightly inferior performance of FIS in the first test section, and the peak in the second test section, which SVR fails to estimate.

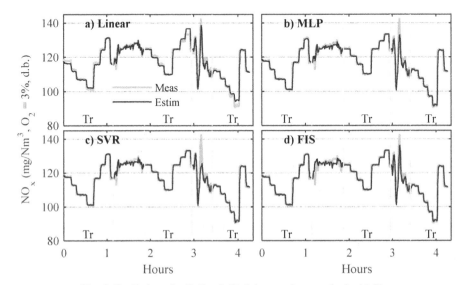

Fig. 6. Predictions for Boiler 3. Training sections marked with Tr.

The linear model performs extraordinarily on the test run with Boiler 3 as seen in Fig. 7. Both, MLP and SVR perform even better, but underestimate the NO_x level at around five hours in the run. The performance of FIS is surprisingly poor. There is a significant bias at all power and O_2 levels.

The performances of the methods are compared by the rmse (root mean squared error). The rmse values of all distinct sections in the trials are collected in Table 1. All nonlinear models have similar performance with the training data, whereas the linear regression is clearly inferior. The rmse values of the test data show that the linear and MLP have most consistent performance. The SVR has significantly higher rmse on the test set of Boiler 3. The performance of the FIS is the best in training sets. However, it fails very much in all test sets, especially in the test sections of Boiler 2.

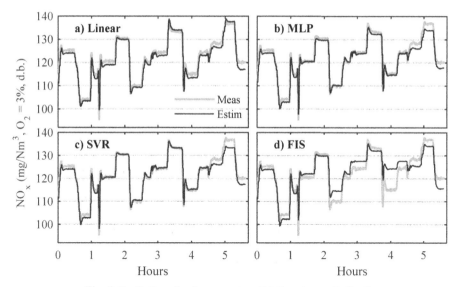

Fig. 7. Predictions for the separate validation day on Boiler 3.

Table 1. The rmse values represent the performance of the methods.

| | Boiler 2 | | Boiler 3 | | |
	Train	Test	Train	Test	Test day
Linear	2.20	2.63	2.03	2.65	2.27
MLP	0.84	2.17	1.71	2.90	2.21
SVM	0.84	2.16	1.71	3.57	2.25
FIS	0.83	7.52	1.70	3.42	3.89

5   Conclusions and Future Work

This paper presents models for estimating NO$_x$ emission in natural gas fired hot water boilers. The models utilize online process measurements that are always available even in relatively simple set-ups. Linear multivariate regression was compared with three nonlinear regression methods: MLP, SVR and FIS.

The linear regression and MLP network had the most consistent performance in the test cases. MLP was the best when other boilers were also in operation with Boiler 2. Despite the cross validation, the FIS was over fitted, performing very well with training sets but failing in tests. It could most probably be fine-tuned to perform better. However, that does not comply with the requirement of easy maintainability.

The estimation models must be fitted separately for each boiler, and retrained occasionally for robust performance. Additional tests are required to test the repeatability of the estimates. Based on preliminary tests only two main variables were used in this study. The effect of additional variables will also be studied.

Acknowledgements. The work was conducted in Measurement, Monitoring and Environmental Assessment (MMEA) program managed by CLEEN Ltd and primarily financed by Tekes, which are gratefully acknowledged. Moreover, project partners, especially Helen, are gratefully acknowledged.

References

1. Industrial Emissions Directive (2010/75/EU) by European Union (2010)
2. Iliyas, S.A., Elshafei, M., Habib, M.A., Adeniran, A.A.: RBF neural network inferential sensor for process emission monitoring. Control Engineering Practice **21**, 962–970 (2013)
3. Ikonen, E., Najimb, K., Kortela, U.: Neuro-fuzzy modelling of power plant flue-gas emissions. Engineering Applications of Artificial Intelligence **13**, 705–717 (2000)
4. Li, K., Peng, J., Irwin, G.W., Piroddi, L., Spinelli, W.: Estimation of NOX emissions in thermal power plants using eng-genes neural networks. In: IFAC WC 2005, Prague, pp. 115–120 (2005)
5. Lv, Y., Liu, J., Yang, T., Zeng, D.: A Novel Least Squares Support Vector Machine Ensemble Model for NOx Emission Prediction of a Coal-Fired Boiler. Energy **55**, 319–329 (2013)
6. Liukkonen, M., Hiltunen, T.: Adaptive monitoring of emissions in energy boilers using self-organizing maps: An application to a biomass-fired CFB (circulating fluidized bed). Energy **73**, 443–452 (2014)
7. Ferretti, G., Piroddi, L.: Estimation of NOx Emissions in Thermal Power Plants using Neural Networks. J. Eng.Gas Turbines Power **123**(2), 465–471 (2001)
8. Draper, N.R., Smith, H.: Applied regression analysis, 3rd edn. Wiley, New York (1998)
9. Haykin, S.: Neural Networks. McMillan, New York (1994)
10. NNSYSID Toolbox - for use with MATLAB. http://www.iau.dtu.dk/research/control/nnsysid.html
11. Sugeno, M.: Industrial Applications of Fuzzy Control. Elsevier, New York (1985)
12. Jang, J.-S.R.: ANFIS: Adaptive-Network-based Fuzzy Inference Systems. IEEE Transactions on Systems, Man, and Cybernetics **23**(3), 665–685 (1993)
13. Cristianini, N., Shawe-Taylor, J.: An Introduction to Support Vector Machines and Other Kernel-based Learning Methods. Cambridge University Press (2000)
14. Schölkopf, B., Smola, A., Williamson, R.C., Bartlett, P.L.: New support vector algorithms. Neural Computation **12**, 1207–1245 (2000)
15. Hsu, C.-W., Chang, C.-C., Lin, C.-J.: A Practical Guide to Support Vector Classification. http://www.csie.ntu.edu.tw/~cjlin/papers/guide/guide.pdf
16. Chang, C.-C., Lin, C.-J.: LIBSVM: a library for support vector machines. ACM Transactions on Intelligent Systems and Technology **2**, 27:1–27:27 (2011). http://www.csie.ntu.edu.tw/~cjlin/libsvm

Neural Network Approaches to Solution of the Inverse Problem of Identification and Determination of the Ionic Composition of Multi-component Water Solutions

Sergey Dolenko[1(✉)], Alexander Efitorov[1,2], Sergey Burikov[1,2], Tatiana Dolenko[1,2], Kirill Laptinskiy[1,2], and Igor Persiantsev[1]

[1] D.V. Skobeltsyn Institute of Nuclear Physics,
M.V. Lomonosov Moscow State University, Moscow, Russia
dolenko@srd.sinp.msu.ru
[2] Physical Department, M.V. Lomonosov State University, Moscow, Russia

Abstract. The studied inverse problem is determination of ionic composition of inorganic salts (concentrations of up to 10 ions) in multi-component water solutions by their Raman spectra. The regression problem was solved in two ways: 1) by a multilayer perceptron trained on the large dataset, composed of spectra of all possible mixing options of ions in water; 2) dividing the data set into compact clusters and creating regression models for each cluster separately. Within the first approach, we used supervised training of neural network, achieving good results. Unfortunately, this method isn't stable enough; the results depend on data subdivision into training, test, and out-of-sample sets. In the second approach, we used algorithms of unsupervised learning for data clustering: Kohonen networks, k-means, k-medoids and hierarchical clustering, and built partial least squares regression models on the small datasets of each cluster. Both approaches and their results are discussed in this paper.

Keywords: Inverse problems · Multi-component solutions · Ionic composition · Identification · Clusterization · Raman spectroscopy

1 Introduction

The problem of determination of concentrations of substances dissolved in water is very important for ecological monitoring and control of mineral and waste waters. It is required to be solved in non-contact express mode with acceptable precision.

The method of Raman spectroscopy complies with all these requirements. Principle opportunity of using Raman spectra for diagnostics of solutions results from high sensitivity of their characteristics to types and concentrations of ions present in water. In [1,2] it was suggested to use Raman spectra of complex ions (such as bands of NO_3^-, SO_4^{2-}, PO_4^{3-}, CO_3^{2-} anions in the area about 1000 cm^{-1}) to determine types and concentrations of salts in water. Anion type can be determined by the position of the corresponding band, its concentration by its intensity. However, this method can be used only for analysis of substances having proper Raman bands. In [3,4], the concentration of the

© Springer International Publishing Switzerland 2015
L. Iliadis and C. Jayne (Eds.): EANN 2015, CCIS 517, pp. 109–118, 2015.
DOI: 10.1007/978-3-319-23983-5_11

single salt in the solution was determined by water Raman valence band. In [5,6], a method for determination of concentrations of salts in multi-component solutions by water Raman valence band was suggested and elaborated. This was possible due to use of neural networks (NN) to solve this inverse problem (IP).

In this study, a method of identification and determination of both complex and simple ions is used. The method was first suggested for salts by the authors in [7] and developed in [8-10]. Presence of complex ions is determined in the easiest way by presence of their valence bands in the low-frequency region of Raman spectrum, and their concentrations can be determined by the dependence of the intensity of these bands on concentration, taking into account the influence of other ions. Recognition and determination of concentration of simple ions is performed by the change of shape and position of water Raman valence band in presence of all ions. Simultaneous determination of concentrations of a number of ions present in water and their identification are provided by use of a NN performing simultaneous analysis of both regions of Raman spectrum (valence and low-frequency).

The first approach to the solution of the studied IP is training a NN with the number of inputs equal to the number of channels in raw or pre-processed spectra, and the number of outputs equal to the maximum number of ions in the solution. The amplitude at an output of a trained NN is proportional to the concentration of the ion corresponding to this output. Thus, the task of determination of the component composition of the solution and the inverse problem of determination of the ions concentrations are solved simultaneously.

However, this approach was developed for diagnostics of mainly natural waters - seawater, river waters, mineral waters. In contrast to natural water, ionic composition and ion concentrations in technology waters vary within significantly larger ranges. This means that a method that uses an ANN trained to recognize certain ions, may turn ineligible. Such situation causes the necessity of prior determination of ionic composition of the solution, e.g. by solving the problem of clusterization of data array with subsequent correlation of each cluster to a definite ionic composition.

Kohonen neural networks and self-organizing maps (SOM) [11] have been successfully used for a long time to solve problems of clusterization and visualization of multi-dimensional data in different domains of human activity [12-14]. However, at present time the authors didn't succeed in finding any references to studies of other groups on application of Kohonen ANN and SOM to solve the problem of determination of composition of multi-component mixtures.

The second approach provides the solution of the complex problem of determining concentrations of 5 salts dissolved in water in two stages. The first stage was the division of initial training dataset into small subsets by unsupervised learning clustering methods: k-means, k-means with centers selected by hierarchical clustering, k-medoids and Kohonen ANN [15, 16]. At the second stage, PLS regression models [17] were formed separately for each cluster using only the samples of this cluster. Hence, the application of this system is also performed in two stages: first the new sample is classified by the formed clustering model, and then the PLS regression model for the corresponding cluster determines the concentrations of salts.

2   Experimental

Excitation of Raman spectra was performed by argon laser (wavelength 488 nm, output power 450 mWt). Raman spectra of water solutions of inorganic salts were measured in 90° geometry using monochromator (resolution 2 cm^{-1}) and CCD-camera (Jobin Yvon). The temperature of samples during experiment was stabilized at 22.0±0.2 °C. Spectra were normalized to laser radiation power, to spectrum-channel sensitivity of the detector and to spectrum accumulation time.

Spectra were measured in two regions: 300-2300 cm^{-1} and 2300-4000 cm^{-1} for every sample. Raman spectra of water and water solutions in the region 300-1800 cm^{-1} have lines corresponding to valence vibrations of complex anions. Intensities of the bands depend on concentrations of the corresponding ions. The shape of water Raman valence band (2600-4000 cm^{-1}) depends on types of ions and their concentrations. When ion concentration increases, water Raman valence band in most cases shifts towards higher frequencies, its halfwidth decreases, the intensity of its high-frequency part rises and the intensity of low-frequency part falls [4].

3   Approach 1: Solution of the Inverse Problem by a Single Perceptron

3.1   Data Preparation

The objects of research were water solutions of salts with significant content in nature waters – KF, KHCO$_3$, LiCl, LiNO$_3$, MgSO$_4$, Mg(NO$_3$)$_2$, NaCl, NaHCO$_3$, NH$_4$F, (NH$_4$)$_2$SO$_4$. The total concentration of salts in the solutions changed in the range from 0 to 1.5 M, concentration of each salts changed in the range from 0 to 1.5 M with concentration step 0.15 – 0.25 M.

Initially, each of the two bands of Raman spectrum was recorded in the range 1824 channels wide, in the frequency range 565...4000 cm^{-1}. The procedure of data preparation and pre-processing is described in [7,8]. Due to complexity of the object, there is no adequate physical model that would allow numerically obtaining the dependence of water Raman spectrum on concentrations of the dissolved salts. Therefore, in this study, the "experiment-based" approach to IP solution [18] was used.

The obtained data array (1824 features, 4445 samples) was randomly divided into training set (for NN training), test set (training was stopped by minimum error on this set, to prevent overtraining), and examination set (for out-of-sample testing) in the ratio of 70:20:10. This procedure was repeated 5 times to check the stability of the solution against data representativity.

In all computational experiments within this approach, the problem was solved by a multi-layer perceptron (MLP) with a linear transfer function in the output layer (which often performs better for regression-type tasks), and logistic transfer function in the hidden layers. The following parameters were used: learning rate 0.01; moment 0.5; stop training criterion – 1000 epochs after minimum of the error on test data set. In every experiment, 5 identical neural networks with different initial weight approximations were trained, and the results were averaged, to eliminate the influence of the initial MLP weights choice.

It should be noted that the ratio of initial number of input features (1824) and the number of samples in the training set (4445*70% ≈ 3100) was unfavorable. In [19-21], several methods of dimensionality reduction have been tested, including correlation analysis and PCS. It was demonstrated, that the best method to reduce the number of input features was aggregation of adjacent channels. In order to improve the quality of solution by MLP in this research, the investigation of the effect of application of different degrees of aggregation to the data set was carried out.

3.2 Results

At the first stage of the work, we needed to choose the best architecture of MLP. 8 ANN architectures were checked: 3 MLPs with a single hidden layer (number of neurons in the hidden layer: 32, 64, 80), 3 MLPs with two hidden layers (number of neurons in the first and second hidden layers: 32-16, 64-32, and 80-40, respectively), 2 MLPs with three hidden layers (number of neurons in the hidden layers: 64-32-16 and 48-24-12, respectively). Figure 1 presents the values of the coefficient of multiple determination R^2 (averaged over outputs), which characterizes the accuracy of ions concentration determination, for various architectures used. The best result was demonstrated by the largest MLP with 3 hidden layers, thus this ANN architecture was chosen for further experiments.

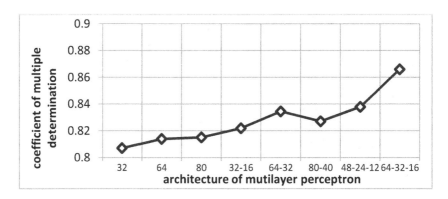

Fig. 1. The dependence of the coefficient of multiple determination R^2 on the examination set (averaged over ANN outputs) on the architecture of a multilayer perceptron (the number of neurons in hidden layers are listed in the signatures of the horizontal axis).

The results of application of different degrees of aggregation to the data set are presented in Figure 2. Only 2-fold aggregation improved the average result of determination of concentrations. Particularly, the most significant improvement 12.5% was achieved for NH_4^+; for HCO_3^-, the improvement was greater than 9,9%, for NO_3^-, F, SO_4^{2-} - greater than 7%. However, for Li^+ and Na^+, the results degraded by 9.2% and 2.1%, respectively.

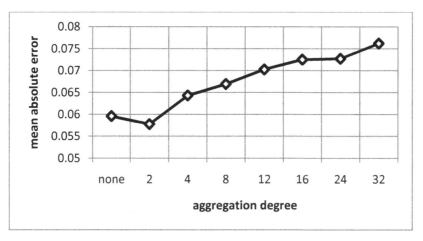

Fig. 2. The dependence of mean absolute error of determination of ions concentrations (averaged over ions, the examination set was used) on aggregation degree (the number of aggregated adjacent channels).

Finally, the variability of the MLP results for determination of the ion concentration for various ions and for 5 random initializations of processing data sets was evaluated. Figure 3 demonstrates, that complex ions are determined much better than simple ones, and that the mean absolute error (MAE) of determination varies quite significantly with data representativity set by the samples comprised in the training set. For HCO_3^- and SO_4^{2-}, the MAE range makes more than 15% of the MAE value.

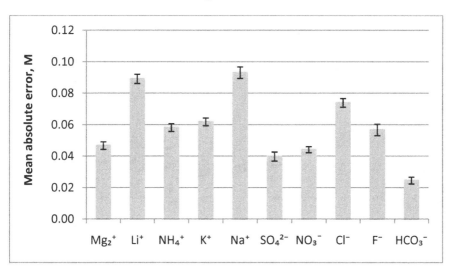

Fig. 3. Mean absolute error of determination of the concentration of ions by MLP averaged over 5 random initializations of the processing data sets. Standard deviation presented by the error bars.

4   Approach 2: Determination of the Ionic Composition of the Solution by Data Clusterization

4.1   Data Preparation

The studied object were the water solutions of KI, NH_4F, $NaNO_3$, $MgSO_4$, $AlCl_3$ inorganic salts. The concentration of each salt in the solutions was changed in the range from 0 to 3-4 M step 0.05 – 0.1 M for all salts, so that the total concentration would not exceed 4 M. As all the 10 ions making the 5 studied salts were different, we can speak of concentrations of salts as well as of concentrations of ions.

The experimentally obtained data array consisted of 807 spectra in 1954 channels. 8-fold spectra aggregation was implemented as the most favorable method of dimension reduction [19-21]. Since the investigated solutions contained from 0 to 5 salts, it was possible to discriminate $2^5=32$ "physical" data classes comprising various combinations of salts present in the solutions belonging to a given class. The initial data array was divided into training and examination sets for each of the 32 classes with 80:20 ratio, to create representative sets for objective checking of methodology.

In [22] it has been demonstrated that the linear method of PLS-regression [17] solved a similar problem better than MLP, provided that the data set was divided into small subsets, and its own regression model was constructed for each subset.

In order to increase the stability of the IP solution, unsupervised learning clustering methods (Kohonen NN, k-means, k-medoid, k-means with centers selected by hierarchical clustering) were used for dataset division along with "physical" division into 32 classes. The number of the obtained clusters varied from 2 to 40. Then a PLS regression model was constructed for each cluster, using only samples from this cluster.

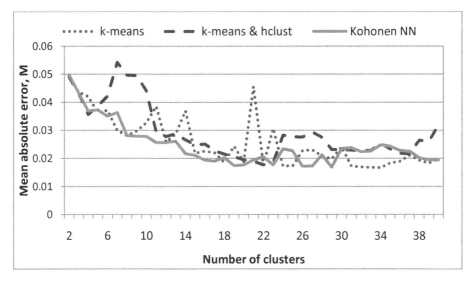

Fig. 4. The dependence of mean absolute error (averaged of salts) of IP solution by PLS-regression models on number of clusters (k-medoids was excluded).

4.2   Results

The dependence of MAE (averaged over all salts) of determination of salt concentrations on the number of clusters is presented in Figure 4. K-medoids was excluded because of the large value of the error. Remember that salts concentrations were calculated by PLS regression models, created on the training samples that fell to each cluster. Table 1 presents the best results of the IP solution. The "physical" division of the dataset was added as a reference point to compare the solution quality.

Table 1. The mean absolute error of salt concentrations determination by PLS regression models created on subsets after division by different clustering methods.

	"physical" classes	k-means	k-means & hclust	Kohonen ANN	k-medoids
No of subsets	32	34	22	29	9
KI	0.0134	0.0146	0.0168	0.0148	0.0347
NH$_4$F	0.0158	0.0250	0.0202	0.0217	0.0325
NaNO$_3$	0.0147	0.0156	0.0193	0.0142	0.0305
MgSO$_4$	0.0086	0.0133	0.0160	0.0158	0.0247
AlCl$_3$	0.0121	0.0150	0.0164	0.0179	0.0255

Obviously, this "physical" division is the best one. To describe the "physicality" of the obtained cauterizations, we introduce the following quantitative index:

$$I_{phys} = \frac{(class'\,samples)}{\sum(cluster's\ samples)}$$

The numerator is the number of samples of the corresponding "physical" class; the denominator is the total number of samples in clusters containing any samples of this class. Special shifting and normalization procedures were applied to achieve uniform variation of the coefficient within [0, 1] for each class. Hence, $I_{phys} = 0$ for some class, if every cluster contains some samples of this class. $I_{phys} = 1$ for some class if there are no samples belonging to other classes in the clusters, containing samples of this class.

Based on Figure 5, we can conclude, that increasing the number of clusters should improve the quality of IP solution. However this assumption is not in agreement with the results presented above (Fig. 4). Since we are using the mathematical PLS regression models to solve the IP, we need to evaluate not only "physicality" of the obtained clustering, but its mathematical quality too. Unfortunately, there is no universal criterion to assess clusterization quality, since different clustering methods "optimize" different criteria, and the "real" clustering is unknown by definition. In our case we need some index, which behavior depends on the number of clusters as well as on the quality of solution of the IP. We considered more than 40 indices available in cluster-Crit package [23]. Finally, two indices were selected: Generalized Dunn's Index (between-cluster distance: single linkage; within-cluster distance measures as mean distance between 2 points belonging to the same cluster) [24] and Xie-Beni index [25]. Figure 6 demonstrates an example of the behavior of the indices and MAE of salt concentrations determination in case of the k-means implementation with centers selected by hierarchical clustering.

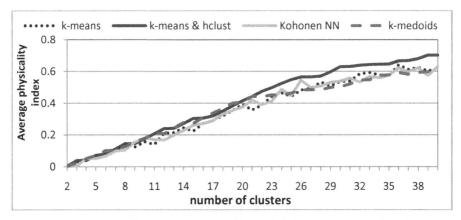

Fig. 5. The dependence of I_{phys} (averaged over 32 physical classes) on the number of clusters.

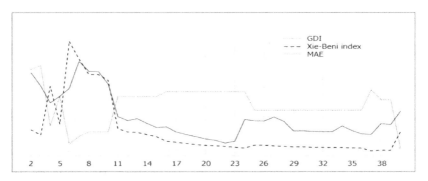

Fig. 6. The dependence of Generalized Dunn's index, Xie-Beni index and mean absolute error of salt concentrations determination on the number of clusters in case of the k-means implementation with centers selected by hierarchical clustering.

5   Conclusion

Complex inverse problems of determining partial concentrations of ions in 10-component water solutions and of salts in 5-component water solutions by Raman spectra have been solved within the "experiment-based" approach using different adaptive supervised and unsupervised learning methods.

In the first approach, the problem of determining concentrations of ions was solved using a multilayer perceptron. The MLP with 3 hidden layers (having 64, 32, 16 neurons, respectively) has demonstrated the best results in comparison with other smaller architectures. The attempts of improving the quality of solution using aggregation of spectra channels demonstrated that only 2-fold aggregation produced a positive result (the averaged over salts improvement of 3.4%). Finally, the variability of the MLP results for determination of the ion concentration for various ions and for 5 random initializations of processing data sets was evaluated. Expectedly, complex ions were determined much better than simple ones, and the error of determination varied significantly with data representativity set by the samples comprised in the training set.

In the second approach, various clustering methods used for solving the problem in two stages were compared. The system based on Kohonen ANN demonstrated the best average result of determining concentrations of 5 salts. Unfortunately, no methods achieved results similar to the supervised partitioning based on salts composition in classes. It has been demonstrated that clustering techniques can be used to determine the chemical composition of some solutions. A criterion named "physicality" has been proposed for evaluation of the obtained clustering. The mathematical quality of clustering, which affects the accuracy of determining the concentrations of salts, can be best evaluated by Generalized Dunn's index and Xie-Beni index.

Acknowledgements. This study was supported by the Russian Scientific Foundation grant no. 14-11-00579 (Approach 1) and by the Russian Foundation for Basic Research grant no. 13-01-00897-a (Approach 2).

References

1. Baldwin, S.F., Brown, C.W.: Detection of Ionic Water Pollutants by Laser Excited Raman Spectroscopy. Water Research **6**, 1601–1604 (1972)
2. Rudolph, W.W., Irmer, G.: Raman and Infrared Spectroscopic Investigation on Aqueous Alkali Metal Phosphate Solutions and Density Functional Theory Calculations of Phosphate-Water Clusters. Applied Spectroscopy **61**(12), 274A–292A (2007)
3. Furic, K., Ciglenecki, I., Cosovic, B.: Raman Spectroscopic Study of Sodium Chloride Water Solutions. J. Molecular Structure **6**, 225–234 (2000)
4. Dolenko, T.A., Churina, I.V., Fadeev, V.V., Glushkov, S.M.: Valence Band of Liquid Water Raman Scattering: Some Peculiarities and Applications in the Diagnostics of Water Media. J. Raman Spectroscopy **31**, 863–870 (2000)
5. Burikov, S.A., Dolenko, T.A., Fadeev, V.V., Sugonyaev, A.V.: New Opportunities in the Determination of Inorganic Compounds in Water by the Method of Laser Raman Spectroscopy. Laser Physics **15**(8), 1–5 (2005)
6. Burikov, S.A., Dolenko, T.A., Fadeev, V.V., Sugonyaev, A.V.: Identification of Inorganic Salts and Determination of Their Concentrations in Water Solutions from the Raman Valence Band Using Artificial Neural Networks. Pattern Recognition and Image Analysis **17**(4), 554–559 (2007)
7. Burikov, S.A., Dolenko, S.A., Dolenko, T.A., Persiantsev, I.G.: Neural network solution of the inverse problem of identification and determination of partial concentrations of inorganic salts in multi-component water solution. In: Proceedings of the XIIth All-Russian scientific and technical conference on Neuroinformatics-2010, part 2, pp. 100-110. MEPhI, Moscow (2010). (In Russian)
8. Burikov, S.A., Dolenko, S.A., Dolenko, T.A., Persiantsev, I.G.: Use of adaptive neural network algorithms to solve problems of identification and determination of concentrations of salts in multi-component water solution by Raman spectra. Neurocomputers: development, application, No. 3, 55-69 (2010). (In Russian)

9. Burikov, S.A., Dolenko, S.A., Dolenko, T.A., Persiantsev, I.G.: Application of Artificial Neural Networks to Solve Problems of Identification and Determination of Concentration of Salts in Multi-Component Water Solutions by Raman spectra. Optical Memory and Neural Networks (Information Optics) **19**(2), 140–148 (2010)
10. Dolenko, S.A., Burikov, S.A., Dolenko, T.A., Persiantsev, I.G.: Adaptive Methods for Solving Inverse Problems in Laser Raman Spectroscopy of Multi-Component Solutions. Pattern Recognition and Image Analysis **22**(4), 551–558 (2012)
11. Kohonen, T.: Self-Organizing Maps, 3rd edn. Springer, Berlin (2001)
12. Deboeck, G., Kohonen, T.: Visual explorations in finance with self-organizing maps. Springer-Verlag London Limited (1998)
13. Seiffert, U., Jain, L.C.: Self-Organizing neural networks: recent advances and applications. Physica-Verlag, Heidelberg (2002)
14. Self-Organizing Maps - Applications and Novel Algorithm Design (2011). http://www.intechopen.com/books/self-organizing-maps-applications-and-novel-algorithm-design
15. Wehrens, R.: Chemometrics with R, p. 286. Springer-Verlag, Heidelberg (2011)
16. Zhao, Y.: R and Data Mining: Examples and Case Studies, p. 256. Academic Press, Elsevier (2012)
17. Wold, S., Geladi, P., Esbensen, K., Öhman, J.: Multi-way principal components-and PLS-analysis. J. of Chemometrics **1**(1), 41–56 (1987)
18. Gerdova, I.V., Dolenko, S.A., Dolenko, T.A., Persiantsev, I.G., Fadeev, V.V., Churina, I.V.: New opportunity solutions to inverse problems in laser spectroscopy involving artificial neural networks. Izv.AN SSSR. Seriya Fizicheskaya **66**(8), 1116–1124 (2002)
19. Dolenko, S., Burikov, S., Dolenko, T., Efitorov, A., Persiantsev, I.: Methods of input data compression in neural network solution of inverse problems of spectroscopy of multi-component solutions. In: Proceedings of the 11th International Conference on Pattern Recognition and Image Analysis: New Information Technologies (PRIA-11-2013), Samara, September 23–28, 2013, vol. II, pp. 541–544. IPSI RAS, Samara (2013)
20. Dolenko, S., Dolenko, T., Burikov, S., Fadeev, V., Sabirov, A., Persiantsev, I.: Comparison of input data compression methods in neural network solution of inverse problem in laser raman spectroscopy of natural waters. In: Villa, A.E., Duch, W., Érdi, P., Masulli, F., Palm, G. (eds.) ICANN 2012, Part II. LNCS, vol. 7553, pp. 443–450. Springer, Heidelberg (2012)
21. Dolenko, S., Burikov, S., Dolenko, T., Efitorov, A., Gushchin, K., Persiantsev, I.: Neural network approaches to solution of the inverse problem of identification and determination of partial concentrations of salts in multi-component water solutions. In: Wermter, S., Weber, C., Duch, W., Honkela, T., Koprinkova-Hristova, P., Magg, S., Palm, G., Villa, A.E.P. (eds.) ICANN 2014. LNCS, vol. 8681, pp. 805–812. Springer, Heidelberg (2014)
22. Efitorov, A.O., Burikov, S.A., Dolenko, T.A., Persiantsev, I.G., Dolenko, S.A.: Comparison of the Quality of Solving the Inverse Problems of Spectroscopy of Mult-Component Solutions with Neural Network Methods and with the Method of Projection to Latent Structures. Optical Memory and Neural Networks (Information Optics) **24**(2), 93–101 (2015)
23. Desgraupes, B.: Clustering Indices. (University of Paris Ouest - Lab Modal'X), p. 34 (2013). http://www.r-project.org
24. Dunn, J.: A Fuzzy Relative of the ISODATA Process and Its Use in Detecting Compact Well-Separated Clusters. Cybernetics and Systems **3**(3), 32–57 (1973)
25. Xie, X.L., Beni, G.: A validity measure for fuzzy clustering. IEEE Transactions on Pattern Analysis and Machine Intelligence **13**(4), 841–847 (1991)

Prediction of Soil Nitrogen from Spectral Features Using Supervised Self Organising Maps

Xanthoula Eirini Pantazi[1(✉)], Dimitrios Moshou[1], Antonios Morellos[1], R.L. Whetton[2], J. Wiebensohn[3], and A.M. Mouazen[2]

[1] Laboratory of Agricultural Engineering, Faculty of Agriculture, Aristotle University of Thessaloniki, Univ. Box 275 54124, Thessaloniki, Greece
renepantazi@gmail.com, dmoshou@agro.auth.gr
[2] Cranfield Soil and AgriFood Institute, Cranfield University, Bedfordshire MK43 0AL, UK
[3] Faculty of Agricultural and Environmental Sciences, Rostock University, Justus-Von-Liebig-Weg 6 18051, Rostock, Germany

Abstract. Soil Total Nitrogen (TN) can be measured with on-line visible and near infrared spectroscopy (vis-NIRS), whose calibration method may considerably affect the measurement accuracy. The aim of this study was to compare Principal Component Regression (PCR) with Supervised Self organizing Maps (SSOM) for the calibration of a visible and near infrared (vis-NIR) spectrophotometer for the on-line measurement of TN in a field in a German farm. A mobile, fiber type, vis-NIR spectrophotometer (AgroSpec from tec5 Technology for Spectroscopy, Germany) mounted in an on-line sensor platform, comprising of measurement range of 305–2200 nm was utilized so as to obtain soil spectra in diffuse reflectance mode. Both PCR and SSOM calibration models of TN were validated with independent validation sets. The obtain root mean square error (rmse) was equal to 0.0313. The component maps of SSOM allow for a visualization of different correlations between spectral components and nitrogen content.

Keywords: Precision farming · Remote sensing · Neural networks · Machine learning · Optical sensing

1 Introduction

Environmental pollution, intensive farming and the following deposition of acidifying substances especially nitrogen and sulfur compounds have boosted soil acidification. This phenomenon can lead to an imbalanced nutrient supply and a lower nutrient uptake by crops [1]. Subsequently, it is important to achieve optimum yields and consistent soil quality aiming to maintain the optimum pH level in the top layer of the field. If soil acidity is not controlled it can result in poor crop growth and enormous yield losses. Additionally the waste of lime is costly and can lead problems in the availability of some micronutrients [2].

Advances in proximal soil sensing methods have proven that on-line sensors can give trustful and data of high resolution on some essential soil properties. Compared to available practices, the visible and near infrared (vis-NIR) spectroscopy appear to

© Springer International Publishing Switzerland 2015
L. Iliadis and C. Jayne (Eds.): EANN 2015, CCIS 517, pp. 119–126, 2015.
DOI: 10.1007/978-3-319-23983-5_12

be the most suitable technology for on-line classification and quantification in internal field variation of soil properties [3, 4, 5, 6, 7, 8, 9]. Linear PLSR analysis is the most common technique for spectral calibration and prediction [10]. Nonlinear architectures e.g. artificial neural networks (ANN) have not drawn the proper attention and were scarcely explored for analyzing soil spectra. Mouazen et al. [11] have demonstrated substantial improvement in the vis-NIR prediction accuracy of key soil properties. This was achieved by combining PLSR and Back- propagation neural networks. Li et al. [12] have presented hybrid modeling of PLSR and ANN aiming to quantify lunar surface minerals. Other nonlinear calibration techniques applied in soil sciences include regression trees [13, 14, 10], multivariate adaptive regression splines [15, 10], support vector machine regression (SVM) [16, 10], and penalized-spline signal regression [16]. ANN was one of the most used methods.

Limited examples in literature concern the application of the ANN architecture for soil analysis with NIR spectroscopy. Fidêncio et al. [17] have applied radial basis function networks (RBFN) in the NIR region (1000–2500 nm). Additionally, Daniel et al. [18] have utilized ANN in the vis-NIR region (400–1100 nm). Mouazen et al. [19] and Viscarra Rossel and Behrens [10] used PLSR with ANN and discrete wavelet transform (DWT) with ANN, respectively. Besides the wide range of non-linear calibration methods demonstrated, ANN appears to be one of the most fitting techniques to classify on-line collected vis-NIR spectra [20]. This is due to its ability to give stable models and to the provision of commercial software to perform the analyses. Concerning the prediction of soil TN from on-line collected vis-NIR spectra it has been shown in Kuang et al.[4] that is possible to predict TN with an accuracy of 0.08 root mean square error (rmse). Until now no literature describes the use of the Supervised SOM to classify on-line collected vis-NIR soil spectra. The objective of this study is to investigate the performance of the SSOM modeling technique for on-line prediction of TN in one field in a German farm.

2   Materials and Methods

The on-line soil measurement with the vis-NIR sensor was carried out in Field 2 in Germany at the location of Nebeliner Weg, in Premslin area, using the on-line sensing platform of CU. Measurements were performed on 03rd August, 2013. During the measurements 140 soil samples were collected and delivered to CU. Scanning of soil samples in the laboratory was made in CU using the same vis-NIR spectrophotometer as that used during the on-line measurement. The laboratory physico-chemical analyses were carried out in CU soil laboratory and in NRM a commercial laboratory analysis provider (www.nrm.uk.com) in the UK. The on-line soil measurement system designed and developed by Mouazen [21] was utilized so as to measure the experimental field in Germany. It comprised of a subsoiler, which was capable of penetrating the soil to the essential depth by making a trench. The trench bottom was smoothed by the vertical forces acting on the subsoiler. The optical probe was enclosed in a steel lens holder. This was mounted to the backside of the subsoiler chisel and obtains diffuse reflectance spectra from a fiber pointing at the bottom of the trench. The subsoiler was combined with the optical unit mounted on a frame. The whole structure was installed on the three

point linkage of the tractor [22] An AgroSpec mobile, fiber type, vis-NIR spectrophotometer (Tec5 Technology for Spectroscopy, Germany) having a measurement range of
305–2200 nm was utilized to provide more precision information about the systems can
be found in [11]. 140 soil samples were gathered during the on-line measurement procedure. The bottom of the trench was capable of collecting the soil samples. For each
sample, around 200 g soil was gathered at a depth of about 15–20 cm. They were closed
and put in a refrigerator at 4 °C to retain the field moisture. Each soil sample was split
into two parts, : one part used to carry out the laboratory reference measurement of soil
TN and the other utilized for optical scanning.

2.1   Optical Measurement in the Laboratory and Feature Extraction

Fresh soil samples were placed into glass containers and a mixer after big stones and
plant residue were taken out [7]. Soil from each sample was put into three Petri dishes.
Soil samples were measured by the same AgroSpec portable spectrophotometer (Tec5
Technology for Spectroscopy, Germany), utilized during the on-line measurement.
Firstly, the range of these spectra was reduced to 371–2150 nm aiming to remove the
noise at both edges of each spectrum. Additional preprocessing of the spectra involved
Savitzky–Golay smoothing, maximum normalization and first derivation [23]. Principal
component regression is a regression analysis technique that is based on principal component analysis (PCA). It considers regressing the outcome (also known as the response
or, the dependent variable) on a set of covariates (also known as predictors or, explanatory variables or, independent variables) based on a standard linear regression model,
but uses PCA for estimating the unknown regression coefficients in the model.

2.2   Supervised Self-Organizing Map with Embedded Output

When SOM is extended with output weights $\mathbf{w}_s^{(out)}$ it is capable of actually learning in
a supervised manner the mapping $\mathbf{y} = f(\mathbf{x})$ when it is presented with training input
and output data. This association is demonstrated schematically in Fig. 1.

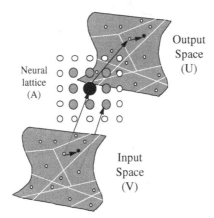

Fig. 1. Construction of input-output mapping by utilizing a Self-Organizing Map during the
learning phase

In order to train an extended map in a self-organizing manner (as opposed to supervised) is to modulate the self-organization procedure to an arbitrarily selected input part, which will rule the self-organizing process during the training period. This method has been introduced by Moshou et al.[24]. The output part can be driven by a small weighting factor for example a=0.005. In this way the input pattern will drive the self-organization due to the best-matching unit (BMU) assessment which is found from the unconstrained input part only. At the operational phase a new test input unforeseen by the map, with a void output part to the SOM. SOM will obtain the BMU and subsequently find the associated output. The algorithm that learns weight updates is given in Eq. (1).

$$\begin{bmatrix} \Delta \mathbf{w}_s^{(in)} \\ \Delta \mathbf{w}_s^{(out)} \end{bmatrix} = \varepsilon h \left(\begin{bmatrix} \mathbf{x} \\ a * \mathbf{y} \end{bmatrix} - \begin{bmatrix} \mathbf{w}_s^{(in)} \\ \mathbf{w}_s^{(out)} \end{bmatrix} \right)$$

(1)

Training parameters ε and h represent the learning rate and neighborhood shape respectively. During the retrieval procedure a new input is introduced to the input map. The neuron in the input map that becomes selected based on the smallest Euclidean distance to the input vector is the winning neuron. Through the mapping that has been learned during the training phase the associated output is retrieved. The whole retrieval process is illustrated in Fig. 2.

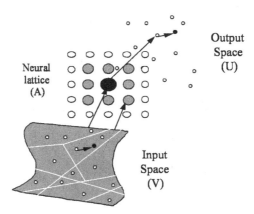

Fig. 2. The retrieval of a learned function from the output space is derived from the matching of the most proximal neuron in the input space.

In the case that the output module is utilized for class prediction, a binary or multiple-valued indicator can be utilized as augmented output on top of the training set. SSOM may be considered as a filtered version of the classical discrete label association scheme. The embedded output module can be utilized for function prediction defined between high-dimensional vector spaces. In this case, the output space data can be characterized as associated output data. Additionally, examples of these data can be used as augmented output in the training set. The prediction of an output target vector following the competitive selection of unit (i) is grounded on the output component $\mathbf{w}_s^{(out)}$ of the neuron. These output components are not fed into the winner neuron search, but they are learned together with the other components as in Eq. (1). In the same way as in classification, in the case of prediction of augmented output data, SSOM may be seen as a smoothed version of the discrete labeling scheme expanded with high dimensional metadata.

3   Results and Discussion

In order to train the SSOM, 106 spectra were used for training and 24 were used for testing. The first two principal components explaining the 93% of variance were used. After training the associated nitrogen layer is retrieved by determining the best matching unit from the incoming spectra. In this way unknown spectra can be used so as to retrieve the nitrogen level. The root mean square error (rmse) was equal to 0.0313, while in the case of PCR it was equal to 0.008. The R^2 was not used due the fact that the small size of the testing set would give misleading results. The advantage of using supervised layer for the SOM is that by encapsulating the output with reducing capability to affect the clustering it is possible to lead the Self Organization to function properly, while at the same time achieve an association with the output values. In the current case study, has been shown that spectral components from soil spectrophotometer can been fed as an input to train the Self Organizing Map and construct an association with nitrogen levels by automatically mining the predicted nitrogen content derived from soil spectra. By training a larger SSOM, it is possible to have a more sensitive visualization of the principal components and how these vary in comparison with nitrogen content. Main aim for this approach was to show the association between the principal components and the values of nitrogen and not to make an accurate regression model from SSOM. So the value of TN retrieved per SSOM neuron, is actually discrete for all input vectors that activate this neuron. This explains the lower value of the rmse, comparing to PCR. As it can be shown in Fig. 3, the second principal component demonstrates slightly higher correlation with the nitrogen distribution. However, it is evident that both principal components have to be combined in order to explain the TN variance. The prediction of TN is considered of high importance for agronomic reasons that are closely related to the growth and development of plants. Nitrogen is a critical component in chlorophyll and it is also vital for photosynthesis procedure and crop protein formation. However, over-use of N fertilizers leads to financial losses but also possible ground water degradation. Determination of within field variability of soil TN at high resolution sampling could assist management of this property in agronomic and environmental systems. The spatial distribution of TN as predicted in this paper, can enable the site specific application of fertilizer in the places were the predicted TN is deficient, while at the same time avoiding over-fertilization in TN rich areas.

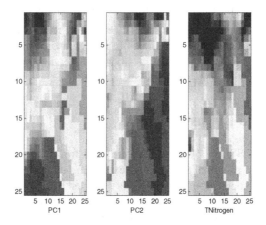

Fig. 3. The component maps of the trained SSOM (25 x 25 neurons) showing the correlation between the first principal component and the nitrogen content.

4   Conclusions

The aim of this study was to propose Supervised Self organizing Maps (SSOM) in order to predict total nitrogen (TN) based on spectral features obtained with a visible and near infrared (vis-NIR) spectrophotometer. A mobile fiber-type vis-NIR spectro-photometer (AgroSpec tec5 Technology for Spectroscopy, Germany) mounted in an on-line sensor platform, comprising of measurement range of 305–2200 nm was utilized so as to obtain soil spectra in diffuse reflectance mode. A trained SSSOM could predict TN and was validated with independent validation sets. The obtained root mean square error (rmse) was equal to 0.0313.The component maps of SSOM allow for a visualization of different correlations between spectral components and nitrogen content. By training a larger SSOM it is possible to have a more sensitive visualization of the principal components and how these variate in comparison with nitrogen content.

Acknowledgements. The presented research was carried out in project FARMFUSE ICT AGRI 2 ERANET.

References

1. Ingerslev, M.: Effects of liming and fertilization on growth, soil chemistry and soil water chemistry in a Norway spruce plantation on a nutrient-poor soil in Denmark. Forest Ecol. Manag., 55–66 (1997)
2. Department for Environment, Food and Rural Affairs (DEFRA, UK). Fertilizer Manual (RB209). Published by TSO (The Stationery Office) (2010). http://www.tsoshop.co.uk, (accessed on October 10, 2013)

3. Bricklemyer, R.S., Brown, D.J.: On-the-go VisNIR: potential and limitations for mapping soil clay and organic carbon. Comput. Electron. Agric. **70**, 209–216 (2010)
4. Kuang, B., Mouazen, A.M.: Non-biased prediction of soil organic carbon and total nitrogen with vis-NIR spectroscopy, as affected by soil moisture content and texture. Biosyst. Eng. **114**(3), 249–258 (2013)
5. Kuang, B., Mouazen, A.M.: Effect of spiking strategy and ratio on calibration of on-line visible and near infrared soil sensor for measurement in European farms. Soil Till. Res. **128**, 125–136 (2013)
6. Kweon, G., Maxton, C.: Soil organic matter sensing with an on-the-go optical sensor. Biosyst. Eng. **115**, 66–81 (2013)
7. Mouazen, A.M., Maleki, M.R., De Baerdemaeker, J., Ramon, H.: On-line measurement of some selected soil properties using a VIS-NIR sensor. Soil Till. Res. **93**, 13–27 (2007)
8. Shi, Z., Cheng, J.L., Huang, M.X., Zhou, L.Q.: Assessing reclamation levels of coastal saline lands with integrated stepwise discriminant analysis and laboratory hyperspectral data. Pedosphere **16**(2), 154–160 (2006)
9. Shibusawa, S., Imade Anom, S.W., Sato, S., Sasao, A., Hirako, S.: Soil mapping using the real-time soil spectrophotometer. In: Grenier, G., Blackmore, S. (eds.) ECPA, Third European Conference on Precision Agriculture, vol. 1, pp. 497–508. Agro Montpellier, Montpellier (2001)
10. Rossel, R.A.V., Behrens, T.: Using data mining to model and interpret soil diffuse reflectance spectra. Geoderma **158**, 46–54 (2010)
11. Mouazen, A.M., Kuang, B., De Baerdemaeker, J., Ramon, H.: Comparison between principal components: partial least squares and artificial neural network analyses for accuracy of measurement of selected soil properties with visible and near infrared spectroscopy. Geoderma **158**, 23–31 (2010)
12. Li, S., Lin, L., Milliken, R., Song, K.: Hybridization of partialleast squares and neural network models for quantifying lunar surface minerals. Icarus **221**, 208–225 (2012)
13. Brown, D.J., Shepherd, K.D., Walsh, M.G., Mays, M.D., Reinsch, T.G.: Global soil characterization with VNIR diffuse reflectance spectroscopy. Geoderma **132**, 273–290 (2006)
14. Vasques, G.M., Grunwald, S., Sickman, J.O.: Comparison of multivariate methods for inferential modeling of soil carbon using visible/near-infrared spectra. Geoderma **146**(1–2), 14–25 (2008)
15. Shepherd, K.D., Walsh, M.G.: Development of reflectance spectral libraries for characterization of soil properties. Soil Sci. Soc. Am. J. **66**, 988–998 (2002)
16. Stevens, A., Udelhoven, T., Denis, A., Tychon, B., Lioy, R., Hoffmann, L., van Wesemael, B.: Measuring soil organic carbon in croplands at regional scale using airborne imaging spectroscopy. Geoderma **158**, 32–45 (2010)
17. Fidêncio, P.H., Poppi, R.J., Andrade, J.C.: Determination of organic matter in soils using radial basis function networks and near infrared spectroscopy. Anal. Chim. Acta **453**, 125–134 (2002)
18. Daniel, K.W., Tripathi, N.K., Honda, K.: Artificial neural network analysis of laboratory and in situ spectra for the estimation of macronutrients in soils of Lop Buri (Thailand). Aust. J. Soil Res. **41**(1), 47–59 (2003)
19. Mouazen, A.M., Kuang, B., De Baerdemaeker, J., Ramon, H.: Comparison between principal components: partial least squares and artificial neural network analyses for accuracy of measurement of selected soil properties with visible and near infrared spectroscopy. Geoderma **158**, 23–31 (2010)

20. Stenberg, B.: Effects of soil sample pretreatments and standardised rewetting as interacted with sand classes on Vis-NIR predictions of clay and soil organic carbon. Geoderma **158**(1–2), 15–22 (2010)
21. Mouazen, A.M.: Soil Survey Device. International publication published under the patent cooperation treaty (PCT). World Intellectual Property Organization, International Bureau. International Publication Number: WO2006/015463; PCT/BE2005/000129; IPC: G01N21/00; G01N21/00 (2006)
22. Mouazen, A.M., De Baerdemaeker, J., Ramon, H.: Towards development of on-line soil moisture content sensor using a fibre-type NIR spectrophotometer. Soil Till. Res. **80**, 171–183 (2005)
23. Martens, H., Naes, T.: Multivariate Calibration, 2nd edn. John Wiley & Sons Ltd, Chichester (1989)
24. Moshou, D., Deprez, K., Ramon, H.: Prediction of spreading processes using a supervised Self-Organizing Map. Mathematics and Computers in Simulation **65**(1–2), 77–85 (2004)

Learning-Algorithms

Multithreaded Local Learning Regularization Neural Networks for Regression Tasks

Yiannis Kokkinos and Konstantinos G. Margaritis[✉]

Parallel and Distributed Processing Laboratory, Department of Applied Informatics, University of Macedonia, 156 Egnatia Str., P.O. Box 1591 54006 Thessaloniki, Greece
kmarg@uom.gr

Abstract. We explore four local learning versions of regularization networks. While global learning algorithms create a global model for all testing points, the local learning algorithms use neighborhoods to learn local parameters and create on the fly a local model specifically designed for any particular testing point. This approach delivers breakthrough performance in many application domains. Usually however the computational overhead is substantial, and in some cases prohibited. For speeding up the online predictions we exploit both multithreaded parallel implementations as well as interplay between locally optimized parameters and globally optimized parameters. The multithreaded local learning regularization neural networks are implemented with OpenMP. The accuracy of the algorithms is tested against several benchmark datasets. The parallel efficiency and speedup is evaluated on a multi-core system.

Keywords: Local learning · Regularization · Parallel processing · Neural networks

1 Introduction

Regression modeling via neural networks often involves trying multiple networks with different architectures and training parameters in order to achieve acceptable model accuracy. Hence, analysis of large amounts of data with neural networks always poses new theoretical and experimental challenges to neural network, machine learning and data mining research communities. In the same time this is a fundamental problem in real life applications. Regularization Networks [1] [2] [3] [4] are a well known family of feed-forward neural networks with one hidden layer, derived from approximation theory. They are viewed as approximation techniques for reconstructing input-output mappings in high-dimensional spaces. They have very good theoretical background [1] [2] [3] [4], simple training and belong to kernel methods. Like any global learning algorithm Regularization Networks are trained by using all the training points to create a global model that serves as a predictor for all the testing points. On the other hand, there are different learning approaches, like the recently emerged local learning approach (see [5–18]). A local learning algorithm builds on the fly a different model for a different testing point using only a local list of neighbour training points closer to each testing point. We explore here this local learning approach applied in four different cases of locally optimized Regularization Networks.

© Springer International Publishing Switzerland 2015
L. Iliadis and C. Jayne (Eds.): EANN 2015, CCIS 517, pp. 129–138, 2015.
DOI: 10.1007/978-3-319-23983-5_13

Bottou and Vapnik [5] proposed the local learning idea which uses neighbourhoods to learn the parameters and create the model at runtime after each testing point is known, by using only the k nearest training points, or the training points lying inside a user defined region, of which the centre is the testing point. This approach was proved effective in many application domains and inspires a variety of algorithms [5-18] that are based on neighbourhoods. However, local learning algorithms as formulated in [5] fall in the class of lazy learning that have great overhead on the testing phase. Hence, some times the computational cost is significant, and in many cases prohibited. To this end, we study four cases of local learning Regularization Networks and their multithreaded parallelization for regression tasks implemented for multi-core systems.

The paper is organized as follows. Section 2 surveys related local learning algorithms. Section 3 describes the local regularization network training and operation. Section 4 outlines the basic principles of the multithreaded implementation. Section 5 provides experimental results and section 6 presents conclusions and summaries.

2   Related Work in Local Learning Algorithms

After the local learning idea in [5] was introduced in the neural network community it has been adapted for various tasks [6]. A local learning neural network studied in [7] explores the concept of sequential learning and the effectiveness of global and local neural network learning algorithms. The work in [8] proposes the Local adaptive subspace regression which localizes the linear regressions in order to approximate nonlinear functions by means of linear models. This method works successfully in high dimensional spaces. In this spirit, the Locally Weighted Projection Regression presented in [9] is an algorithm that achieves nonlinear function approximation in high dimensional spaces with redundant and irrelevant input dimensions.

The Transductive Classification [10] formulation, which learns from both labeled and unlabeled data, has been studied via Local Learning Regularization in [11]. The Local Learning Projection method for Dimensionality reduction was proposed in [12] and it reveals that the projection of a point can be well estimated based on its neighbors in the same class. Two Local Support Vector Machine classifiers where independently proposed in [13] for remote sensing images and in [14] for visual recognition tasks. The main idea of Local SVM classifier is to build at prediction time an example-specific maximal marginal hyperplane based on the set of k-neighbours. An analogous approach of adaptive Local hyperplane classification is given in [15].

A Local Learning Approach for unsupervised clustering tasks is formulated in [16] and achieves good performance. The same clustering objective is also combined with a global label smoothness regularizer to obtain a method for clustering with local and global regularization that was proposed in [17]. From the view of Bayesian confidence the work in [18] has shown how a Local Learning Probabilistic Neural Network classifier that uses few k-nearest neighbour neurons can maintain the confidence ratio of the correctly classified samples and by intrinsically optimizing the number of k-nearest neighbours it reduces substantially the operation cost for large datasets.

3   Local Learning Regularization Networks

3.1   Regularization Network Basics

The Regularization Network algorithm was proposed in [1]. In this seminal paper Poggio and Girosi have shown that for the problem of reconstructing a real function of several variables from a finite number of measurements the regularized solution is a neural network, called Regularization Network (RN). An individual RN [1][2][3][4] has an input layer, a hidden layer, and an output layer. All the training points are loaded into the hidden neurons to form the centers of the kernel functions. The global model $f()$ is the weighted linear combination of these kernel functions [1][2][3][4].

3.2   Local Learning Regularization Networks

While a global learning algorithm creates one model $f(\mathbf{x})$ using all the training data, a local learning algorithm [5] builds on the fly a different local model $f_k(\mathbf{q})$ for each testing point $\mathbf{q}$ using only the k-NN list of k nearest training points closest to $\mathbf{q}$. Hence, a local learning Regularization Network is defined for each $\mathbf{q}$ by the k-NN list which gives the local training set $\{\mathbf{x}_j, y_j\}_{j=1}^{k}$ where all $\mathbf{x}_j$ become the centers of the kernel functions. The output of a local RN is the weighted linear combination of k kernel functions. These are strictly positive radially symmetric functions for which the Gaussian kernel $\exp(-\|\mathbf{x}_i - \mathbf{x}_j\|^2/\sigma^2)$ is commonly used, and the output is given by:

$$f_k(\mathbf{q}) = \sum\nolimits_{j=1}^{k} w_j \exp(-\|\mathbf{q} - \mathbf{x}_j\|^2/\sigma^2) \qquad (1)$$

where k is the number of the k-nearest neighbours and w_j are the weights of the kernel functions. Given the k-NN list of $\mathbf{q}$, the training of the local model $f_k()$ finds the weight vector $\mathbf{w}$ by solving in Reproducing Kernel Hilbert Space H_K a minimization problem for a regularized functional:

$$\arg\min_{f \in H_K}\left\{\frac{1}{k}\sum\nolimits_{i=1}^{k} (y_i - f_k(\mathbf{x}_i))^2 + \frac{\lambda}{k}\|f_k\|^2\right\} \qquad (2)$$

From eq. 2 the weight vector $\mathbf{w}$ is given by solving the system $\mathbf{w} = (\mathbf{K} + \lambda\mathbf{I})^{-1}\mathbf{y}$, where $\mathbf{I}$ is the identity matrix, $\mathbf{K}$ is the local kernel matrix, λ is the regularization parameter and $\mathbf{y}$ is the target vector. The local kernel matrix $\mathbf{K}$ (of size $k\times k$) has entries $\mathbf{K}_{i,j} = \exp(-\|\mathbf{x}_i - \mathbf{x}_j\|^2/\sigma^2)$ and is formed from the local training set. The two hyper-parameters of the model are the regularization parameter λ and the Gaussian width σ which can be locally optimized during model selection via cross-validation. This is usually done by monitoring the virtual leave-one-out squared error [19] given by:

$$E_{loo} = (1/k)\sum\nolimits_{j}^{k} e_j^2 \quad \text{with} \quad e_j = w_j / diag[(\mathbf{K} + \lambda\mathbf{I})^{-1}]_{jj} \qquad (3)$$

where the notation $[]_{jj}$ designates the j-th diagonal element.

3.3   Interplay Between Locally Optimized and Globally Optimized Parameters

Although local learning reduces training cost, the operation cost becomes larger. Learning the best local RN model for each testing point $\mathbf{q}$ requires to locally find the best pair of hyper-parameters $\{\lambda, \sigma\}$ that minimizes the local error E_{loo}. This strategy which we call Case 1 has a large cost. By testing 20 values for λ, 20 values for σ, then the local learning phase must build on the fly 400 different local models.

There exists an interesting interplay between locally optimized parameters and globally optimized parameters. One can try to reduce the operation cost by using a single global parameter, either σ or λ, and only optimize locally the other parameter for each network. Table 1 illustrates the hyper-parameters needed to be optimized from each local regularization network (RN) case used in the experimental comparisons. While a global parameter is optimized during training by using all training data points, a local parameter is optimized during operation for every testing point.

Hence, beyond case 1 there are other cases to consider also. Case 2 uses one globally optimized width σ and optimize locally only the regularization parameter λ. Then, case 3 uses one global λ and optimizing a local σ for each model and case 4 exclusively uses a global λ and a global σ.

Table 1. The hyper-parameters needed to be optimized for each algorithm.

Algorithm	Parameters to be optimized during training (off-line)	Parameters to be optimized during operation (on-line)
Local RN case 1	global k	local σ, local λ
Local RN case 2	global k, global σ	local λ
Local RN case 3	global k , global λ	local σ
Local RN case 4	global k , global λ, global σ	

We explore all these cases in the experimental section.

The online operation of the local learning RN for any of the four cases is given by algorithm 1 in table 2.

Table 2. Online operation phase of local regularization network.

Algorithm 1: local learning RN operation
1 For each testing query point $\mathbf{q}$ find the k-nearest neighbor training points
2 Use this k-NN list as a local training set in order to optimize the local parameters: (either case 1 {local σ, local λ}, or case 2 {local λ}, or case 3 {local σ}, or case 4 none)
3 Predict the output $f_k(\mathbf{q})$

The training of the Local Regularization Networks must find the best number k of nearest neighbor training points. Fortunately, given a training set $\{\mathbf{x}_n, y_n\}_{n=1}^{N}$, the local approach permits a natural leave-one-out cross-validation for the training points, since every $\mathbf{x}_n$ uses its neighbors (and not itself) to train the local model and predict its own target value y_n. Algorithm 2 in table 3 is the training phase of a local Regularization Network for case 1 (the other cases are similarly processed).

Table 3. Training phase of a local regularization network.

Algorithm 2: local RN training for case 1
1 For each training point $\mathbf{x}_n$ ($n = 1,\dots,N$) find its k-nearest training points $\{\mathbf{x}_j,y_j\}_{j=1}^{k}$
2   For each candidate σ_m value
3   Build the local kernel matrix $\mathbf{K}$ with entries $\mathbf{K}(i,j)= \exp(- \|\mathbf{x}_i - \mathbf{x}_j\|^2/\sigma_m^2)$
4   Compute the vector $\mathbf{h}$ with $\mathbf{h}(j)= \exp(- \|\mathbf{x}_n - \mathbf{x}_j\|^2/\sigma_m^2)$ and $j = 1,\dots,k$
5   For each candidate λ_l value
6   Solve the local weights $\mathbf{w} = (\mathbf{K} + \lambda_l\mathbf{I})^{-1}\, \mathbf{y}$
7   Store the residual error $r_n(k,\ \sigma_m,\ \lambda_l) = (y_n - \mathbf{h}\,\mathbf{w})^2$
8   Compute and store $E_{loo}(k,\ \sigma_m,\lambda_l)$ using eq. 3
9   Use the weight's vector with the lowest $E_{loo}(k,\ \sigma_m,\lambda_l)$ to predict $f_k(\mathbf{x}_n) = \mathbf{h}\,\mathbf{w}$
10   update the training error $e(k) = e(k) + (y_n - f_k(\mathbf{x}_n))^2$
11 Select the best k number with minimum training error $e(k)$.

We use a lot of caching to speed up the process. Note that the training must also search the best k number. We search for the best k number in the grid $\{\delta L, 2\delta L,\dots,,L\}$ where L is the maximum candidate k number of neighbors. A local distance matrix maintains the cached distances between the L neighbour points. Based on this matrix a cached local kernel matrix is created once for every candidate σ_m value. Since a local matrix $\mathbf{K} + \lambda_l\mathbf{I}$ is symmetric positive definite, an appropriate solver is through Cholesky factorization which is progressive. Thus for each L, σ_m and λ_l value only one Cholesky is computed for the kernel matrix of the L-NN list of nearest points. Then, progressively the Cholesky back substitution solves for the local weights $\mathbf{w}_k$ of all the k candidate values. All four local RN cases can use the residual errors $r_n(k,\ \sigma_m,\ \lambda_l)$ to find the best global parameters.

3.4   Multithreaded Implementation

The computational cost of the Local RN operation is an issue. A direct way to satisfy the computational demands of such local learning models is to use parallel computing [20] and divide the problem into a number of smaller problems that can be solved concurrently. To this end we examine a shared memory multi-core CPU platform in which the multithreaded implementations divide the computational load into many threads. Parallelisation on multi-core platforms is based on the thread programming model which uses Pthreads, OpenMP, Intel Cilk++ or Intel TBB. We utilize Open Message Passing (OpenMP), which is an API based on fork-join operations where the program enters into a parallel region. All threads can exhibit both task and data-level parallelism and can independently execute code within a same region.

The multithreaded implementation of case 1 (the other cases are similarly defined) exploits the fact that the local Regularization Networks use the k-NN list of each query point q, in order to optimize either the local parameter λ or the local width σ or both of them. There is a candidate parameter set. For the optimum regularization parameter λ we define this set to be composed of M values and the search is conducted into the grid $\{\lambda_1, \lambda_2, \dots, \lambda_M\}$. The same holds for the optimum local width.

Hence, a local RN predictor needs to build, on the fly, M candidate local models in the case it locally optimizes σ, M candidate local models when it locally optimizes λ, and M^2 candidate local models in the case it must optimize locally both parameters. These calculations are the most time-consuming part of the code. The multi-core architecture supports task-level parallelism where each core can asynchronously execute separate threads on separate data regions. We assign every thread to execute a different candidate local model. The model with the lower virtual leave-one-out error is selected as most suitable to compute $f_k(q)$.

4   Experimental Simulations

The first set of simulations aims to compare the four cases of the local Regularization Network. We compare the error rates and the training and testing times. The next set of simulations present the speedup of the multithreaded implementations.

The benchmark datasets for regression are listed in Table 4. California Housing, White wine quality, Red wine quality and Housing can be found in the UCI repository (http://www.ics.uci.edu/). Kinematics, Computer activity and Puma dynamic were downloaded from the Delve Repository (http://www.cs.toronto.edu/delve). The rest benchmark datasets like Ailerons, Delta Ailerons, Delta Elevators can be found in the Torgo's site (http://www.dcc.fc.up.pt/~ltorgo/Regression).

Table 4. Benchmark datasets for regression

Dataset	Instances	Features
California Housing	20640	8
Ailerons	13750	39
Delta Elevators	9517	6
Computer activity	8192	12
Kinematics	8192	8
Delta Ailerons	7129	5
Puma dynamic (8NH)	8193	8
White wine quality	4898	11
Red wine quality	1599	11
Housing	506	13

Each dataset is randomly splitted into 50% training set and 50% test set. All the input features and targets have been normalized into the range [0,1]. The Root Mean Squared Error (RMSE) is used to measure the prediction error on the test set. The RMSE are averaged over 20 runs.

Parameter settings are essential. The optimum number k of nearest neighbours is found globally by searching the grid {10, 20, ..., 90} with step 10. The optimum width parameter σ as well as the optimum regularization parameter λ are found by searching into the grid {0.05, 0.1, 0.15,..., 5.0} with varying step 0.05, using 25 candidate values for each parameter.

4.1   Comparison Results

Table 5 illustrates the averaged root mean squared errors from the comparisons of the four local RN cases {local σ, local λ}, {global σ, local λ}, {local σ, global λ} and {global σ, global λ}. For each dataset the lowest RMSE is shown in boldface. From the last column of table 5 we can see that the fourth case performs better.

Table 5. Root Mean Squared Errors of the test set, from the four cases of the local learning regularization networks, averaged over 20 runs

Dataset	Local σ Local λ	Global σ Local λ	Local σ Global λ	Global σ Global λ
California Housing	0.11641	0.11494	0.11596	**0.11450**
Ailerons	0.04763	**0.04601**	0.04681	0.04612
Delta Elevators	0.05407	0.05332	0.05346	**0.05304**
Computer activity	0.02987	0.02922	0.02954	**0.02901**
Kinematics	0.06716	0.06738	0.06590	**0.06386**
Delta Ailerons	0.03928	0.03874	0.03901	**0.03852**
Puma dynamic	0.15575	0.15322	0.15399	**0.15115**
White wine quality	0.12171	0.11882	0.11816	**0.11778**
Red wine quality	0.13352	0.13041	0.13201	**0.12924**
Housing	0.08374	0.08056	0.08342	**0.08019**

The time measurements are helpful in order to find out which local algorithm is the fastest. Table 6 illustrates the corresponding training time and testing time for the four local RN cases. We expect from the training phase of all the four cases to have computational complexity of the order $O(NM^2k^3)$ and thus to scale well with the number N of the training set, when the number k of nearest neighbors is small enough. During the testing phase the local RN cases that need less local parameters to optimize are favored more. So, the last case, that uses global parameters is the fastest.

Table 6. Training time (tr) and testing time (te) in seconds

Dataset	Local σ Local λ		Global σ Local λ		Local σ Global λ		Global σ Global λ	
	tr	te	tr	te	tr	te	tr	te
California	5752	3584	5738	194	5717	283	3050	11
Ailerons	4095	3676	4037	195	4043	212	2189	12
Delta Elevators	2706	2418	2690	117	2691	138	1424	40
Computer activity	2573	1825	2482	76	2589	105	1372	33
Kinematics	2260	2039	2257	105	2261	114	1201	39
Delta Ailerons	2201	1757	2172	88	2175	89	1180	32
Puma dynamic	836	734	858	40	869	45	460	1.5
White wine	991	705	991	18	992	13	527	0.5
Red wine	323	256	325	13	324	14	171	0.5
Housing	104	62	103	3	104	4	55	0.1

Table 7 presents the comparison of the Standard Deviations of the testing errors when averaged over 20 runs. The standard deviations of the errors are small and similar through out the four cases of the local learning regularization networks.

Table 7. Comparison of the Standard Deviations of the testing errors from the four cases of the local learning regularization networks, averaged over 20 runs

Dataset	Local σ Local λ	Global σ Local λ	Local σ Global λ	Global σ Global λ
California Housing	0.001384	0.001220	0.001156	0.001185
Ailerons	0.000407	0.000305	0.000346	0.000358
Delta Elevators	0.000372	0.000336	0.000334	0.000320
Computer activity	0.000512	0.000356	0.000495	0.000387
Kinematics	0.000986	0.000968	0.001060	0.000902
Delta Ailerons	0.000691	0.000614	0.000636	0.000610
Puma dynamic	0.001664	0.001658	0.001522	0.001629
White wine quality	0.001644	0.001661	0.001456	0.001628
Red wine quality	0.003713	0.002818	0.003061	0.002875
Housing	0.010258	0.009199	0.008988	0.008839

Table 8 illustrates for each dataset and each local RN case the best global parameters that were found during their training phase.

Table 8. Best k number of neighbors and best global parameters σ and λ.

Dataset	Local σ Local λ	Global σ Local λ		Local σ Global λ		Global σ Global λ		
	k	k	σ	k	λ	k	σ	λ
California	76	77	0.7500	88	0.0197	78	0.7916	0.0032
Ailerons	90	90	5.0000	88	0.0058	90	5.0000	0.0071
Delta Elevators	89	87	3.5454	89	0.0036	86	4.1818	0.0131
Computer activity	81	75	2.4090	83	0.0031	72	2.4090	0.0031
Kinematics	90	90	1.6750	90	0.0051	90	1.2500	0.0031
Delta Ailerons	84	85	2.7545	81	0.0050	82	2.9250	0.0035
Puma dynamic	79	78	2.2500	79	0.0187	80	2.1000	0.0100
White wine	67	41	2.4000	28	0.2540	25	1.0375	0.0943
Red wine	74	77	2.5625	72	0.5790	74	1.3875	0.1750
Housing	65	64	1.9625	68	0.0098	62	2.1625	0.0050

4.2   Parallel Speedup Measurements

Experimental simulations for the multithreaded implementations are carried out using OpenMP for the multi-core local learning regularization network. The experiments were performed on a system consisted of two CPU's, AMD opteron(tm) Processor 6128 HE, with eight cores each, 800MHz clock speed and 16GB RAM, under Ubuntu Linux 10.04 operating system. The OpenMP version 3.0 was used. We simulate case 1 and we measure the total parallel execution time versus the number of cores and the speedup S / P, where S the sequential run time in a single core and P the time that simulates the network in parallel.

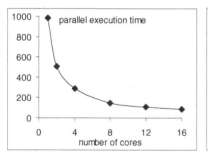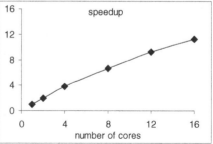

Fig. 1. (a) Reduction of execution time versus the number of threads, (b) Speedup

Figure 1(a) illustrates the execution time reduction on increasing the number of threads. Using 16 cores the parallel execution time was reduced by 92%. Fig. 1(b) presents the speedup ratio that assists in the scalability analysis.

5 Conclusions

This work explores four local learning approaches of regularization networks that use the k-nearest neighbor training points to learn local parameters and create on run time a locally optimized model, exclusively for predicting a particular testing point. For speeding up the online predictions we exploit both multithreaded implementations as well as interplay between locally optimized parameters and globally optimized parameters. Simulations show that the local regularization network parallelizes well in multi-core systems. In order to tackle with the extra computational overhead during the network operation that must optimize locally the width parameter σ and the regularization parameter λ, we also investigate which of the hyper-parameters can be optimized globally and off-line. The resulting four cases use {local σ, local λ}, {global σ, local λ}, {local σ, global λ} and {global σ, global λ} with the last case being the fastest during off-line training as well during on-line operation. In addition, the last case (case 4) delivers more accurate results. Throughout their off-line training all four cases scale well with the number N of the training set when the number k of nearest neighbors is kept small enough (to give a scalability example, for the California dataset they compute and validate 10000 local models per second). Although such results seem promising more extensive experiments could be considered in the feature for establishing other trends. The regularization network solution is very closely related to other kernel-based algorithms that combine Tikhonov regularization, like kernel ridge regression, regularized least-squares classification, kernel least mean squares, gaussian process regression and least-squares support vector machine. These methods use a kernel width parameter and a regularization parameter. Hence the present study can have straightforward extensions to such methods as well.

References

1. Poggio, T., Girosi, F.: Regularization algorithms for learning that are equivalent to multilayer networks. Science **247**, 978–982 (1990)
2. Girosi, F., Jones, M., Poggio, T.: Regularization theory and neural networks architectures. Neural Computation **7**, 219–269 (1995)
3. Evgeniou, T., Pontil, M., Poggio, T.: Regularization Networks and Support Vector Machines. Advances in Computational Mathematics **13**, 1–50 (2000)
4. Poggio, T., Smale, S.: The mathematics of learning: Dealing with data. Notices of the American Mathematical Society **50**(5), 537–544 (2003)
5. Bottou, L., Vapnik, V.: Local learning algorithms. Neural Computation **4**(6), 888–900 (1992)
6. Vapnik, V., Bottou, L.: Local Algorithms for Pattern Recognition and Dependencies Estimation. Neural Computation **5**(6), 893–909 (1993)
7. Robins, A., Frean, M.: Local Learning Algorithms for Sequential Tasks in Neural Networks. Advanced Computational Intelligence **2**, 221–227 (1998)
8. Vijayakumar, S., Schaal, S.: Local Adaptive Subspace Regression. Neural Processing Letters **7**(3), 139–149 (1998)
9. Vijayakumar, S., Schaal, S.: Locally weighted projection regression: an O(n) algorithm for incremental real time learning in high dimensional space. In: Proceedings of Seventeenth International conference on Machine Learning (ICML 2000), pp. 1079–1086 (2000)
10. Zhou, D., Bousquet, O., Lal, T.N., Weston, J., Schölkopf, B.: Learning with local and global consistency. Advances in Neural Information Processing Systems **16** (2004)
11. Wu, M., Schölkopf, B.: Transductive classification via local learning regularization. In: Proceedings of the Eleventh International Conference on Artificial Intelligence and Statistics (2007)
12. Wu, M., Yu, K., Yu, S., Schölkopf, B.: Local learning projections. In: Proceedings of the Twenty-Fourth International Conference on Machine Learning (2007)
13. Blanzieri, E., Melgani, F.: An adaptive SVM nearest neighbor classifier for remotely sensed imagery. In: Proceedings of IEEE International Conference on Geoscience and Remote Sensing Symposium (IGARSS 2006), pp. 3931–3934 (2006)
14. Zhang, H., Berg, A.C., Maire, M., Malik, J.: SVM-KNN: discriminative nearest neighbor classification for visual category recognition. In: Proceedings of IEEE Computer Society Conference on Computer Vision and Pattern Recognition, vol. 2, pp. 2126–2136 (2006)
15. Yang, T., Kecman, V.: Adaptive local hyperplane classification. Neurocomputing **71**, 3001–3004 (2008)
16. Wu, M., Schölkopf, B.: A local learning approach for clustering. In: Advances in Neural Information Processing Systems, vol. 19 (2006)
17. Wang, F., Zhang, C., Li, T.: Clustering with local and global regularization. In: Proceedings of the Twenty-Second AAAI Conference on Artificial Intelligence (2007)
18. Kokkinos, Y., Margaritis, K.: Parallel and local learning for fast probabilistic neural networks in scalable data mining. In: ACM Proceedings of 6th Balkan Conference in Informatics, (BCI 2013), pp. 47–52 (2013)
19. Rifkin, R.M., Lippert, R.A.: Notes on regularized least squares. Technical report, MIT Computer Science and Artificial Intelligence Laboratory (2007)
20. Buyya, R. (ed.): High Performance Cluster Computing: Programming and Applications, vol. 2. Prentice Hall, New Jersey (1999)

Self-Train LogitBoost for Semi-supervised Learning

Stamatis Karlos[1(✉)], Nikos Fazakis[2], Sotiris Kotsiantis[1], and Kyriakos Sgarbas[2]

[1] Department of Mathematics, University of Patras, Patras, Greece
{stkarlos,kotsiantis}@upatras.gr
[2] Department of Electrical and Computer Engineering, University of Patras, Patras, Greece
fazakis@ece.upatras.gr, sgarbas@upatras.gr

Abstract. Semi-supervised classification methods are based on the use of unlabeled data in combination with a smaller set of labeled examples, in order to increase the classification rate compared with the supervised methods, in which the total training is executed only by the usage of labeled data. In this work, a self-train Logitboost algorithm is presented. The self-train process improves the results by using the accurate class probabilities for which the Logitboost regression tree model is more confident at the unlabeled instances. We performed a comparison with other well-known semi-supervised classification methods on standard benchmark datasets and the presented technique had better accuracy in most cases.

Keywords: Semi-supervised learning · Logitboost · Classification method · Labeled and/or unlabeled data

1 Introduction

Supervised machine learning algorithms need a large number of labeled data to assign an unlabeled example to a class. As a consequence, this characteristic demands too much effort from a specialist, as the stage of labeling all the instances, is necessary. On the contrary, semi-supervised techniques are more automated, since their needs for labeled data are dramatically reduced and can be easily applied in a variety of fields, such as text mining, image or speech classification etc. [2].

Sun [11] reviews theories developed to understand the properties of multi-view learning and gives a taxonomy of approaches according to the supervised and semi-supervised machine learning mechanisms involved. In this work, a self-training method that combines the power of Logitboost and regression trees for semi-supervised tasks is proposed. We performed a comparison with other well-known semi-supervised classification methods on standard benchmark datasets and the presented technique had better accuracy in most cases.

2 Semi-supervised Techniques

The main idea of the self-training technique is the tuning of more than one classifiers, so as to achieve a better classification accuracy [1]. The whole procedure is split into

© Springer International Publishing Switzerland 2015
L. Iliadis and C. Jayne (Eds.): EANN 2015, CCIS 517, pp. 139–148, 2015.
DOI: 10.1007/978-3-319-23983-5_14

several phases. During the first one, a classifier of the user's choice is trained using the small set of labeled examples. After this phase has been completed, unlabeled examples are examined and classified according to the knowledge that our learner has acquired from the previous stage. When the second phase has been finished, all the instances that were previously unlabeled and for which the learner's prediction exceeds a trust-threshold, are added to the training set along with their predicted labels. Next steps include the re-training of the classifier, under the assumption that the new training set provides us more useful information, until all the stopping criteria are met.

Co-Training is based on the hypothesis that the attribute space can be split into two disjoint subsets and that each subset may contribute to correct classification [4]. According to this guess, a single learner is trained on each subset. To begin with, both learners are trained only on labeled data. The next phase comprises the classification of usually a small number of unlabeled examples by both previous learners. During this phase, the most confident predictions of each one learner are added to the training set of the other one. This procedure continues to be repeated for a number of times, until a stopping criteria to be satisfied. Didaci et al [16] evaluated co-training performance as a function of the size of the labeled training set. It seems that the concept of using co-training can work even with very few instances per class. However, Du et al [20] made a number of experiments and came to the conclusion that relying on small labeled training sets, verification of both the sufficiency and independence assumptions of splitting single view into two views are unreliable.

Sun and Jin [10] proposed robust co-training, in which the predictions of co-training on the unlabeled data are tested through Canonical correlation analysis (CCA) with the intention to enlarge the training set, adding only these instances whose predicted labels are consistent with the outcome of CCA. Wang et al [5] proposed to combine the probabilities of class membership with a distance metric between unlabeled instances and labeled instances. If two instances have the same class probability value, the one with the smallest distance will have larger chance to be selected. Xu et al [6] proposed DCPE co-training algorithm. For each classifier, if unlabeled instances have the same prediction labels and the highest class probability values differences between this classifier and the other one, then these unlabeled examples are added into the training set of the classifier. COTRADE [18] uses a number of predicted labels with higher confidence of either learner are passed to the other one, if a number of constraints are imposed to avoid introducing noise.

Li and Zhou [7] proposed Co-Forest algorithm. According to this algorithm, a number of Random Trees are trained on bootstrap sample data from the data set. Then each Random Tree is refined with a small number of unlabeled instances during the training process and the final prediction is produced by majority voting. Deng and Guo [12] proposed a new Co-Forest algorithm named ADE-Co-Forest which uses a data editing technique to identify and discard probably mislabeled instances during the iterations. Co-training by committee has been proposed by Hady and Schwenker [8]. In their work, an initial committee was built with the labeled data set. Three ensemble methods were used: Bagging, AdaBoost and Random Subspace and these semi-supervised learning algorithms were named as CoBag, CoAdaBoost and CoRSM, respectively. Liu and Yuen [22] also proposed a boosted co-training algorithm.

Tri-training algorithm has been proposed by Zhou and Li [3]. In each round of tri-training algorithm, an unlabeled instance is labeled for a learner if the other two learners agree on the labeling. Guo and Li [17] proposed improved tri-training algorithm (im-tri-training) that addresses some issues existed in tri-training such as unsuitable error estimation. Democratic co-learning [9] also uses multiple classifiers. Initially, each classifier is trained with the same data. The classifiers are then used to label the unlabeled data. Each instance is then labeled with the majority voting, and the labeled instance is added to the training set of the classifier whose prediction disagree with the majority. Sun and Zhang [19] proposed an ensemble of classifiers to be trained from each view, and the consensus prediction of the ensemble to be used to select confident labeled instances from the unlabeled data to teach the other ensemble from the other view.

3   Proposed Algorithm

Regression trees are obtained using a fast divide and conquer greedy algorithm that recursively partitions the given training set into smaller subsets. The most well-known regression tree inducer is the M5 [14]. In spite of their advantages regression trees are also known for their instability, since a small change in the training data can lead to a different choice when building a node, which in turn can produce a dramatic change in the regression tree, particularly if the change occurs in top level nodes. A well-known technique to improve the accuracy of tree-based classifiers is the boosting procedure. The idea of boosting is to combine the prediction of many simple classifiers to form a powerful 'committee'. The simple classifiers are trained on reweighted versions of the training data, such that training instances that have been misclassified by the classifiers built so far, receive a higher weight and the new classifier can concentrate on these hard instances.

Additive logistic regression algorithm: Logitboost [13] is based on the observation that boosting is in essence fitting an additive logistic regression model to the training data. An additive model is an approximation to a function $F(x)$ of the form:

$$F(x) = \sum_{m=1}^{M} c_m f_m(x) \tag{1}$$

where the c_m are constants to be determined and f_m are basis functions. It must be mentioned that m is number of classifiers and is set equal to 10 in our implementation.

If we assume that $F(x)$ is the mapping that we seek to fit as our strong aggregate hypothesis, and $f(x)$ are our weak hypotheses, then it can be shown that the two-class boosting algorithm is fitting such a model by minimizing the criterion:

$$J(F) = E(e^{-yF(x)}) \tag{2}$$

where y is the true class label in $\{-1,1\}$. Logitboost minimises this criterion by using Newton-like steps to fit an additive logistic regression model to directly optimise the binomial log-likelihood:

$$-\log(1 + e^{-2yF(x)})\qquad(3)$$

Friedman et al [13] used Logitboost algorithm with one level decision tree as base learner for solving supervised classification problems. A very useful property of this classification method is that it directly yields class conditional probability estimates that are crucial for constructing self-train classifiers. Regression trees are known for their simplicity and efficiency when dealing with domains with large number of features and instances. In this work, we propose a self-training method that combines the power of Logitboost and regression trees for semi-supervised tasks. The proposed algorithm (Self-Logit-M5) is presented in Figure 1. The self-train process produces good results by using the more accurate class probabilities of Logitboost regression tree model for the unlabeled instances.

Input: An initial set of labeled instances L and a set of unlabeled instances U
Initialization:
- Train Logitboost using M5 as base model on L

Loop for a number of iterations (40 in our implementation)
- Use Logitboost classifier to select the most confident predictions, remove them from U and add them to L. In each iteration about 2 instances per class are removed from U and added to L.

Output: Use Logitboost trained on L to predict class labels of the test cases.

Fig. 1. The Self-Logit-M5 Algorithm

For the implementation, it must be mentioned that we made use of the free available code of Weka [24] and KEEL [25].

4   Experiments

The experiments are based on standard classification datasets taken from the KEEL-dataset repository [23]. These data sets have been partitioned using the 10-fold cross-validation procedure. For each generated fold, a given algorithm is trained with the examples contained in the rest of folds (training partition) and then tested with the current fold. Each training partition is divided into two parts: labeled and unlabeled examples. In order to study the influence of the amount of labeled data, we take two different ratios when dividing the training set: 10% and 20%.

For the experiments, the proposed method has been compared with other state of the art algorithms integrated into the KEEL tool [23] such as Self-Training (C45) [1], Self-Training (SMO) [27], Co-Bagging (C45) [8], TriTraining (C45) [3], Democratic-Co [9], CoForest [7] and Co-Training (C45) [4]. It must be mentioned that the default parameters of KEEL were used for all the tested algorithms. The classification accuracy of each tested algorithm using 10% and 20% as labeled ratio is presented in Table 1 and Table 2 respectively. As it concerns the values in bold style, they actually point the best accuracy value in each row among the different algorithms.

Table 1. Classification accuracy (Labeled Ratio 10%)

Datasets	Alg1	Alg2	Alg3	Alg4	Alg5	Alg6	Alg7	Alg8
appendicitis	**0.862**	0.832	0.754	0.805	0.805	0.822	0.823	0.832
australian	0.828	0.828	0.796	0.828	**0.845**	**0.845**	0.841	0.835
banana	0.877	0.848	**0.896**	0.855	0.848	0.842	0.527	0.848
breast	0.672	0.722	0.691	0.725	0.722	0.729	**0.734**	0.677
bupa	0.606	0.539	**0.633**	0.612	0.574	0.510	0.585	0.574
chess	0.956	0.954	0.896	0.954	**0.958**	0.920	0.944	0.952
coil2000	0.922	0.937	0.572	0.935	0.936	0.932	0.930	**0.940**
contraceptive	**0.500**	0.489	0.450	0.483	0.481	0.436	0.485	0.446
crx	0.801	**0.866**	0.832	0.850	0.856	0.850	0.821	0.816
dermatology	0.863	0.856	0.840	0.876	0.882	0.876	**0.905**	0.843
ecoli	0.641	0.647	0.622	0.656	**0.659**	0.637	0.628	0.577
flare	**0.721**	**0.721**	0.516	0.714	0.716	**0.721**	0.402	0.574
german	0.713	0.706	0.592	0.711	**0.717**	0.716	0.686	0.690
glass	0.553	0.485	0.496	0.490	0.492	0.487	**0.559**	0.450
haberman	0.650	0.705	0.601	0.712	0.709	0.716	0.601	**0.719**
heart	0.763	0.678	0.770	0.704	0.715	**0.800**	0.693	0.700
hepatitis	0.793	**0.834**	**0.834**	**0.834**	**0.834**	**0.834**	0.811	**0.834**
housevotes	0.921	**0.941**	0.876	0.920	0.916	0.890	0.922	0.823
iris	0.900	0.840	**0.940**	0.800	0.727	0.913	0.933	0.847
led7digit	**0.686**	0.614	0.568	0.564	0.604	0.616	0.634	0.514
lymphography	**0.737**	0.631	0.549	0.595	0.612	0.490	0.646	0.573
magic	**0.849**	0.822	0.839	0.832	0.825	0.784	0.844	0.820
mammographic	0.809	0.803	0.782	0.809	**0.818**	0.796	0.794	0.807
monk-2	0.953	**0.973**	0.783	0.966	0.966	0.908	0.939	**0.973**
mushroom	0.996	**0.997**	0.992	0.995	0.996	0.993	0.908	**0.997**
nursery	**0.962**	0.906	0.815	0.901	0.904	0.895	0.381	0.903
page-blocks	0.953	0.952	0.940	0.957	0.956	0.908	**0.959**	0.949
penbased	0.965	0.892	**0.976**	0.905	0.903	0.947	0.955	0.896
phoneme	**0.829**	0.777	**0.829**	0.789	0.777	0.787	0.801	0.765
pima	0.680	0.664	0.608	0.634	0.656	**0.697**	0.663	0.670
ring	0.904	0.840	**0.970**	0.858	0.854	0.874	0.882	0.837
saheart	0.645	0.652	0.613	0.650	0.678	**0.682**	0.656	0.636
satimage	**0.870**	0.805	0.825	0.821	0.822	0.846	0.860	0.806
segment	**0.931**	0.890	0.907	0.916	0.900	0.903	0.903	0.902
sonar	0.686	0.643	0.669	0.701	0.702	0.601	**0.755**	0.582
spambase	0.917	0.867	0.851	0.895	0.881	0.878	**0.919**	0.888
spectfheart	0.749	0.682	0.656	0.757	0.757	0.738	**0.775**	0.724
splice	**0.938**	0.827	0.541	0.825	0.825	0.898	0.507	0.831
texture	0.929	0.831	**0.963**	0.850	0.852	0.894	0.907	0.829
thyroid	0.987	**0.992**	0.932	0.991	0.992	0.939	0.986	**0.992**
tic-tac-toe	**0.741**	0.711	0.627	0.704	0.709	0.690	0.597	0.693
titanic	0.780	0.775	0.776	**0.784**	0.777	0.776	0.707	0.778
twonorm	0.949	0.814	**0.973**	0.860	0.862	0.965	0.899	0.809
vehicle	**0.633**	0.579	0.586	0.603	0.619	0.502	0.612	0.575
vowel	0.495	0.424	0.463	0.456	0.453	0.416	**0.522**	0.438
wine	0.854	0.741	**0.955**	0.787	0.820	0.949	0.859	0.808
wisconsin	0.945	0.909	0.952	0.928	0.931	**0.965**	0.936	0.906
yeast	**0.506**	0.462	0.458	0.477	0.491	0.489	0.456	0.489
zoo	0.841	0.679	0.493	0.746	0.719	**0.931**	0.909	0.636

[1]Alg1: Self-training (Logitboost), Alg2: Self-Training (C45), Alg3: Self-Training (SMO), Alg4: Co-Bagging (C45), Alg5: TriTraining (C45), Alg6: Democratic-Co, Alg7: CoForest, Alg8: Co-Training (C45)

Table 2. Classification accuracy (Labeled Ratio 20%)

Datasets	Alg1	Alg2	Alg3	Alg4	Alg5	Alg6	Alg7	Alg8
					Algorithms[2]			
appendicitis	**0.887**	0.851	0.705	0.832	0.832	0.860	0.877	0.850
australian	**0.859**	0.858	0.830	0.836	0.852	0.858	0.842	0.838
banana	0.886	0.880	**0.899**	0.880	0.874	0.879	0.541	0.874
breast	0.687	0.716	0.636	0.722	0.723	0.718	**0.727**	0.723
bupa	**0.641**	0.606	0.611	0.603	0.620	0.550	0.602	0.606
chess	**0.978**	**0.978**	0.948	**0.978**	**0.978**	0.939	0.951	0.976
coil2000	0.923	**0.940**	0.817	0.934	0.937	0.933	0.927	**0.940**
contraceptive	**0.515**	0.478	0.471	0.481	0.488	0.484	0.494	0.504
crx	0.833	0.857	**0.860**	0.851	0.854	0.859	0.841	0.850
dermatology	**0.924**	0.910	0.910	0.919	0.921	0.930	**0.924**	0.879
ecoli	**0.774**	0.720	0.670	0.732	0.721	0.724	0.748	0.733
flare	0.720	**0.728**	0.585	0.718	0.727	0.740	0.416	0.690
german	0.714	0.697	0.618	0.712	0.684	**0.729**	0.684	0.699
glass	0.628	0.509	0.551	0.563	0.610	0.480	**0.634**	0.535
haberman	0.686	0.709	0.660	**0.735**	0.709	0.722	0.640	**0.735**
heart	0.789	0.752	0.789	0.748	0.793	**0.822**	0.730	0.756
hepatitis	0.747	0.843	0.834	0.820	0.843	0.834	**0.861**	0.843
housevotes	0.931	**0.959**	0.888	0.953	**0.959**	0.902	0.945	0.936
iris	0.893	0.893	0.913	0.867	0.880	**0.953**	0.940	0.893
led7digit	**0.692**	0.678	0.644	0.658	0.676	0.662	0.678	0.644
lymphography	**0.817**	0.706	0.649	0.747	0.755	0.461	0.693	0.775
magic	**0.855**	0.830	0.842	0.841	0.834	0.802	0.852	0.832
mammographic	0.823	0.823	0.793	**0.825**	0.816	0.808	0.787	0.823
monk-2	0.977	**0.980**	0.869	0.973	0.973	0.944	0.979	**0.980**
mushroom	**1.000**	0.999	0.997	0.999	0.999	0.998	0.910	0.999
nursery	**0.984**	0.924	0.576	0.924	0.926	0.911	0.384	0.926
page-blocks	**0.965**	0.960	0.957	0.961	0.961	0.912	0.959	0.961
penbased	0.979	0.924	**0.983**	0.937	0.929	0.963	0.966	0.925
phoneme	**0.851**	0.784	0.848	0.824	0.798	0.806	0.831	0.802
pima	**0.734**	0.681	0.646	0.708	0.694	0.732	0.711	0.687
ring	0.928	0.866	**0.971**	0.892	0.880	0.897	0.893	0.863
saheart	0.667	0.652	0.602	0.686	0.678	0.697	0.671	**0.704**
satimage	**0.886**	0.824	0.857	0.841	0.835	0.861	0.872	0.826
segment	**0.953**	0.924	0.945	0.928	0.925	0.931	0.934	0.929
sonar	0.725	0.664	0.691	0.659	0.663	0.647	**0.745**	0.634
spambase	0.926	0.891	0.900	0.896	0.894	0.894	**0.931**	0.891
spectfheart	**0.802**	0.719	0.671	0.783	0.749	0.731	0.791	0.761
splice	**0.954**	0.883	0.578	0.888	0.884	0.912	0.521	0.879
texture	0.960	0.867	**0.982**	0.884	0.893	0.918	0.935	0.863
thyroid	0.992	**0.994**	0.939	**0.994**	**0.994**	0.942	0.988	0.993
tic-tac-toe	**0.779**	0.757	0.672	0.729	0.751	0.733	0.607	0.721
titanic	0.777	0.782	0.782	0.781	**0.783**	0.780	0.723	0.782
twonorm	0.955	0.817	**0.973**	0.874	0.867	0.971	0.903	0.828
vehicle	**0.679**	0.649	0.675	0.655	0.662	0.483	0.655	0.649
vowel	0.644	0.530	**0.705**	0.557	0.555	0.503	0.640	0.530
wine	0.848	0.837	**0.961**	0.826	0.842	0.954	0.842	0.786
wisconsin	0.951	0.934	0.952	0.936	0.925	**0.964**	0.936	0.934
yeast	**0.567**	0.526	0.502	0.520	0.549	0.549	0.489	0.549
zoo	0.891	0.831	0.601	0.823	0.826	0.890	**0.897**	0.734

In the sequel, in Table 3 and Table 4 the results of Friedman test [13] together with a statistical test [13] are presented, which are used in order to conduct comparisons among all algorithms considered in the study and the proposed algorithm for both situations that have already been examined.

---

[2] Alg1: Self-training (Logitboost), Alg2: Self-Training (C45), Alg3: Self-Training (SMO), Alg4: Co-Bagging (C45), Alg5: TriTraining (C45), Alg6: Democratic-Co, Alg7: CoForest, Alg8: Co-Training (C45)

Table 3. Average rankings of the algorithms in 10% labeled ratio (Friedman) and Holm / Hochberg (alpha=0.05)

Algorithm	Friedman Ranking	p	Holm/Hochberg Test
Self-training (Logitboost)	3.0408163265306114	-	-
TriTraining (C45)	3.9693877551020407	0.060602	0.05
CoForest	4.265306122448979	0.013347	0.025
Co-Bagging (C45)	4.316326530612245	0.009953	0.01666
Democratic-Co	4.591836734693876	0.001723	0.0125
Self-Training (C45)	4.979591836734695	8.94E-5	0.01
Self-Training (SMO)	5.336734693877552	3.49E-6	0.00833
Co-Training (C45)	5.500000000000001	6.72E-7	0.00714

Table 4. Average rankings of the algorithms in 20% labeled ratio (Friedman) and Holm / Hochberg (alpha=0.05)

Algorithm	Friedman Ranking	p	Holm/Hochberg Test
Self-training (Logitboost)	2.724489795918366	-	-
TriTraining (C45)	4.469387755102042	4.22E-4	0.05
CoForest	4.510204081632654	3.08E-4	0.025
Co-Bagging (C45)	4.5306122448979576	2.63E-4	0.01666
Democratic-Co	4.530612244897959	2.63E-4	0.0125
Co-Training (C45)	4.969387755102043	5.72E-6	0.01
Self-Training (C45)	5.030612244897958	3.16E-6	0.00833
Self-Training (SMO)	5.2346938775510194	3.93E-7	0.00714

As a result, the proposed algorithm gives statistical better results than all the tested algorithms. Better probability-based ranking and high classification accuracy could select the high-confidence predictions in the selection step of self-training and therefore, the proposed method improved the performance of self-training. The proposed approach performs better than the tested state of the art algorithms in the tested datasets.

5   Tool Presentation

In this section a short presentation of a tool that implements the semi-supervised proposed algorithm is provided. First of all, its display is illustrated in Figure 2. Secondly, it is easily observable that it is simple enough, as it contains only three buttons through which the user can interact with. Moreover, because it is developed using JAVA, it is platform independent. The only restriction is the hardware requirement of 64-bit CPU combined with 4GB RAM.

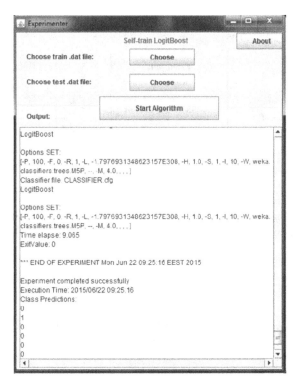

Fig. 2. Illustration of the corresponding tool for Self-train LogitBoost Algorithm.

To begin with the description, anyone who wants to run the proposed algorithm has to follow the standard steps that semi-supervised theory includes. Particularly, a train set has to be chosen through the first button Choose. It must be mentioned that the train set has to consist of both labeled and unlabeled instances. The provided sets contain 10% labeled instances, but the user can adjust the percentage of unlabeled instances over his train set. After the completion of this step, the user can choose the test set. The difference in comparison with the train set is that the column which refers to the class of each instance simply does not exist. After that, selecting the Start Algorithm button, a file named as predictions.txt will be created in the same folder with the Experimenter program, which finally will contain the class prediction of each tested instance.

As it considers the data sets that are used by this tool, are managed by plain ASCII text files, with the .dat extension. Each data file is composed by header (basic meta-data describing the data set) and data (content of the dataset). The corresponding link for downloading this tool and some basics instructions is the following: http://www. math.upatras.gr/~sotos/SelfLogitBoost-Experiment.zip.

6   Conclusion

It is promising to develop techniques that use both labeled and unlabeled instances in classification tasks. The limited availability of labeled examples makes the learning process difficult, as supervised learning methods cannot produce a classifier with good generalization performance.

The main difficulty in self-training task is to find a set of high confidence predictions of unlabeled instances. Although for a lot of domains, decision tree classifiers produce accurate classifiers, they provide poor probability estimates [28-29]. The reason is that the sample size at the leaves is small in most of the time, and all instances at a leaf get the same probability. In addition, the probability estimate is the proportion of the majority class at the leaf of a pruned decision tree.

In this work, a self-train Logitboost algorithm is presented. The self-train process improves the results by using the accurate class probabilities for which the Logitboost regression tree model is more confident at the unlabeled instances. We performed a comparison with other well-known semi-supervised classification methods on standard benchmark datasets and the presented technique had better accuracy in most cases. In spite of these results, no general method will work always.

In the near future, we will try to improve the results of proposed method combining the probabilities of class membership with a distance metric between unlabeled instances and labeled instances [21].

References

1. Rosenberg, C., Hebert, M., Schneiderman, H.: Semi-supervised self-training of object detection models. In: 7th IEEE Workshop on Applications of Computer Vision, pp. 29–36 (2005)
2. Friedhelm, S., Edmondo, T.: Pattern classification and clustering: A review of partially supervised learning approaches. Pattern Recognition Letters **37**, 4–14 (2014)
3. Zhou, Z.-H., Li, M.: Tri-Training: Exploiting Unlabeled Data Using Three Classifiers. IEEE Trans. on Knowledge and Data Engg. **17**(11), 1529–1541 (2005)
4. Chapelle, O., Schölkopf, B., Zien, A.: Semi-supervised learning. MIT Press, Cambridge (2006)
5. Wang, S., Wu, L., Jiao, L., Liu, H.: Improve the performance of co-training by committee with refinement of class probability estimations. Neurocomputing **136**, 30–40 (2014)
6. Xu, J., He, H., Man, H.: DCPE co-training for classification. Neurocomputing **86**, 75–85 (2012)
7. Li, M., Zhou, Z.: Improve computer-aided diagnosis with machine learning techniques using undiagnosed samples. IEEE Trans. Syst. Man Cybernet, 1088–1098 (2007)
8. Hady, M., Schwenker, F.: Co-training by committee: a new semi-supervised learning framework. In: Proceedings of the IEEE International Conference on Data Mining Workshops, pp. 563–572 (2008)
9. Zhou, Y., Goldman, S.: Democratic co-learning. In: Ictai, 16th IEEE International Conference on Tools with Artificial Intelligence (ICTAI 2004), pp. 594–202 (2004)
10. Sun, S., Jin, F.: Robust co-training. Int. J. Pattern Recognit. Artif. Intell. **25**, 1113–1126 (2011)

11. Sun, S.: A survey of multi-view machine learning. Neural Computing and Applications **23**(7–8), 2031–2038 (2013)
12. Deng, C., Guo, M.Z.: A new co-training-style random forest for computer aided diagnosis. Journal of Intelligent Information Systems **36**, 253–281 (2011)
13. Friedman, J., Hastie, T., Tibshirani, R.: Additive logistic regression: a statistical view of boosting. Ann. Statist. **28**(2), 337–407 (2000)
14. Torgo, L.: Inductive learning of tree-based regression models. AI Communications **13**(2), 137–138 (2000)
15. Jiang, Z., Zhang, S., Zeng, J.: A hybrid generative/discriminative method for semi-supervised classification. Knowledge-Based Systems **37**, 137–145 (2013)
16. Didaci, L., Fumera, G., Roli, F.: Analysis of co-training algorithm with very small training sets. In: Gimel'farb, G., Hancock, E., Imiya, A., Kuijper, A., Kudo, M., Omachi, S., Windeatt, T., Yamada, K. (eds.) SSPR&SPR 2012. LNCS, vol. 7626, pp. 719–726. Springer, Heidelberg (2012)
17. Guo, T., Li, G.: Improved tri-training with unlabeled data. In: Wu, Y. (ed.) Software Engineering and Knowledge Engineering: Vol. 2. AISC, vol. 115, pp. 139–148. Springer, Heidelberg (2012)
18. Zhang, M.-L., Zhou, Z.-H.: CoTrade: Confident co-training with data editing. IEEE Trans. Syst. Man Cybernet, Part B: Cybernetics **41**(6), 1612–1626 (2011)
19. Sun, S., Zhang, Q.: Multiple-View Multiple-Learner Semi-Supervised Learning. Neural Process. Lett. **34**, 229–240 (2011)
20. Du, J., Ling, C.X., Zhou, Z.-H.: When. does cotraining work in real data? IEEE Trans. on Knowledge and Data Engg. **23**(5), 788–799 (2011)
21. Zhu, X., Goldberg, A.: Introduction to semi-supervised learning. Synthesis Lectures on Artificial Intelligence and Machine Learning. Morgan & Claypool (2009)
22. Liu, C., Yuen, P.C.: A boosted co-training algorithm for human action recognition. IEEE Trans. on Circuits and Systems for Video Technology **21**(9), 1203–1213 (2011). 5739520
23. Alcalá-Fdez, J., Fernandez, A., Luengo, J., Derrac, J., García, S., Sánchez, L., Herrera, F.: KEEL Data-Mining Software Tool: Data Set Repository, Integration of Algorithms and Experimental Analysis Framework. Journal of Multiple-Valued Logic and Soft Computing **17**(2–3), 255–287 (2011)
24. Hall, M., Frank, E., Holmes, G., Pfahringer, B., Reutemann, P., Witten, I.: The WEKA Data Mining Software: An Update. SIGKDD Explorations **11**(1) (2009)
25. Triguero, I., Garca, S., Herrera, F.: Self-labeled techniques for semi-supervised learning: taxonomy, software and empirical study. Knowledge and Information Systems **42**(2), 245–284 (2015)
26. García, S., Fernández, A., Luengo, J., Herrera, F.: Advanced nonparametric tests for multiple comparisons in the design of experiments in computational intelligence and data mining: Experimental analysis of power. Inf. Sciences **180**(10), 2044–2064 (2010)
27. Keerthi, S.S., Shevade, S.K., Bhattacharyya, C., Murthy, K.R.K.: Improvements to Platt's SMO Algorithm for SVM Classifier Design. Neural Computation **13**(3), 637–649 (2001)
28. Mease, D., Wyner, A.J., Buja, A.: Boosted classification trees and class probability/quantile estimation. J. Mach. Learn. Res. **8**, 409–439 (2007)
29. Provost, F.J., Domingos, P.: Tree induction for probability based ranking. Mach. Learn. **52**, 199–215 (2003)

Scalable Digital CMOS Architecture for Spike Based Supervised Learning

Shruti R. Kulkarni and Bipin Rajendran[(✉)]

Department of Electrical Engineering, IIT Bombay, Mumbai, India
{shrutirk,bipin}@ee.iitb.ac.in

Abstract. Supervised learning algorithm for Spiking Neural Networks (SNN) based on Remote Supervised Method (ReSuMe) uses spike timing dependent plasticity (STDP) to adjust the synaptic weights. In this work, we present an optimal network configuration amenable to digital CMOS implementation and show that just 5 bits of resolution for the synaptic weights is sufficient to achieve fast convergence. We estimate that the implementation of this optimal network architecture in 65 nm and a futuristic 10 nm digital CMOS could result in systems with close to 0.85 and 30 Million Synaptic Updates Per Second (MSUPS)/Watt.

Keywords: Spiking neural networks · Supervised learning · Bit-precision · Digital neuromorphic architecture

1 Introduction

Neurons in the brain communicate with each other by transmitting short duration impulse like signals, known as spikes through synapses [1]. Spikes from upstream neurons propagate to downstream neurons based on the effective conductance of the synapses connecting them. The biological frequencies of neuronal computation is about 10 Hz to 100 Hz [2], so a significant reduction in power can be achieved if such spike based computational methods can be developed and implemented to solve real-world problems.

A simple scheme of computation using spikes is shown in Fig. 1. The information is carried and processed using the precise timings of the spikes [3], the output could be used as an input signal for controlling a motor in neuroprosthetics [4] or in robotics [5] or even as an intermediate signal in cognitive processing tasks. Hence, it is highly desirable to design and develop general SNNs capable of reproducing desired spike trains depending on the nature of input excitation.

This paper is organized as follows. Section 2 presents some of the common learning algorithms from the literature inspired by neurobiology. Section 3 gives a background of one particular supervised learning algorithm called the Remote Supervised Method (ReSuMe) by Ponulak [4], [6]. We then present a scheme for mapping this algorithm onto a digital neuromorphic hardware, inspired by existing schemes [7,8]. Sections 4 and 5 present the results of the network and synaptic weight optimization studies, respectively. We then present a design scheme for realizing this network onto a digital hardware, in section 6, with the

© Springer International Publishing Switzerland 2015
L. Iliadis and C. Jayne (Eds.): EANN 2015, CCIS 517, pp. 149–158, 2015.
DOI: 10.1007/978-3-319-23983-5_15

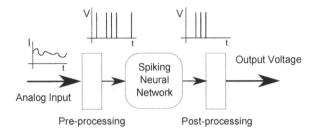

Fig. 1. High level scheme showing spike based information processing. The input pre-processing layer receives analog input and converts it into a stream of spikes, which is processed by the spiking neural network. The output spikes can be either converted back into appropriate signals for interfacing to other systems or can be used as intermediate signals for higher levels within the network.

area and power estimates for the design on an existing 65 nm node based on preliminary simulations and synthesis, and present their scaling projections at 10 nm, based on the trends in [9].

2 Learning in Neural Networks

Learning involves extracting key features or information content in previously observed samples of the input data and using that to make inferences or predictions about new input samples. This information is stored in links connecting two neurons, the synapse (also called weights), that determine the fraction of input to be delivered to the downstream neuron. This in turn, controls the time at which neuron's membrane potential exceeds the threshold, causing a spike. Among the several mechanisms proposed to understand the weight update from neurobiological studies, the most prominent was proposed by Donald O. Hebb [10], who postulated that the strength of the synaptic connection between two neurons is proportional to their spiking rates or activities. However, a drawback to this rule was that there was no mechanism to bound the weights. Recently, a modification to this rule was proposed, called Spike Timing Dependent Plasticity (STDP) [11] wherein, the weights get updated according to the precise timings of the pre- (t_{pre}) and post- (t_{post}) synaptic neurons, in a specific learning window.

3 Supervised Learning in SNN

Supervised learning aims to determine the optimal network that can produce a desired output spike stream when excited by a specified input spike stream (Fig. 2). In the ReSuMe algorithm, [4],[6], the activity of the teacher neuron (desired spike train) along with the input spike train influences the changes to the synaptic weights of the output neurons, which then get trained to generate the spike patterns close to the desired ones. Similar to supervised learning, in biology hetero-synaptic plasticity has been observed, where the synaptic plasticity is more dependent on the activity of the post-synaptic neuron and an external signal. It has also been shown to stabilize the synaptic dynamics [12].

3.1 Weight Update Rule

ReSuMe algorithm is inspired by the STDP rule as the weight adjustment of the synapse depends on precise timing of the spikes at the input neurons (pre-synaptic) and the output or the teacher neurons (post-synaptic). The cost to be minimized is the time difference between the desired (S^d) and the observed output (S^o) spikes by adjusting the weights. As described in Ponulak's work [4], the precise formula for weight update is given by the relation,

$$\frac{d}{dt}w(t) = S^d(t)\left[a_d + \int_0^\infty W^d(s^d).S^{in}(t-s^d)ds^d\right]$$
$$+S^o(t)\left[a_o + \int_0^\infty W^o(s^o).S^{in}(t-s^o)ds^o\right] \qquad (1)$$

where, $W(s)$, the learning window is given by,

$$W^d(s^d) = \begin{cases} +A^d_+ exp\left(\frac{-s^d}{\tau_+}\right) & \text{if } s^d > 0 \\ 0 & \text{if } s^d \leq 0 \end{cases} \qquad (2)$$

$$W^o(s^o) = \begin{cases} -A^o_+ exp\left(\frac{-s^o}{\tau_-}\right) & \text{if } s^o > 0 \\ 0 & \text{if } s^o \leq 0 \end{cases} \qquad (3)$$

and s is given as, $s = t_{post} - t_{pre}$.

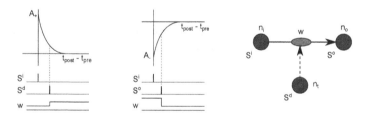

Fig. 2. Basic scheme of weight update in ReSuMe [4]. A simple case of training with one pre-synaptic (input) neuron (n_i), one post-synaptic (output) neuron (n_o) and one teacher neuron (n_t), generating the spike streams S^i, S^o and S^d, respectively.

3.2 ReSuMe: Network Architecture and Weight Training Process

The network (Fig. 3(a)) consists of input neurons which apply the input spike streams, a recurrently connected Neural Micro-Circuit (NMC) which generates a rich spatio-temporal pattern of spikes (Fig. 4(c)), and the output neurons that receive the inputs from the NMC neurons through adjustable synapses. All the connections in the network are stochastic in nature, allowing for generalizing the architecture for any application. The liquid state machine network, [13], [4], acts as the NMC block. Fig. 3(b) shows a typical adjacency matrix of the NMC, and it can be seen that connectivity is sparse with high probability of connection between the neighboring neurons.

Dedicated hardware has been demonstrated to implement SNNs in power-efficient neuromorphic systems ([7], [8], [14]). These chips implement neuron and synaptic dynamics and use SRAM or DRAM cells to store the synaptic weights. These chips consist of large crossbar arrays of cores that can be connected to each other through routing networks, enabling arbitrary connectivity between neurons and operate in an event-driven manner. There have been hardware implementations of ReSuMe, [15], though they were limited to FPGAs. We will use the generic architecture used in the custom chips [14], [16] as the framework for our work to understand the trade-offs and optimization strategies to implement supervised learning algorithm such as ReSuMe. Our proposed high level scheme for placement of a sample NMC block with 800 neurons and the manner of partitioning into neurosynaptic cores is shown in Fig. 3(c).

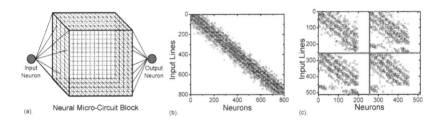

Fig. 3. (a) Network architecture for realization of the ReSuMe algorithm. The dots within the 3D block represent the neurons within the Neural Micro Circuit (NMC). (b) A sample adjacency matrix showing the connectivity within the NMC consisting of 800 neurons. (c) Actual placement of synapses per block. The size of crossbar array in each block is kept fixed at 256×256, though the number of neurons per block is 200.

The neurons in our implementation of the ReSuMe algorithm are modeled by the classical Leaky Integrate-and-Fire (LIF) model, where the membrane potential, V is determined by the differential equation

$$C\frac{dV(t)}{dt} = -g_L(V(t) - E_L) + I_{app}(t) + I_{syn}(t) \tag{4}$$

C is the membrane capacitance, g_L is the leak conductance and E_L is the resting potential. When the membrane potential (V) exceeds a threshold (V_T), a spike is issued and V is reset to E_L, and held at E_L for the next t_{ref} ms, which mimics the refractory period of biological neurons. I_{app} and I_{syn} are the total currents due to external stimulus and due to spikes arriving on all the interconnecting synapses respectively and are modeled by an exponential kernel (as used in [4]).

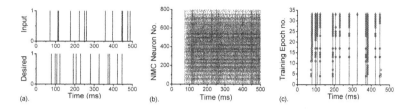

Fig. 4. (a) An exemplary input and a desired spike pattern used in our study. (b) Spike raster plot of the neurons of the NMC when excited by the input in (a). (c) Output spike training raster plot. The lines are the desired spike times, while the circles show the output spikes at every epoch. The network trained in 33 epochs.

An exemplary input and a desired spike pattern, each of 500 ms duration, for a training task, which was simulated using MATLAB, is shown in Fig. 4(a). The input and the desired spike streams are Poisson processes, with an average spiking rate of 20 Hz, close to the observed biological spike rate. The output spikes can be seen evolving to the desired spikes during the training (Fig. 4(c)) within an accuracy of 2.5 ms.

4   Network Optimization for ReSuMe

In order to use the algorithm in real-world applications, and also to port to power-efficient hardware, it is necessary to determine the optimal size of the NMC network and its dependency on the duration of the desired spike train, T_d. We study the performance of the network by first determining the number of training iterations necessary to achieve convergence with different number of inputs (NMC connections) to the output neuron. This experiment is also repeated for different spike stream durations, viz. 500 ms, 1000 ms and 1500 ms and results are presented for 100 runs each.

From Fig. 5 we can see an improvement in training speed with increasing NMC size, which in turn, shifts to the right with increase in spike duration. The results presented here serve as a guideline for scaling the network for longer spike streams. For the later part of our analysis, the lengths of the spike streams are kept at 500 ms, and the results and the proposed architecture are presented for the network consisting of the NMC with 800 neurons, which is also an optimal choice for $T_d = 1000$ ms.

5   Synaptic Parameter Optimization

The learning algorithm proposed in ReSuMe does not impose any restriction on the minimum and maximum values that the synaptic conductance can attain during training. Studies on deep learning networks [17] have shown that successful classification can be achieved even with limited precision of data representation. We studied the performance of ReSuMe, first by limiting the range of

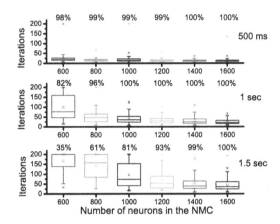

Fig. 5. Learning performance with length of spike streams for different sizes of the NMC. The percentage above each box-plot shows the training success rates for the NMC sizes. At least 800 and 1200 neurons are required in the NMC for reproducing 1 s and 1.5 s spike trains.

conductance that a synapse can attain during the course of training and then as a function of the number of discrete levels available. From Fig. 6 it can be seen that the histograms of the trained synaptic weights (for 100 runs), without any constraints, approximately obey a log-normal distribution. To quantize the weights, the optimal on-off ratio (maximum to minimum) required for faster training was studied by running the algorithm at different on-off ratios. The best possible on-off ratio was found to be 100 (Fig. 7(a)).

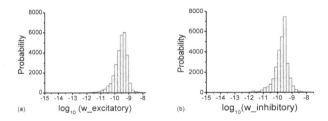

Fig. 6. Histogram of the logarithm of the absolute values of trained weights for 100 runs. It can be seen from the histograms that though the on-off ratio is of the order of 10^5, significant fraction of the weights lie only in the range 10^{-11} to 10^{-9}. (a) excitatory synaptic weights, with $\mu = -9.57$ and $\sigma = 0.47$, and (b) inhibitory weights, with $\mu = -9.72$ and $\sigma = 0.47$ on normalized $\log_{10}$ scale.

To study the effect of discretization of the synaptic weights on the network's learning performance, the weights' on-off ratios are kept fixed. The neuronal membrane potentials are also discretized by choosing 20 bits for representation, following a recent implementation [14]. For discretization, we apply the transformations $V_q(t) = (V(t) - E_L)/(V_{th} - E_L)$ and $w = w_0 \times w_q$, for membrane potential and weights (w_q represents the quantized weight value), respectively. Equation 4 can then be written as a difference equation as,

$$V_q(t+1) = V_q(t) + \frac{dt}{C}\left[\frac{w_0}{Vth - Em}\sum_{i=0}^{N_{fanin}} w_{q_i}\exp(-t/\tau) - \frac{V_q(t)}{R}\right] \quad (5)$$

Fig. 7(b) shows that an on-off ratio of 100 with 5 bits of precision is optimal for weights' representation, to obtain a good network training performance.

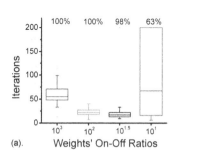
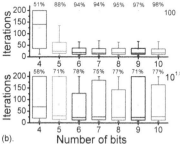

Fig. 7. (a) Variation of number of training iterations required with the on-off (min-max) ratios of the weights. 100% convergence is obtained even at an on-off ratio of 100, but further reduction lowers the convergence. (b) Variation of training iterations with number of bits for representing the weights. The top and bottom charts show the case with on-off ratio of 100 and $10^{1.5}$ respectively.

6   Hardware Architecture for ReSuMe

Our proposed design for the network of 800 NMC neurons, consists of three major components (Fig. 8). The input block, that applies the spike streams to the network, has a mean fan-out of 240 and consists of a register file maintaining the list of the addresses (10-bit) of the fan-out NMC neurons. The output block consists of the output neuron (having a fan-in of 560), an array of registers storing the synaptic weights, and a learning module, which computes change in the conductance values during training. The neurosynaptic cores (inspired by [8], [14], [16]) form the third component of this architecture, housing the NMC, consist of the crossbar arrays (as seen in section 3), with 6T SRAM cells as the storage units for the synaptic weights, and the neuron circuitry.

The synaptic connections within the NMC are programmed only once. However, the weights of the synapses connecting the output neuron get updated during the training. Each output synapse is associated with a 5-bit down-counter

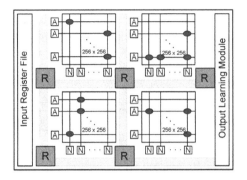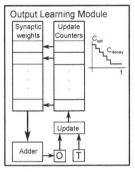

Fig. 8. ReSuMe implementation scheme. The dark circles in the core are the active synapses. Lines labeled as 'A' are the axon (input) lines and 'N' are the neurons. The input fan-out information is stored in the input register file. The learning mechanism is implemented in the output block (right). Block labeled 'O' represents the output neuron and 'T', the teacher neuron. However, in actual implementation the spikes from the input and teacher neurons would be applied as an external input. Both the input and output blocks are connected to the cores via the on-chip mesh network (shown in gray).

(as in [16]), which is initialized to C_{set} whenever the corresponding pre-synaptic neuron spikes, and decremented by a fixed value C_{decay} at each time step to linearly approximate the exponentially decaying learning window (Fig. 2) [15]. Whenever the output neuron spikes or there is a spike from the desired signal, the synapses which received a recent spike, get incremented or decremented, depending on the selector update block.

The power consumed by the NMC cores during the operation is composed of two components – for communicating synaptic conductance values to neurons and for updating the neuronal membrane potential [8], [9]. The average spike rate within the NMC is close to 100 Hz. Using a time step of 0.1 ms, we get the probability of a spike issue for a neuron in any emulation time step to be ~ 0.01. For the hardware implementation, we accelerate the dynamics by a factor of 1000, so the power analysis is done by assuming a hardware clock running at $f_{avg} = 10$ MHz. Tables 1 and 2 list the power consumed and area required, respectively, by the different blocks in the design, including the four cores, at 65 nm scaled from our simulation results of a single LIF neuron (fixed-point implementation), the STDP counter and a single 6T SRAM cell. The tables also list the numbers for 10 nm, scaled as per the trends reported in [9].

The power for inter core communication via the on-chip mesh network has three components: spike communication between different NMC cores, from input block to the NMC neurons and from the NMC neurons to the output neuron. It depends on the number of hops (N_{hops}) a spike packet has to travel to reach destination neurons, the number of cores (N_{cores}) a neuron communicates with and the hop distance (L_{hops}). N_{hops} and N_{cores} for an NMC neuron is 1. For the input and output blocks the N_{hops} and N_{cores} is 2. The voltage swing (V_{sw}) and V_{dd} for 10 nm node was 0.25 V [9]; we can see from Table 3 that the communication power is reduced by a factor of ~ 64 from 65 nm (where $V_{dd} = V_{sw} = 1\,\mathrm{V}$).

Table 1. Power estimates in μW for 65 nm (based on Cadence simulations and RTL synthesis) and 10 nm (scaling projections based on values in [9])

Technology node	NMC cores	Input Block	Output Block	Total
65 nm	24, 500	46	724	25, 270
10 nm	730	1.43	17.5	748.93

Table 2. Area estimates (in μm^2) for 65 nm (RTL synthesis) and 10 nm (scaling projections based on values in [9])

Technology node	NMC cores	Input Block	Output Block	Total
65 nm	1, 810, 000	1, 800	66, 700	1, 878, 500
10 nm	111, 000	200	3, 800	115, 000

Table 3. Communication Power in μW for 65 nm and 10 nm node

Technology node	NMC to NMC	Input to NMC	NMC to Output	Total
65 nm	30	257	600	887
10 nm	0.5	4.0	9.5	14.0

To quantify the learning capability of our design, we compute the metric Synaptic Updates Per Second (SUPS) per Watt from the relation,

$$SUPS/W = \frac{\text{Avg no. of weight updates per synapse} \times \text{No. of synapses}}{T_d \times \text{Power consumed}} \quad (6)$$

The average number of synaptic updates for training $T_d = 500\,\mathrm{ms}$ spike streams was ~ 20. With 560 learning synapses, we estimate that the 65 nm CMOS implementation (total power of 26.2 mW) to be capable of 0.85 MSUPS/Watt and the corresponding projected value for the 10 nm (with total power of 0.76 mW) node to scale to about 30 MSUPS/Watt.

7   Conclusion

The supervised learning algorithm ReSuMe has been analyzed in terms of its learning speed and the network size required. The results presented here give an indication of the size of the network to be used for longer spike streams. It was seen that an on-off ratio of just 100 is sufficient to represent the weights, with a precision of 5 bits. Based on this analysis, we have presented a high level scalable architecture for on-chip learning, inspired by the recently demonstrated SNN implementations. At an acceleration of 1000 our design is projected to scale to support 30 MSUPS/Watt at 10 nm node. The architecture proposed in this work thus presents a baseline for designing larger spike based learning systems.

Acknowledgments. This work was supported partially by Intel Technology India Pvt Ltd.

References

1. Gerstner, W., et al.: Neuronal dynamics: From single neurons to networks and models of cognition. Cambridge University Press (2014)
2. Bean, B.P.: The action potential in mammalian central neurons. Nature Reviews Neuroscience **8**(6), 451–465 (2007)
3. Gabbiani, F., Metzner, W.: Encoding and processing of sensory information in neuronal spike trains. Journal of Experimental Biology **202**(10), 1267–1279 (1999)
4. Ponulak, F., Kasinski, A.: Supervised learning in spiking neural networks with resume: sequence learning, classification, and spike shifting. Neural Computation **22**(2), 467–510 (2010)
5. Bora, A., Rao, A., Rajendran, B.: Mimicking the worman adaptive spiking neural circuit for contour tracking inspired by c. elegans thermotaxis. In: 2014 International Joint Conference on Neural Networks (IJCNN), pp. 2079–2086. IEEE (2014)
6. Kasinski, A., Ponulak, F.: Experimental demonstration of learning properties of a new supervised learning method for the spiking neural networks. In: Duch, W., Kacprzyk, J., Oja, E., Zadrożny, S. (eds.) ICANN 2005. LNCS, vol. 3696, pp. 145–152. Springer, Heidelberg (2005)
7. Gehlhaar, J.: Neuromorphic processing: a new frontier in scaling computer architecture. ACM SIGPLAN Notices **49**(4), 317–318 (2014)
8. Merolla, P., et al.: A digital neurosynaptic core using embedded crossbar memory with 45pj per spike in 45nm. In: 2011 IEEE Custom Integrated Circuits Conference (CICC), pp. 1–4. IEEE (2011)
9. Rajendran, B., et al.: Specifications of nanoscale devices and circuits for neuromorphic computational systems. IEEE Transactions on Electron Devices **60**(1), 246–253 (2013)
10. Hebb, D.O.: The organization of behavior: A neuropsychological theory. Psychology Press (2005)
11. Bi, G.Q., Poo, M.M.: Synaptic modification by correlated activity: Hebb's postulate revisited. Annual Review of Neuroscience **24**(1), 139–166 (2001)
12. Lee, C.M., et al.: Heterosynaptic plasticity induced by intracellular tetanization in layer 2/3 pyramidal neurons in rat auditory cortex. The Journal of Physiology **590**(10), 2253–2271 (2012)
13. Maass, W., Natschläger, T., Markram, H.: Real-time computing without stable states: A new framework for neural computation based on perturbations. Neural Computation **14**(11), 2531–2560 (2002)
14. Merolla, P.A., et al.: A million spiking-neuron integrated circuit with a scalable communication network and interface. Science **345**(6197), 668–673 (2014)
15. Kraft, M., Kasinski, A., Ponulak, F.: Design of the spiking neuron having learning capabilities based on fpga circuits. Discrete-Event System Design **3**, 301–306 (2006)
16. Seo, J., et al.: A 45nm cmos neuromorphic chip with a scalable architecture for learning in networks of spiking neurons. In: 2011 IEEE Custom Integrated Circuits Conference (CICC), pp. 1–4. IEEE (2011)
17. Gupta, S., et al.: Deep learning with limited numerical precision (2015). arXiv preprint arXiv:1502.02551

Enhanced KNNC Using Train Sample Clustering

Hamid Parvin[1(✉)], Ahad Zolfaghari[2], and Farhad Rad[2]

[1] Department of Computer Engineering, Mamasani Branch,
Islamic Azad University, Mamasani, Iran
`parvin@iust.ac.ir`
[2] Department of Computer Engineering, Yasouj Branch, Islamic Azad University, Yasouj, Iran
`rad@comp.iust.ac.ir`

Abstract. In this paper, a new classification method based on k-Nearest Neighbor (kNN) lazy classifier is proposed. This method leverages the clustering concept to reduce the size of the training set in kNN classifier and also in order to enhance its performance in terms of time complexity. The new approach is called Modified Nearest Neighbor Classifier Based on Clustering (MNNCBC). Inspiring the traditional lazy k-NN algorithm, the main idea is to classify a test instance based on the tags of its k nearest neighbors. In MNNCBC, the training set is first grouped into a small number of partitions. By obtaining a number of partitions employing several runnings of a simple clustering algorithm, MNNCBC algorithm extracts a large number of clusters out of those partitions. Then, a label is assigned to the center of each cluster produced in the previous step. The assignment is determined with use of the majority vote mechanism between the class labels of the patterns in each cluster. MNNCBC algorithm iteratively inserts a cluster into a pool of the selected clusters that are considered as the training set of the final 1-NN classifier as long as the accuracy of 1-NN classifier over a set of patterns included the training set and the validation set improves. The selected set of the most accurate clusters are considered as the training set of proposed 1-NN classifier. After that, the class label of a new test sample is determined according to the class label of the nearest cluster center. While kNN lazy classifier is computationally expensive, MNNCBC classifier reduces its computational complexity by a multiplier of 1/k. So MNNCBC classifier is about k times faster than kNN classifier. MNNCBC is evaluated on some real datasets from UCI repository. Empirical results show that MNNCBC has an excellent improvement in terms of both accuracy and time complexity in comparison with kNN classifier.

Keywords: Edited nearest neighbor classifier · kNN · Combinatorial classification

1 Introduction

One of the most important goals of artificial intelligence is to design models with high recognition rates [30-33]. In pattern recognition, the input space is mapped into the high dimensional feature space, and in the feature space it is tried to determine the

© Springer International Publishing Switzerland 2015
L. Iliadis and C. Jayne (Eds.): EANN 2015, CCIS 517, pp. 159–168, 2015.
DOI: 10.1007/978-3-319-23983-5_16

optimal hyperplane(s), so that the mapped function better approximates the main function for each unseen data. The mapping that is named classification is an interesting subject in machine learning and data mining communities with a lot of studies around it [16-18].

Despite the simplicity, k-Nearest Neighbor (kNN) classifier is one of the most fundamental classifiers. It is also the simplest classifier. When there is a little or no prior knowledge about the data distribution, kNN classifier could be automatically the first choice for a classification study.

In a lot of recent emerging applications, such as text categorization [21-22], multiple classifier systems (MCS) [23-24], intrusion detection field [25] (an intrusion detection problem is first converted to a text categorization problem; then it is treated as a text categorization problem), medical systems such as diagnosis of diabetes diseases [26], thyroid diseases [27] and myocardial infarction [28], and image classification [29] and etc., kNN classifier has been successfully applied and proves its effectiveness.

The kNN classifies a test sample x by assigning it the label most frequently represented among the k nearest samples; in other words, a decision is made by examining the labels on the k-nearest neighbors and taking a majority vote mechanism. kNN classifier was developed from the need to perform discriminant analysis when reliable parametric estimates of the probability densities are unknown or difficult to determine. In 1951, Fix and Hodges introduced a non-parametric method for pattern classification that is known the k-nearest neighbor rule [1] and [15]. Later in 1967, some of the formal properties of the k-nearest neighbor rule have been worked out; for instance it was shown that for $k=1$ and $n \to \infty$ the kNN classification error is bounded above by twice the Bayes error rate [2]. Once such formal properties of kNN classification were established, a long line of investigation ensued including new rejection approaches [3], refinements with respect to Bayes error rate [4], distance weighted approaches [5-6], soft computing [7] methods and fuzzy methods [8-9].

Some advantages of kNN include: the simplicity to use, the robustness to learn in a noisy training data (especially if it uses the inverse square of weighted distances as the "distance metric"), and finally the effectiveness in learning at a large scale training dataset. Although kNN has the mentioned advantages, it has some drawbacks such as: high computational cost (because it needs to compute distance of each query instance to all training samples); large memory consumption (in proportion with the size of training set); ineffectiveness in multidimensional datasets; sensitivity to well-setting of parameter k (number of effective nearest neighbors); sensitivity to the used distance metric; and finally ignoring a weighting mechanism for features [10].

The High computational cost of the nearest neighbor algorithm, in both space (storage of prototypes) and time (distance computation) has received a great deal of analysis. Suppose we have N labeled training samples in d dimensions, and seek to find the closest to a test point x ($k = 1$). In the most naive approaches we inspect each stored point in turn, calculate its Euclidean distance to x, retaining the identity only of the current closest one. Each distance calculation is $O(d)$, and thus this search is $O(dN^2)$ [10].

2   Related Work

ITQON et al. in [11] proposed a classifier, TFkNN, aiming at upgrading of distinction performance of kNN classifier and combining plural KNNs using testing characteristics. Their method not only upgrades distinction performance of kNN but also brings an effect of stabilizing variance of recognition ratio.

Parvin et al. proposes two modified versions of kNN named MKNN1 and MKNN2. The main idea of their works is to assign the class label of the data according to the k nearest neighbor data samples that are previously validated. On the other side, the validity of all data samples in the training set is computed. Then, a weighted kNN is performed on any test samples [18].

A comparative analysis of kNN and decision tree classifiers for the Irish National Forest Inventory has been done and the results were assessed based on two evaluation metrics: the estimation errors (primarily using the Root Mean Square Error) and relative mean deviation (bias); in terms of both metrics it has been proved that the performance of kNN is better than decision tree [19].

Input:
    PTrS: patterns of train set
    PTeS: patterns of test set
    c: the number of classes
    clus_alg: clustering algorithm
    pq: the threshold for assigning a label to a
        cluster center
    Maxiteration: the maximum of the allowed
        iterations
    Condition: a condition for decision
Output:
    accuracy: accuracy of MNNCBC over *PTeS*
$PC=\{\}$
partition *PTrS* into two clusters: *TS* and *VS*
For $i = 1$ to *Maxiteration*
   1. $P=clus\_alg(TS,k)$ where $2*c \geq k \geq c$;
     resulting cluster centers P_i
   2. For each $p \in P$
     if *condition(cvr, pq)*, then p_i will be added to
     the set *PC;* where *cvr* is the maximum
     number of the *pq* nearest patterns of P_i in
     PTrS that have consensus vote to an
     identical label
$SPC=\{\}$
$cur\_acc=0$
For each $q \in PC$
   1. $TTS=SPC \cup q$
   2. $acc=1\text{-}NN(TTS, PTrS)$
   3. if$(acc>cur\_acc)$ $SPC=TTS$
$acc=1\text{-}NN(TTS, PTeS)$

Fig. 1. Pseudo-code of training phase of the nearest cluster classifier algorithm

Su has proposed a genetic weighted kNN classifier. A genetic algorithm has been used to train an optimal weight vector for features; meanwhile, an unsupervised clustering algorithm has been applied to reduce the number of instances in the dataset, in order to shorten training and execution time, as well as to promote the system's overall performance. He has shown that more instances of the dataset can be replaced by a more meaningful subset of them [20]. In this paper inspiring from the method proposed by Su, clustering is employed as a preprocessing phase of kNN. While Su has not proposed how to cluster the data, it is explored in this paper. The proposed kNN can augment the performance of original kNN classifier in terms of the accuracy, the time complexity and the memory complexity. This method which is called that named Modified Nearest Neighbor Classifier Based on Clustering (MNNCBC), applies the clustering techniques to reduce the number of training prototypes. Despite of reducing train samples, the clustering will guide us to find a more significant sub-sample out of the data.

3   MNNCBC: Modified Nearest Neighbor Classifier Based on Clustering

In all experiments the 10-fold cross validation is employed to obtain the performance of the MNNCBC algorithm. In 10-fold cross validation the dataset is randomly partitioned into 10 clusters. Considering each partition as test set, *PTeS*, and the other data as train set, *PTrS*, the MNNCBC algorithm reaches 10 experiments. Averaging the performances of the MNNCBC algorithm over all these 10 test sets, the final performance of the MNNCBC algorithm is obtained. In each experimentation, the train set, *PTrS*, is divided into two sub-sets, train sub-set, *TS*, and evaluation (validation) subset, *VS*. The main idea of the proposed approach is assigning data to the nearest cluster who is naturally consisted the neighbor points. To implement the idea, first, the samples of the train sub-set, *TS*, are clustered into k clusters where k is a number equal to or greater than the number of real classes, c, and equal to or less than $2*c$. The clustering is done with use of a simple rough clustering algorithm, *clus_alg*. Then, the obtained cluster centers are considered as new points and their labels are determined based on the simple majority vote mechanism. Indeed the label of a cluster is obtained based on a kNN classifier over train set where k is equal to an input parameter, denoted by *pq*. The *condition* is computed as equation 1.

$$condition(cvp, pq) = \begin{cases} cvp \geq pq \\ cvp \geq \dfrac{pq}{2} \\ cvp \leq \dfrac{pq}{2} \end{cases} \tag{1}$$

For example, if $condition(cvp, pq) = cvp > \frac{pq}{2}$; it means that each cluster center that is labeled with less than $\frac{pq}{2}$ votes in kNN classifier (using *PTrS* as train set) is immediately omitted; otherwise it is added in a pool of clusters, *PC*. This procedure is iterated as many as *Maxiteration*. So finally the MNNCBC algorithm has a pool of clusters with $Maxiteration \times c^3$ ($Maxiteration \times k \times k \times \frac{k}{2}$) clusters at most.

The quality of obtained clusters is evaluated employing a 1-NN with considering *PTrS* as test set. Each pattern of the *PTrS* is used as a test sample; with specifying the nearest cluster, its label is assigned to the sample. After that, in comparison with the ground true labels of data, the accuracy of the obtained classifier is derived. So, after obtaining the pool of clusters, *PC*, the MNNCBC algorithm iteratively selects them into a subset of pool of clusters (*SPC*) if the accuracy of a 1-NN using *SPS* as train set over *PTrS* is improved.

In the test phase of the MNNCBC algorithm, a new test sample is assigned to the label of the nearest cluster center. The pseudo code of training phase of the Nearest Cluster Classifier algorithm is shown in Fig. 1. Until here, the training of the MNNCBC is finished. After here, any test samples are classified using the trained classifier. In Fig. 1, training and test sets, number of classes, the used clustering algorithm, the threshold for assigning a label to a cluster center, the maximum of the allowed iterations and the condition for decision are fed to the proposed algorithm. Because of the need of MNNCBC algorithm to validation set, it first divides the training set into two subsets: training and test subsets. In the first Iteration of the algorithm, the following parts are done as many as *Maxiteration*. In each iteration training subset is partitioned into k clusters where k is a random value ranging from a to $2*a$, where a is the real number of classes. Then by testing the cluster centers using kNN with the k parameter equal to *pq*, we assign a label to any cluster center; also the number of agreements to assign a label to the cluster center is denoted by *cvr*. If the condition in equation 1 is not satisfied for a cluster center using its *cvr* and *pq*, the cluster center will be dropped; otherwise, it will be inserted into the set *PC*. Finally we will select a cluster center in the set *PC* if it increases the accuracy of 1-NN (with train set of selected cluster centers so far) in a greedy mechanism.

Now we explain how the class labels are used to specify the labels of the cluster centers which are explanatory points of the clusters. There are some combining methods to aggregate the class labels of the cluster members. When the individual votes of classifiers are crisp (not soft/fuzzy), the Simple Majority Vote approach is the common logical way that votes a pattern x to a class j if a little more than n/c of classifiers (here cluster members) assigns to class j [12], where n and c stand for the number of cluster members and the number of classes, respectively. In this paper, the majority vote mechanism is used to assign a class label to cluster centers. It means that *pq* nearest neighbors of a cluster center in the *PTrS* make a consensus decision about its label.

There are many methods to evaluate the clustering result. They may use whether external indices, whether internal indices or relative indices [13]. External index needs further information to evaluate the clusters. In this work, the *PTrS* is used to measure the performance of the different clusterings. It is a kind of external index usage. First, the MNNCBC algorithm is trained on the *TS*. Then, by executing the trained classifier on the *PTrS*, the accuracy of this method is obtained using the ground true class labels of the *PTrS*.

As it is shown in Fig. 1, the steps 1, 2 and 3 are repeated *Maxiteration* times. In this method there is a procedure to select a set of satisfactory good cluster centers from several times of performing clustering techniques; however, the cluster centers obtained from any iteration can be considered as the solution. The method enhances both the accuracy and robustness of the kNN classifier algorithm, significantly; however, it needs less time and memory in testing phase. Based on empirical result, it can be induced that, usually the best results may be obtained when the *SCP* size is chosen near to the value of number of classes, c. Since each cluster center has d dimensions, examining each test sample needs to $O(cd)$. In the worst case the time complexity is $O(c^3d)$. It shows that the proposed combinational method can be performed with less order than the kNN classifier method which is $O(dkN)$.

4   Experimental Study

This section discuses the experimental results and compares the performance of the MNNCBC algorithm with original kNN methods.

The proposed method is examined over 9 different standard datasets and one artificial dataset. It has been tried to keep diversity in their numbers of features, samples, and true classes. a large variety of the used datasets can more validate the final results. Detailed information about the used datasets is available is available in [14]

Note that datasets which are marked with star (*) in paper are normalized. The experiments are made on the normalized features in the stared dataset. It means each feature is normalized with the mean of 0 and the variance of 1, N(0, 1). The artificial HalfRing dataset is depicted in Fig. 2. The HalfRing dataset is considered as one of the most challenging dataset for the proposed MNNCBC algorithm.

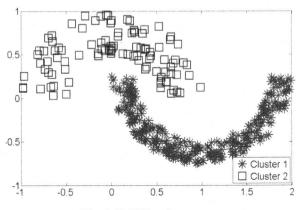

Fig. 2. Half Ring dataset.

All experiments are reported on 10-fold cross validation. Parameters pq and *Maxiteration* are respectively set to 5 and 10 in all of experiments experimentations. Parameter *clus_alg* is k-means. It means that the k-means clustering algorithm is considered as clustering algorithm. The maximum allowed iterations in the k-means clustering algorithm is equal to 2 in order to obtain the rough and unlike partitions out of data. The number of partitions that is requested from k-means is a random value between c and $2*c$. Validation set, *VS*, is 22.22% of train set, *PTrS*, through all experiments.

Table 1 shows final accuracies of the MNNCBC algorithm using three different conditions. In each column, the accuracies obtained by the MNNCBC algorithm employing one condition are shown. Next to the accuracy of each MNNCBC algorithm on each dataset, the averaged number of cluster centers in the final 1-NN classifier is presented. It is obvious that the condition $cvp \geq \frac{pq}{2}$ is the best condition among the three used conditions. But there is a hidden rule among the results obtained with the condition $cvp \geq \frac{pq}{2}$ in the MNNCBC algorithm and the results obtained by using the condition $cvp \geq pq$ in the MNNCBC algorithm. First we define a new column in Table 1. It is the ratio of column 3 to column 5. We present this defined column in last column of Table 1. By a detailed considering of the values in the last column of Table 1, it is inferred when the last column for a dataset is lower than the averaged last column over all datasets (depicted at last column and last row in Table 1, it is equal to 1.58), the MNNCBC algorithm with the condition $cvp \geq pq$ is superior to the others; otherwise the MNNCBC algorithm with the condition $cvp \geq \frac{pq}{2}$ is superior.

Table 1. Final results of the MNNCBC algorithm using different conditions.

	$cvp \geq \frac{pq}{2}$		$cvp \geq pq$		$cvp \leq \frac{pq}{2}$			
	MNN CBC	Average # of Proto- types	MNN CBC	Average # of Proto- types	MNN CBC	Average # of Proto- types	All Prototypes	Ratio of Column 3 to 5
1	**97.25**	6.90	96.25	6.20	96.25	9.00	280	1.11
2	95.33	8.60	**96.00**	7.30	**96.00**	7.00	105	1.18
3	96.47	6.20	**97.06**	5.10	95.88	5.00	479	1.22
4	**60.29**	10.40	52.53	2.50	58.24	10.00	242	4.16
5	**71.56**	52.00	67.50	28.90	69.69	141.00	227	1.80
6	**66.67**	28.90	52.86	15.00	65.71	98.00	151	1.93
7	82.86	9.90	**85.71**	6.90	84.00	8.00	246	1.43
8	**69.78**	10.10	65.65	1.60	68.26	9.00	324	6.31
9	95.29	9.20	95.29	7.40	**95.88**	8.00	126	1.24
10	**57.70**	92.40	57.53	67.31	54.53	480.00	1040	1.37
average	**79.32**	23.46	76.64	14.82	78.44	77.50	322	1.58

The hidden rule says when the number of prototypes in the MNNCBC with the condition $cvp \geq pq$ is less than the number of prototypes in the MNNCBC with the condition $cvp \geq \frac{pq}{2}$ by large margin; the dataset is a hard one. So we must turn to the MNNCBC with the best prototypes. It means that the MNNCBC with the condition $cvp \geq pq$ is the best option. It also says when the number of prototypes in the MNNCBC with the condition $cvp \geq pq$ is less than the number of prototypes in the

MNNCBC with the condition $cvp \geq \frac{pq}{2}$ by small margin, the dataset is an easy one. So we can turn to the MNNCBC with the more prototypes to cover total feature space. It means that the MNNCBC with the condition $cvp \geq \frac{pq}{2}$ is the best option.

By using the hidden rule, we can present a combinatorial selective classifier that contains both MNNCBCs and uses each of them depending on the defined ratio for the two classifiers.

Table 2 shows the performances of different kNN classifiers and the proposed combinatorial selective classifier comparatively. MNNCBC is compared with original versions of kNN. The MNNCBC method outperforms some of kNN classifiers in terms of average accuracy. In addition, because of the lower number of stored prototypes, the results of the proposed combinatorial selective classifier are gained while the testing phase of the MNNCBC method has less computational cost in both cases of time and memory rather than the kNN classifiers.

Table 2. Final accuracies of the MNNCBC comparing with the results of different kNN classifiers.

	Different Method								
	1-NN	2-NN	3-NN	4-NN	5-NN	6-NN	7-NN	MNN CBC	Average # of Prototypes
1	100.00	100.00	100.00	100.00	100.00	100.00	100.00	96.25	6.20
2	96.00	95.33	96.67	97.33	97.33	98.00	98.00	96.00	7.30
3	96.91	94.71	97.21	96.76	97.50	96.18	96.62	97.06	5.10
4	63.53	60.00	63.82	62.06	60.59	55.59	58.53	60.29	10.40
5	81.88	82.81	80.63	77.51	75.94	71.25	70.94	71.56	52.00
6	69.05	68.57	69.52	63.80	64.29	63.33	62.86	66.67	28.90
7	86.57	89.43	84.00	86.57	84.86	85.71	83.43	85.71	6.90
8	65.65	66.30	68.70	69.13	65.43	67.17	66.52	69.78	10.10
9	95.29	94.12	94.71	94.71	96.47	94.71	96.47	95.29	7.40
10	52.77	51.55	55.47	55.61	57.91	57.57	57.03	57.53	67.31
average	80.77	80.28	81.07	80.35	80.03	78.95	79.04	79.61	20.16

5   Conclusion and Future Works

In this paper, a new method is proposed to improve the performance of kNN classifier. The proposed method which is called MNNCBC, standing for Nearest Cluster Classifier, improves the kNN classifier in terms of both time and memory order. The MNNCBC algorithm employs clustering technique to find the same groups of data in multi-dimensional feature space.

Despite of reducing training prototypes, the clustering technique can cause to find the natural groups of data. On the other hands, the natural neighborhoods can be successfully recognized by clustering technique. Moreover, unlike the kNN method which classifies any sample without considering the data distribution, only based on exactly k nearest neighbor, in the MNNCBC algorithm, the data is grouped into k clusters unequally, according to the data distribution and the position of data samples in feature space.

MNNCBC method is examined over nine benchmarks from UCI repository and one synthetic dataset, HalfRing. According to the obtained results, it can be concluded that the proposed algorithm is comparatively not worse than kNN classifier. MNNCBC method is even more accurate than kNN classifier in some cases.

References

1. Fix, E., Hodges, J.L.: Discriminatory analysis, nonparametric discrimination: Consistency properties. Technical Report 4, USAF School of Aviation Medicine, Randolph Field, Texas (1951)
2. Cover, T.M., Hart, P.E.: Nearest neighbor pattern classification. IEEE Trans. Inform. Theory, IT **13**, 21–27 (1967)
3. Hellman, M.E.: The nearest neighbor classification rule with a reject option. IEEE Trans. Syst. Man Cybern. **3**, 179–185 (1970)
4. Fukunaga, K., Hostetler, L.: k-nearest-neighbor bayes-risk estimation. IEEE Trans. Information Theory **21**(3), 285–293 (1975)
5. Dudani, S.A.: The distance-weighted k-nearest-neighbor rule. IEEE Trans. Syst. Man Cybern., SMC **6**, 325–327 (1976)
6. Bailey, T., Jain, A.: A note on distance-weighted k-nearest neighbor rules. IEEE Trans. Systems, Man. Cybernetics **8**, 311–313 (1978)
7. Bermejo, S., Cabestany, J.: Adaptive soft k-nearest-neighbour classifiers. Pattern Recognition **33**, 1999–2005 (2000)
8. Jozwik, A.: A learning scheme for a fuzzy k-nn rule. Pattern Recognition Letters **1**, 287–289 (1983)
9. Keller, J.M., Gray, M.R., Givens, J.A.: A fuzzy k-nn neighbor algorithm. IEEE Trans. Syst. Man Cybern., SMC **15**(4), 580–585 (1985)
10. Duda, R.O., Hart, P.E., Stork, D.G.: Pattern Classification. John Wiley & Sons (2000)
11. Itqon, S.K., Satoru, I.: Improving Performance of k-Nearest Neighbor Classifier by Test Features. Springer Transactions of the Institute of Electronics, Information and Communication Engineers (2001)
12. Lam, L., Suen, C.Y.: Application of majority voting to pattern recognition: An analysis of its behavior and performance. IEEE Transactions on Systems, Man, and Cybernetics **27**(5), 553–568 (1997)
13. Jain, A.K., Dubes, R.C.: Algorithms for Clustering Data. Prentice-Hall, Englewood Cliffs (1988)
14. Newman, C.B.D.J., Hettich, S., Merz, C.: UCI repository of machine learning databases (1998). http://www.ics.uci.edu/~mlearn/MLSummary.html
15. Wu, X.: Top 10 algorithms in data mining. Knowledge information, 22-24. Springer-Verlag London Limited (2007)

16. Parvin, H., Minaei-Bidgoli, B., Ghatei, S., Alinejad-Rokny, H.: An Innovative Combination of Particle Swarm Optimization, Learning Automaton and Great Deluge Algorithms for Dynamic Environments. International Journal of the Physical Sciences, IJPS **6**(22), 5121–5127 (2011)
17. Parvin, H., Helmi, H., Minaei-Bidgoli, B., Alinejad-Rokny, H., Shirgahi, H.: Linkage Learning Based on Differences in Local Optimums of Building Blocks with One Optima. International Journal of the Physical Sciences, IJPS **6**(14), 3419–3425 (2011)
18. Parvin, H., Alizadeh, H., Minaei-Bidgoli, B.: Validation Based Modified k-Nearest Neighbor. Book Chapter in IAENG Transactions on Engineering Technologies, II–Special Edition of the World Congress on Engineering and Computer Science (2008)
19. McInerney, D.O., Nieuwenhuis, M.B.: A comparative analysis of kNN and decision tree methods for the Irish National Forest Inventory. International Journal of Remote Sensing **30**(19), 4937–4955 (2009)
20. Su, M.Y.: Using clustering to improve the kNN-based classifiers for online anomaly network traffic identification. Journal of Network and Computer Applications **34**(2), 722–730 (2010)
21. Bi, Y., Bell, D., Wang, H., Guo, G., Guan, J.: Combining multiple classifiers using dempster's rule text caractrization. Applied Artificial Intelligence: An International Journal **21**(3), 211–239 (2007)
22. Tan, S.: An effective refinement strategy for KNN text classifier. Expert Systems with Applications **30**(2), 290–298 (2005)
23. Yan, W.Y., Shaker, A.: The effects of combining classifiers with the same training statistics using Bayesian decision rules. International Journal of Remote Sensing **32**(13), 3729–3745 (2011)
24. Gao, Y., Gao, F.: Edited AdaBoost by weighted kNN. Neurocomputing **73**(16–18), 3079–3088 (2010)
25. Liao, Y., Vemuri, V.R.: Use of K-Nearest Neighbor classifier for intrusion detection. Computers & Security **21**(5), 439–448 (2002)
26. Chikh, M.A., Saidi, M., Settouti, N.: Diagnosis of Diabetes Diseases Using an Artificial Immune Recognition System2 (AIRS2) with Fuzzy K-nearest Neighbor. Journal of Medical Systems (2011) (Online)
27. Liu, D.Y., Chen, H.L., Yang, B., Lv, X.E., Li, L.N., Liu, J.: Design of an Enhanced Fuzzy k-nearest Neighbor Classifier Based Computer Aided Diagnostic System for Thyroid Disease. Journal of Medical Systems (2011) (Online)
28. Arif, M., Malagore, I.A., Afsar, F.A.: Detection and Localization of Myocardial Infarction using K-nearest Neighbor Classifier. Journal of Medical Systems **36**(1), 279–289 (2012)
29. Mejdoub, M., Amar, C.B.: Classification improvement of local feature vectors over the KNN algorithm. Multimedia Tools and Applications (2011) (Online)
30. Qodmanan, H.R., Nasiri, M., Minaei-Bidgoli, B.: Multi objective association rule mining with genetic algorithm without specifying minimum support and minimum confidence. Expert Systems with Applications **38**(1), 288–298 (2011)
31. Parvin, H., Minaei-Bidgoli, B., Alizadeh, H.: Detection of cancer patients using an innovative method for learning at imbalanced datasets. In: Yao, J., Ramanna, S., Wang, G., Suraj, Z. (eds.) RSKT 2011. LNCS, vol. 6954, pp. 376–381. Springer, Heidelberg (2011)
32. Daryabari, M., Minaei-Bidgoli, B., Parvin, H.: Localizing program logical errors using extraction of knowledge from invariants. In: Pardalos, P.M., Rebennack, S. (eds.) SEA 2011. LNCS, vol. 6630, pp. 124–135. Springer, Heidelberg (2011)
33. Parvin, H., Minaei-Bidgoli, B.: Linkage learning based on local optima. In: Jędrzejowicz, P., Nguyen, N.T., Hoang, K. (eds.) ICCCI 2011, Part I. LNCS, vol. 6922, pp. 163–172. Springer, Heidelberg (2011)

Intelligent Telecommunications
Modeling

A Metric for Determining the Significance of Failures and Its Use in Anomaly Detection

Case Study: Mobile Network Management Data from LTE Network

Robin Babujee Jerome[1(✉)] and Kimmo Hätönen[2]

[1] Department of Communication Engineering, Aalto University School
of Electrical Engineering, Espoo, Finland
robin.babujeejerome@aalto.fi
[2] Nokia Networks Research, Espoo, Finland
kimmo.hatonen@nokia.com

Abstract. In big data analytics and machine learning applications on telecom network measurement data, accuracy of findings during the analysis phase greatly depends on the quality of the training data set. If the training data set contains data from Network Elements (NEs) with high number of failures and high failure rates, such behavior will be assumed as normal. As a result, the analysis phase will fail to detect NEs with such behavior. High failure ratios have traditionally been considered as signs of faults in NEs. Operators use well-known Key Performance Indicators (KPIs), such as, e.g., Drop Call Ratio and Handover failure ratio to identify misbehaving NEs. The main problem with these KPIs based on failure ratios is their unstable nature. This paper proposes a method of measuring the significance of failures and its use in training set filtering.

Keywords: Anomaly detection · Pre-processing · Self-Organizing maps · Training set filtering

1 Introduction

Anomalies in telecommunication networks can be signs of errors or malfunctions, which originate from a wide variety of reasons. Huge amount of data collected from Network Elements (NEs) in the form of counters, server logs, audit trail logs etc. can provide significant information about the normal state of the system as well as possible anomalies [1]. Anomaly detection (AD) forms a very important task in telecommunication network monitoring [2] and has been the topic of several research works in the past few years [3] [1]. Since it is very difficult to obtain reference data with labeled anomalies from industrial processes, unsupervised methods are chosen for AD [4]. Among these unsupervised techniques, Self-Organizing Maps (SOMs) [5] is a tool often used for analyzing telecommunication network data [1] [6] [7]; characterized by its high volume and high dimensionality. The key idea of SOM is to map

© Springer International Publishing Switzerland 2015
L. Iliadis and C. Jayne (Eds.): EANN 2015, CCIS 517, pp. 171–180, 2015.
DOI: 10.1007/978-3-319-23983-5_17

high-dimensional data into low-dimensional space by competitive learning and topological neighbourhood so that the topology is preserved [8].

Network traffic data obtained from several sources need to be pre-processed before they can be fed to SOMs or other AD mechanisms. These pre-processing steps could include numerization of log data, cleaning and filtering of training set, scaling and weighting of variables etc. depending on the type of data analyzed and goal of the AD task. SOMs, as well as other neural network models, follow the "garbage in - garbage out" principle. If poor quality data is fed to the SOM, the result will also be of poor quality [4]

The important events that happen in various NEs during the operation of the network are counted, which forms the raw low level data called counters. Since the number of these low level data counters is too large and often unmanageable, several of these counters are aggregated to form high level counter variables called *Key Performance Indicators* (KPIs) [9].

Several well-known KPIs used by telecom operators such as Drop Call Ratio (DCR), Handover failure ratio (HO_Fail), SMS failure ratio (SMS_Fail) *etc.* are typical examples of failure ratio based KPIs. The failure ratio metric (u, n) for u failures out of n attempts, defined in equation (1), does not take the magnitude of the number of attempts into account. Hence one failure out of one attempt, as well as, hundred failures out of hundred attempts, give the same resultant failure ratio.

$$fr(u, n) = (u/n) = (failure\ count/total\ attempts) \tag{1}$$

If there has been a lot of activity and both numerator and denominator of the equation are high, the failure ratio is a meaningful metric. However, if there has not been much activity, both numerator and denominator are low, and the resulting failure ratio metric can be randomly high. Using $fr(u, n)$ as a metric for filtering the training set can have mainly two drawbacks: Firstly, it removes random points from the training dataset and overall quality of the dataset cannot be guaranteed to be high. Secondly, network monitoring personnel can be misguided by such wrong signals and is likely to spend their time analyzing an anomaly which might result due to a high failure ratio and low number of attempts.

It is possible to give thresholds for n and u, above which the failure ratio can be considered as a measure of failure. However, this approach too comes with its own set of problems. Regional, seasonal, daily, weekly and even hourly traffic variations can lead to such rules being unable to detect important anomalies.

In this paper, we introduce more stable *failure significance metric* in Section 2. Sections 3 and 4 shortly introduce an AD method and some training data set filtering techniques that we used. In Section 5 the proposed metric is used for training set filtering for an AD experiment on performance measurement data obtained from a live LTE network. Finally, Section 6 concludes the paper.

2   Failure Significance Metric

The *failure significance metric* (fsm) is a metric that evaluates the significance of a failure based on the number of attempts that have been made. The fsm metric balances the failure ratio so that the failure ratios based on lesser number of attempts are scaled to be of smaller value when compared to the failure ratio based on higher number of attempts. Equation (2) defines a weight function that helps in scaling a value based on the sample size n.

$$f(n) = 2/(1 + e^{\frac{w}{n}}) \tag{2}$$

The term w is a term that can be used to adjust the sensitivity of the function. The value of this parameter could vary from 0 to 1. In cases when w is 0, there is no scaling based on number of attempts as the whole fraction becomes equal to 1. The behavior is opposite at the other end of the range ($w = 1$). Fig. 1 depicts how the value of the scaling function $f(n)$ varies for different values of the sensitivity tuning parameter w.

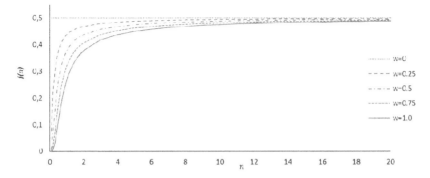

Fig. 1. Variation of scaling function with sensitivity tuning parameter w

As the number of attempts increases, the denominator increases, resulting in high value of $f(n)$. This property of the function can be used to scale the difference of a failure ratio from the average value of failure ratio in the training set. When the difference between the failure ratio of a data point and the average failure ratio of the training set is high, the scaling function $f(n)$ results in a higher scaled value in cases where the number of attempts is high, in contrast to cases where the number of attempts is low. On applying this scaling function to the difference of the failure ratio with the average failure ratio gives a scaled value of failure ratio $fr_{scaled}(u, n)$ as shown in equation (3). Here $fr(u, n)_{avg}$ represents the average failure ratio in the entire dataset.

$$fr_{scaled}(u, n) = \frac{2}{1 + e^{\frac{w}{n}}}\left(fr(u, n) - fr(u, n)_{avg}\right) \tag{3}$$

Let us consider u as the total number of failures that occur in an NE during an aggregation interval, and u_{avg} as the average number of failures that occur for all the NEs during the same aggregation interval in training set. When the number of failures u is high, the impact it has on the network is also high. The value of logarithm of $(u + 1)$ to the base $(u_{avg} + 1)$ (which is also $\log(u + 1) / \log(u_{avg} + 1)$) gives a measure of relative magnitude of failure count when compared to average failure counts. This factor defined here as an impact factor is given in equation (4).

$$i(u) = log\,(u+1)/log\,(u_{avg}+1) \tag{4}$$

Adding one to the numerator and denominator eliminates the need to separately handle the zeroes in the data. The value of $fsm(u,n)$ is further obtained by multiplying the two terms $fr_{scaled}(u,n)$ and $i(u)$ as represented by equation (5). Fig. 2 depicts two views of a three dimensional plot of the obtained fsm function.

$$fsm(u,n) = \frac{2}{1+e^{\frac{w}{n}}}\Big((u/n)-(u/n)_{avg}\Big)*\frac{log(u+1)}{log(u_{avg}+1)} \tag{5}$$

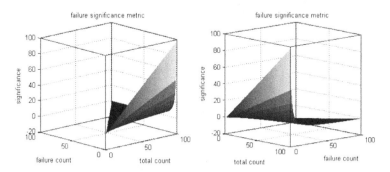

Fig. 2. Graphical representation of the failure significance metric (w=0.5)

2.1   Significance Metric in Training Dataset Filtering

An operator monitoring the network for anomalies will be interested in observing the network elements which have faults corresponding to high fsm. By removing the data points, which correspond to high fsm from the training set, the observations in the analysis dataset which correspond to similar behavior will be detected as anomalies. This is a typical example of a case in which application domain knowledge is used in filtering the training data set. Consider a set of n observations measured from m cells. The fsm based training set filtering approach removes the top k percentile of observations which have the highest value of the fsm metric. For the experiments done as part of this research, the values of k and w were kept at 1% and 0.5 respectively. The value of w was determined through a heuristic approach. The top 1% of the observations in the training set are considered as corresponding to high values of failure significance metric.

3   The Anomaly Detection Mechanism

The AD system used in this research work is based on the method by Höglund et al [10]. This method uses quantization errors of a one or two-dimensional SOM for AD. The basic steps in the algorithm are listed step by step.

1. A SOM is fitted to the reference data of n samples. Nodes without any hits are dropped out.

2. For each of the n samples in the reference data, the distance from the sample to its BMU is calculated. These distances, denoted by $D_1 .... D_n$ are also called as quantization errors.
3. A threshold is defined for the quantization error which is a predefined quantile of all the quantization errors in the reference data.
4. For each of the samples in the analysis data set, the quantization errors are calculated based on the distance D_{n+1} from its BMU.
5. A sample is considered as an anomaly if its quantization error is larger than the threshold calculated in Step 3.

4   Training Dataset Filtering Techniques

This section focuses on some techniques of cleaning up the training set before it is used in the AD process. The presence of outliers in the training set can dominate the analysis results and thus hide essential information. The detection and removal of these outliers from the training set results in improved reliability of the analysis process [11]. The objective for training set filtering is to remove from the training set those data points that have a particular pattern associated with them, so that similar behavior in the analysis dataset is detected as anomalous. The training set filtering techniques evaluated as part of this research are listed in the following sub-sections.

SOM Smoothening Technique: This is one of the widely used training set filtering techniques. This technique is a generic one and does not use any application domain knowledge in filtering the training set. The steps in this process are:

1. The entire training set is used in the training to learn the model of the training data.
2. AD is carried out on the training set to filter out the non-anomalous points using the model generated in Step 1.
3. The non-anomalous points from the training set obtained in Step 2 are used once again to generate a refined model of the training set.
4. This new model is used in the analysis of anomalies in the analysis dataset.

Statistical Filtering Techniques: The general assumption behind this kind of techniques is that extreme values in either side of distribution are representative of anomalies. Hence statistical filtering techniques try to eliminate extreme values from the training set to reduce their impact on scaling variables. Two statistical filtering techniques are evaluated as part of this research: one which uses application domain knowledge and another which does not.

1. Percentile based filtering: This kind of filtering removes $k\%$ of the observations of each of the KPIs from either side of the distribution.
2. Failure ratio based filtering: This kind of filtering removes from the training set $k\%$ of the observations, which correspond to the highest values of a relative KPI. The failure ratio based technique of filtering from the training set can be considered to be a technique that uses the application domain knowledge in filtering the training set as high failure ratios are considered as anomalies in networks.

5   Case Study: LTE Network Management Data

5.1   Traffic Measurement Data

The AD tests done as part of this research were performed on mobile network traffic measurement data obtained from Nokia Serve atOnce Traffica [12]. This component monitors real-time service quality, service usage and traffic across the entire mobile network owned by the operator. It stores detailed information about different real-time events such as call or SMS attempts, handovers etc. Based on the goal of the AD task, as well as the monitored network functionalities, KPIs are extracted out and they form the data on which the AD task is carried out.

The impact of the fsm based training set filtering technique is evaluated by measuring the SMS counters of a group of 11,749 cells over a period of two days. The measurement data from the first day is chosen as the training data set and the data from the subsequent day is chosen as the analysis data set. A set of five KPIs were chosen for this experiment. The five SMS KPIs monitored for this experiment were 1) number of text messages sent from the cell 2) number of text messages received to the cell 3) number of failed text messages 4) number of failures due to core network errors 5) number of failures due to uncategorized errors.

In order to measure the percentage anomaly detection, a set of five anomalous cells (error conditions described in Table 1) are modeled programmatically and synthetic observations corresponding to them are added into the analysis dataset. The performance of a technique is evaluated in 5 realms: 1) number of synthetic anomalies detected, 2) number of synthetic anomalous NEs detected, 3) total number of anomalies detected, 4) total number of anomalous NEs detected and 5) quality of anomalies detected.

Table 1. Synthetic NEs and their error conditions

Synthetic NE Id	Total SMS count	SMS failure %
30000	(10 to 18) times average traffic	(60 to 100)
30001	(5 to 9) times average traffic	(60 to 100)
30002	(2.5 to 4.5) times average traffic	(30 to 60)
30003	(0.5 to 0.9) times average traffic	(60 to 100)
30004	(0.5 to 0.9) times average traffic	(30 to 60)

To measure comparatively the AD capabilities of the fsm based filtering technique with other techniques, three other methods were chosen: 1) SOM smoothening method 2) Percentile based training set filtering and 3) Failure ratio based training set filtering. In each of the cases, *1%* of the observations from the training set were filtered out. Linear scaling technique was used for scaling the KPIs prior to feeding them to the AD system. This kind of scaling divides the variable by its standard deviation as shown in equation (6).

$$x_s = \frac{x}{s_x} \tag{6}$$

A total of 50 neurons in a 5 X 10 grid were allocated to learn the training set. The initial and final learning rates were chosen as 8.0 and 0.1 respectively.

The neighborhood function was chosen to be Gaussian and a rectangular neighborhood shape was used.

5.2   Illustration

Fig. 3 represents the structure of the 2-D SOM (in logarithmic scale) obtained with and without training set filtering techniques. Logarithmic scales are suitable to understand in more detail the structure of SOMs. The darker and bigger dots represent the positions of the neurons and smaller green (or grey) dots represent the data points.

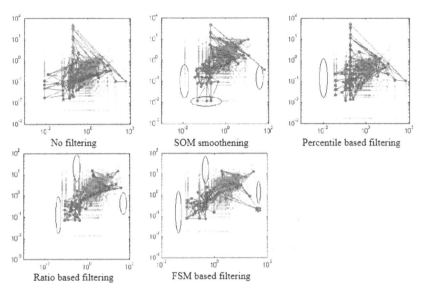

Fig. 3. SOM weight positions (x-axis - weight 1, y-axis - weight 2) in logarithmic scale for different filtering techniques

   The main regions where the structures of the SOMs are different when compared to the case with no filtering are marked with ellipses for easy comprehension. For each of the figures, the ellipses correspond to the region from the training set from where the data points were filtered out. Since, the data points from this region are removed due to the filtering, the neurons which correspond to this behavior are removed as well.

5.3   Quantitative Evaluation

Table 2 summarizes the results in terms of the total number of non-synthetic anomalies and anomalous NEs detected in each of the scenarios. The number of anomalies detected using the fsm based filtering technique is higher when compared to all other techniques. The results clearly show that filtering the training set has a huge impact on the number of anomalies and anomalous NEs detected. This behavior is consistent for synthetic anomalies as well. It is interesting to note that approximately 43% of the synthetic anomalies still remain undetected even using the fsm based filtering technique. In the absence of training set filtering techniques, none of the anomalous NEs nor their anomalous observations could be detected.

Table 2. Statistics of detected anomalies and anomalous NEs

Filtering technique	Non synthetic anomaly count	Non synthetic anomalous NE count	Synthetic anomaly count (out of 120)	Synthetic anomalous NE count (out of 5)
None	121	11	0	0
Percentile	160	17	17	1
SOM smoothening	253	65	31	2
Failure ratio based	525	119	48	2
Fsm based	629	172	68	3

5.4   Qualitative Evaluation

Since, this experiment monitored relatively fewer KPIs, a manual analysis of quality of anomalies was not a tedious task and hence was chosen in this case. Detailed analysis of the entire set of anomalies exhibited ten different kinds of anomalies. The detected typed of anomalies along with their severity are provided in Table 3.

Table 3. Anomaly group id, description and severity

Group Identifier	Anomaly Description	Severity
A1	Sent SMS count = 0, High received SMS count, Failure percentage (~100%), Reason for failure – Uncategorized	Critical
A2	Received SMS count = 0, High sent SMS count, Failure percentage (~100%), Reason for failure – Core network	Critical
A3	Received SMS count = 0, High sent SMS count, Failure percentage (~100%), Reason for failure – Uncategorized	Critical
A4	Moderate/high sent SMS count, Moderate/high received SMS count, Failure percentage (90 - 100%), Reason for failure – Core network	Critical
A5	Moderate/high sent SMS count, Moderate/high received SMS count, Failure percentage (90 - 100%), Reason for failure – Uncategorized	Critical
A6	Moderate/high sent SMS count, Moderate/high received SMS count, Failure percentage (30 - 90%), Reason for failure – Core network	Important
A7	Moderate/high sent SMS count, Moderate/high received SMS count, Failure percentage (30 - 90%), Reason for failure – Uncategorized	Important
A8	Moderate/high sent SMS count, Moderate/high received SMS count, Failure percentage (10 - 30%), Reason for failure – Core network	Moderate
A9	Moderate/high sent SMS count, Moderate/high received SMS count, Failure percentage (10 - 30%), Reason for failure – Uncategorized	Moderate
A10	Moderate/high sent SMS count, Moderate/high received SMS count, Failure percentage (0 - 10%)	Irrelevant

Table 4 presents depicts the number of anomalies and anomalous groups with their criticality levels. Fsm based training set filtering outperforms all other filtering techniques on the basis of the number of critical and important anomalies detected. It is interesting to note here is that, fsm based training set filtering finds approximately four times the number of critical anomalies that are found without any sort of filtering. However, it is also important to note that the fsm based filtering does not outperform other methods in terms of the proportion of critical anomalies detected out of the total number of anomalies detected. Moreover, the number of distinct anomaly groups found by using the application domain knowledge incorporated filtering techniques (failure ratio and fsm) is higher than the cases which do not employ it.

Table 4. Qualitative Analysis: number of anomalies and anomalous groups with their criticality

Filtering technique	Irrelevant		Moderate		Important		Critical	
	#anomalies	#groups	#anomalies	#groups	#anomalies	#groups	#anomalies	#groups
None	0	0	1	1	12	1	108	4
Percentile	3	1	2	1	29	1	143	3
SOM smoothening	35	1	9	1	55	2	185	5
Failure ratio based	14	1	14	2	158	2	387	5
Fsm based	22	1	38	2	228	2	409	5

6   Conclusions

The *fsm* based filtering technique does not aim to replace any of the standard methods of AD in the presence of outliers. This technique should be considered more as a supplement to existing techniques. In cases when there exists some form of prior knowledge about what can be considered as an anomaly, using this information in the training set filtering stage can lead to improved accuracy of AD experiments. The results presented in this paper emphasize the importance of pre-processing techniques in general and training set filtering techniques in particular. Since, this approach is a generic approach for cleaning up training data, there is no dependency on the AD algorithms or mechanisms used commonly.

This paper introduced a novel approach of training set filtering using *fsm* and analysed its impact on the quantity and quality of anomalies detected in an AD experiment performed on network management data obtained from a live LTE network. *Fsm* based training set filtering was found to detect the largest number of synthetic anomalies as well as anomalous NEs. The total number of anomalies and anomalous NEs found were also the highest using this approach. Training set filtering using *fsm* was found to detect the highest percentage of synthetic anomalies and anomalous NEs. In the absence of training set filtering, the results were poor.

References

1. Kumpulainen, P.: Anomaly Detection for Communication Network Monitoring Applications, Doctoral Thesis in Science & Technology, Tampere: Tampere University of Technology (2014)
2. Kumpulainen, P., Hätönen, K.: Anomaly detection algorithm test bench for mobile network management. In: The MathWorks Conference Proceedings of the MathWorks/ MATLAB User Conference Nordic (2008)
3. Chernogorov, F.: Detection of Sleeping Cells in Long Term Evolution Mobile Networks, Master's Thesis in Mobile Technology, University of Jyväskylä, Jyväskylä
4. Kumpulainen, P., Kylväjä, M., Hätönen, K.: Importance of scaling in unsupervised distance-based anomaly detection (2009)
5. Kohonen, T.: Self-Organizing Maps. Springer, Berlin (1997)
6. Laiho, J., Raivio, K., Lehtimäki, P., Hätönen, K., Simula, O.: Advanced Analysis Methods for 3G Cellular Network. IEEE Transactions on Wireless Communications 4(3), 930–942 (2005)
7. Kumpulainen, P., Hätönen, K.: Local anomaly detection for mobile network monitoring. Information Sciences 178(20), 3840–3859 (2008)
8. Yin, H.: The Self-Organizing Maps: Background, Theories, Extensions and Applications. Computational Intelligence: A Compendium: Studies in Computational Intelligence 115, 715–762 (2008)
9. Suutarinen, J.: Performance Measurements of GSM Base Station System. Thesis (Lic.Tech.), Tampere University of Technology, Tampere (1994)
10. Höglund, A.J., Hätönen, K., Sorvari, A.S.: A computer host-based user anomaly detection system using the self-organizing map. IEEE-INNS-ENNS International Joint Conference on Neural Networks (IJCNN) 5, 411–416 (2000)
11. Kylväjä, M., Hätönen, K., Kumpulainen, P., Laiho, J., Lehtimäki, P., Raivio, K., Vehviläinen, P.: Trial Report on Self-Organizing Map Based Analysis Tool for Radio Networks. Vehicular Technology Conference 4, 2365–2369 (2004)
12. Anonymous, Serve atOnce Traffica. Nokia Solutions and Networks Oy. http://networks.nokia.com/portfolio/products/customer-experience-management/serve-atonce-traffica (accessed December 16, 2014)
13. Hätönen, K.: Data mining for telecommunications network log analysis. Doctoral Thesis, Helsinki University, Helsinki (2009)

Fixed-Resolution Growing Neural Gas for Clustering the Mobile Networks Data

Szabolcs Nováczki[1] and Borislava Gajic[2]([✉])

[1] Nokia Networks, Budapest, Hungary
`szabolcs.novaczki@nokia.com`
[2] Nokia Networks, Munich, Germany
`borislava.gajic@nokia.com`

Abstract. An important property of the competitive neural models for data clustering is autonomous discovery of the data structure without a need of a priori knowledge. Growing Neural Gas (GNG) is one of the commonly used incremental clustering models that aims at preserving the topology and the distribution of the input data. Keeping the data distribution unchanged has already been recognized as a problem leading to bias sampling of input data. This is undesired for the use cases such as mobile network management and troubleshooting where is important to capture all the relevant network states uniformly. In this paper we propose a novel incremental clustering approach called Fixed Resolution GNG (FRGNG) that keeps the input data representation at the fixed resolution avoiding the oversampling and undersampling problems of original GNG algorithm. Furthermore, FRGNG introduces a native stopping criteria by terminating the run once the input data is represented with the desired fixed resolution. Additionally, the FRGNG has a potential of the algorithm acceleration which is especially important when large input data set is applied. We apply the FRGNG model to analyze the mobile network performance data and evaluate its benefits compared to GNG approach.

Keywords: Growing Neural Gas · Fixed resolution mobile network clustering

1 Introduction

Different clustering methods are commonly used in the multidimensional data analysis. The topology-preserving clustering algorithms are able to determine the homogeneous groups of input data samples over a whole multidimensional input dataset while keeping the topological structure of the input dataset. The clustering allows reduction in the data set required for analysis by grouping the whole input network data into smaller groups with a similar behavior. The final representation of the multidimensional input space is in the form of the points in space (also with reduced number of dimensions), where the distances between such points in space match the differences between the samples in the input dataset. The Self Organizing Map [1] is the neural network model commonly

© Springer International Publishing Switzerland 2015
L. Iliadis and C. Jayne (Eds.): EANN 2015, CCIS 517, pp. 181–191, 2015.
DOI: 10.1007/978-3-319-23983-5_18

used for clustering, where the clusters are identified with respect to the position of neurons.

The most critical limitation of SOM is the need of a priori knowledge to define the structure of the data. In order to overcome such issue the use of the incremental clustering algorithms such as Growing Neural Gas [2] has been proposed. The Growing Neural Gas algorithm does not require any a priori knowledge about the input data, as it incrementally generates the neurons in order to represent the input data space. The algorithm does not impose the restrictions on the growth of the neural network as a representation of the input space and continues adding new neurons as long as the predefined stopping criteria is not met.

The common stopping criteria is the predefined threshold for the errors accumulated by generated neurons, or the maximum number of generated neurons. Based on the predefined stopping criteria the input data is represented in a certain granularity. Therefore, the definition of the stopping criteria is the critical part of the algorithm since it heavily influences the output of the algorithm.

However, the Growing Neural Gas algorithm has certain drawbacks, as well. One of the issues is uncontrolled proliferation of nodes in the areas of the input dataset that are well presented. There is no stopping criteria defined for terminating the creation of new nodes in the areas that are sufficiently well represented, and the algorithm will continue unnecessary addition of new nodes until the global stopping criteria is not met (e.g. maximum number of generated nodes is reached). *As a consequence the input dataset has an unbalanced representation, i.e. high representation granularity for frequent input samples and very low granularity for rare input samples. Furthermore, as the algorithm continues unnecessary node addition the overall execution time of the algorithm increases.* This characteristic of the GNG algorithm is favorable for the applications where importance of data distribution matching is high. However, for the applications like the anomaly detection in mobile networks the machine learning algorithms do not necessarily need to preserve the information about the frequency of the input data samples. It is more important that the input data is represented with desired granularity. The higher granularity representation can be considered as unnecessary oversampling. The anomaly detection procedure in mobile networks aims at learning all representations/states of the network in normal operation. Such learned normal network states are further on compared with the current network operation. In the case of deviation between normal and current network states the anomaly in network operation is detected.

A solution to the problem of nodes proliferation has been proposed in [4]. The authors present the modified GNG algorithm that suppresses the insertion of new units in already well represented areas. The proposed GNG algorithm minimizes the modifications of the original GNG algorithm by adding only 3 new parameters in order to avoid the oversampling of already well represented areas. Similar solutions that require larger modifications of the original GNG algorithm by adding higher number of new parameters are proposed in [5]. However, such proposals solve the GNG drawbacks only partially, as the problem of unnecessary increase of the execution time still remains while running the algorithm until the global stopping criteria is met. Such global stopping criteria often does not explicitly reflect the granularity of the data representation, and thus can be an incorrect termination point. In this work we propose the Fixed-Resolution Growing Neural Gas algorithm which solves the problems of the original GNG by focusing on following issues:

1. **Uniform Granularity of Input Data Representation** - The main target of the Fixed-Resolution GNG algorithm is to represent all the input data samples uniformly, i.e. with predefined resolution. The resolution is determined by the quantization error between the original input data sample and its corresponding FRGNG representation. Following such an approach the input data oversampling and undersampling is avoided as all the input data samples will be represented with the same granularity.
2. **Native Stopping Criteria** - The FRGNG is terminated once all input data samples are represented with the satisfying resolution. Once the input data sample has satisfied the resolution criteria it is excluded from the further consideration by the algorithm. In such a way the number of input samples to consider in the algorithm run is constantly decreasing during the algorithm run. The algorithm natively stops once all input data samples to be considered are exhausted.

The remainder of this paper is organized as follows: In Section 2 we give the overview of related work. In Section 3 we explain the Growing Neural Gas algorithm, its implementation as well as its extension to suppress the bias representation of input data. In Section 4 we present the novel approach of Fixed Resolution GNG for data clustering. The performance evaluation of our approach based on the real mobile network data is presented in Section 5. Finally, in Section 6 we draw the conclusions from this work.

2 Related Work

The problem of the uncontrolled proliferation of GNG nodes in the sufficiently well represented areas has been addressed in [4] and [5]. The solutions propose addition of new parameters to the original GNG algorithm that foster the insertion of new nodes in the areas that are not well represented which is indicated by the higher quantization error.

The control of the relation between the input data and the weight vector density, i.e. magnification in the neural network algorithms has been addressed in [6]. Such explicit control is important property of neural networks as for different applications is desirable to suppress or emphasize the sparsely covered areas. An idea of growing the neural network only when the input data is not sufficiently well represented rather than after predefined number of algorithm iterations and in the areas where the accumulated error is maximal is presented in [7]. However, the proposed solution is based on the frequency of the nodes being chosen as the BMU, i.e. activity, and the algorithm iterates over *entire input dataset* until all nodes have the satisfying activity.

Another approach for restricting the insertion of new nodes in the GNG model has been proposed in [8]. The algorithm introduces three types of restrictions, based on the radius of nodes, their density and the restriction based on the squared error. The new nodes are added only in the areas that are still not well covered by the model, whereas the coverage threshold is maintained in every node. Furthermore, the modification of GNG by pruning and node relocation in order to improve vector quantization is presented in [9].

All proposed solutions for the GNG unbalanced node proliferation are based on the node level quality measures as a main driving force for the algorithm run, e.g. quantization error, frequency of being selected as a BMU. Our proposal departs from this approach and focuses on the input data samples quality measure, i.e. how satisfying each input data sample is represented. In such a way the accelerations in the algorithm run are possible, as the considered input data set decreases in size as the algorithm run advances.

3   Growing Neural Gas Algorithm and Its Extension to Suppress the Unbalanced Node Proliferation

Growing Neural Gas algorithm performs incremental vector quantization and learning of topology of input data. The output of the GNG algorithm is the graph of neurons representing the topology of the input data. Each neuron generated by the GNG is represented by the weight vector which has the dimensionality of the input space. See Algorithm 1.

As outlined before the GNG algorithm suffers from bias representation of input data samples. In order to suppress the proliferation of the algorithm in the areas with the high density of input data the authors in [4] propose the changes in the error accumulation of the original GNG algorithm. The added value of the accumulated error is attenuated if it is produced by the input data sample that has a lower quantization error than a predefined threshold.

$$accErr = \begin{cases} accErr + \Delta err & \text{if } \Delta err \geq qErr \\ accErr + h * \Delta err & \text{if } \Delta err < qErr \end{cases}$$

Parameter h is in the range $0 \leq h \leq 1$.

Such an attenuation of the accumulated error results in creation of less units in already well represented areas. Consequently the movement of units needs to be adjusted according to the attenuation of the accumulated error increase. The parameter $qErr$ determines the radius of the created units that can overlap to a certain level defined by the parameter sp. Taking into consideration the additional parameters h, $qErr$ and sp the addition of the new unit from original GNG algorithm is modified as described in [4].

Such modifications of the original GNG algorithm solve the problem of oversampling of the high density areas in the input dataset and fosters creation of units in the areas of rare input data samples. In such a way even the uncommon

Algorithm 1. Growing Neural Gas (GNG)

Step 1. Start with two units at random positions ω_a and ω_b in R^n.

Step 2. Generate an input signal ξ according to probability density function $P(\xi)$.

Step 3. Find the nearest unit s_1 and the second nearest unit s_2.

Step 4. Increment the age of all edges starting from the nearest unit s_1.

Step 5. Add the squared distance between the input signal and the nearest unit in input space to a local counter variable $\Delta error(s_1) = \parallel \omega_{s_1} - \xi \parallel$.

Step 6. Move the nearest unit s_1 and its direct topological neighbors towards ξ by fraction ϵ_b and ϵ_n, respectively, of the total distance: $\Delta\omega_{s1} = \epsilon_b(\xi - \omega_{s1})$ and $\Delta\omega_n = \epsilon_n(\xi - \omega_n)$ for all direct neighbors n of s_1.

Step 7. Set the age of the edge between s_1 and s_2 to zero. If the edge does not exist create it.

Step 8. Remove edges with an age larger than a_{max}. If the remaining units have no emanating edges, remove them.

Step 9. If the number of input signals generated so far is an integer multiple of parameter λ:

    a. Insert a new unit r on the half way between the units with the maximal accumulated error (q and f).

    b. Insert the edges to connect r with q and f. Remove the original edges between q and f.

    c. Decrease the error variables of q and f by multiplying them with constant α.

    d. Initialize the error variable of r with the new value of error variable of q.

Step 10. Decrease all errors by multiplying them with the constant d.

Step 11. If the stopping criterion is not fulfilled return to **Step 2**.

samples from the input data are well represented. However, the proposed algorithm has a drawback of creating the invalid units while moving from the areas of high input data density towards the rare input data samples as illustrated in Figure 1. Such invalid units do not correspond to any input data sample but are created as a consequence of the incremental moving of units towards the input data samples. In the mobile network anomaly detection use case such invalid units can have very negative impact as they would misleadingly indicate the network states that are not existing in reality. Therefore, applying such algorithms in mobile network anomaly detection would be suboptimal. Furthermore, due to the lack of the stopping criteria that reflects the quality of the input dataset representation the execution time of the proposed algorithm can still unnecessarily increase.

We propose a novel algorithm Fixed Resolution Growing Neural Gas (FRGNG) algorithm based on the original GNG algorithm that solves the Prototype Proliferation problem as well as the stopping criteria issue from the original GNG and thus is suitable for the mobile networks anomaly detection. As a baseline for our implementation and the evaluation of the novel algorithm proposal FRGNG we used the original GNG algorithm open source R package [3].

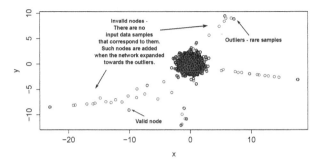

Fig. 1. The output of the modified GNG algorithm from [4] applied on the non-uniform distribution of data points in two-dimensional space. The algorithm suppresses the generation of units on well represented areas and achieves the good representation of rare samples. Invalid units creation in the areas between rare and frequent input data samples is noticeable.

4   Fixed Resolution Growing Neural Gas Algorithm

In this section we present the Fixed Resolution Growing Neural Gas Algorithm. The baseline of our algorithm is in the original GNG algorithm. However, the FRGNG significantly differs from the original GNG.

Instead of accumulated error between the input data and the GNG units being a main driver of the algorithm run we focus on the monitoring of the input data set, especially on the quality of its representation. In order to achieve this we assign the counter to each input data sample initialized to a certain nonnegative value. During the algorithm run a random sample from the set of input samples having counter higher than zero is selected. The algorithm computes the quantization error for the selected sample. If this error is lower than a threshold (RES) than the counter of the input sample is decremented, otherwise it is incremented. Once the counter of the sample reaches a predefined maximum value, a new node is added on the position indicated by the sample vector (the node will represent the sample with zero quantization error).

Furthermore, the edges between the new node and all the nodes that are closer than a certain threshold (maximal edge length) are created. The node adaptation is performed after each step regardless of the resulting quantization error according to the original GNG strategy. The edges that became longer than the maximum edge length will be removed. The stopping criteria is when the active training set is empty, thus the algorithm will add as many nodes are required in order to represent the sample set with the requested resolution, regardless of the sample distribution. In order to collect the statistics on the frequency of input samples the BMU counter can be maintained during the algorithm run. The parameters of the FRGNG algorithm need to be set based on the input data, e.g. its type, range of values, as well as based on the desired granularity of final data representation. See Algorithm 2. Additionally, such approach has potential of accelerating the algorithm as input data samples with counter zero are not any more fed to the algorithm, i.e. well represented input data samples are gradually removed from the input data.

Algorithm 2. Fixed Resolution GNG

Step 0. For a number of input data samples (training samples) associate a counter (ci) to each training sample and initialize it to some integer value.
Step 1. Start by generating one node associated to one of the training samples (randomly selected).
   a. *optionally associate a best matching unit (BMU) counter to the node and initialize it to zero.*
Step 2. Select a new sample which counter is greater than zero and compute its quantization error (e):
   a. if e is larger that the resolution RES then:
      i. increment its counter by *cinc*.
      ii. if the counter of the sample is greater or equal than addition threshold (A) then:
         1. add a new node and initialize it to the sample vector.
         2. connect the new node to all other nodes, which distance is less or equal than a maximum edge length (L).
         3. if BMU counter is used; increment it here.
      iii. if $ci < A$ go to **Step 2c**.
   b. if $e \leq RES$ then:
      i. decrement the samples counter by *cdec*.
      ii. if BMU counter is used; increment it here.
   c. adapt nodes to the sample using GNG strategy but only with a maximum extent (D).
   d. remove edges connecting nodes with distance higher than L.
Step 3. Return to **Step 2** until there are samples with counter greater than zero, STOP is no such sample available.
Step 4. If BMU counters were used, normalize them to get a BMU probability in order to differentiate between rare and common samples.

The Fixed Resolution GNG highlights can be summarized as follows:

- The algorithm aims at representing the whole sample set with a given resolution (RES) regardless of input data sample probability.
- Input data samples which have counter value zero are not fed to the algorithm as they are represented sufficiently well by the model. This gradually reduces the sample size as the algorithm advances.

– Learned information is protected by the D parameter, no extreme deviation towards the input data samples are allowed. No accumulated error and edge aging is used.

5   FRGNG Evaluation

In order to get more insight on the performance of the proposed algorithm we analyze its capabilities applying it on data collected from the real mobile network performance and comparing it with the original GNG. As input data for our experiments we use the mobile network performance indicators data collected over one week. The performance indicators commonly referred as Key Performance Indicators (KPI) or counters are provided by the network elements performing the measurements at regular intervals. As an outcome of such measurements the multi dimensional time-series of KPIs per network element are available.

In our experiments we consider different dimensionality of the input data ranging from 2-dimensional to 6-dimensional datasets taken over one week of cellular network operation. The KPIs considered in our experiments are:

– Number of discarded MAC-hs PDUs due to maximum number of retransmissions
– Number of discarded MAC-hs PDUs due to some other reason than T1 timer or maximum number of retransmissions
– Number of discarded MAC-hs PDUs due to T1 timer
– Number of dropped MAC-d PDUs due to maximum number of retransmissions
– Number of dropped MAC-d PDUs due to other reason
– The sum of sampled values for the number of configured HSUPA users with 10 ms TTI

On each of the multidimensional input datasets we run the Fixed Resolution GNG algorithm and compare it with the original GNG. Running the both algorithms on the same dataset illustrates the drawbacks of the original GNG algorithm and the FRGNG solution. Figure 2 shows obtained results from experiments with 4-dimensional input data (feeding first 4 KPIs that are listed above in FRGNG and GNG algorithms, and plotting first two of them). The FRGNG algorithm succeeds in representing all the input data samples including the outliers. The GNG algorithm does not have the same capabilities as it fails in representing the outliers and it generates the nodes that do not correspond to any input data. As discussed before this effect is a consequence of gradually moving from the high density areas towards outliers.

Another target of the evaluation study is to investigate the impact on the execution time due to the changed concept implemented in the FRGNG which drives the algorithm run based on the value of the input samples counter. Once the input data counter drops below zero and the algorithm starts adding the

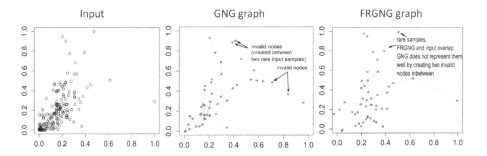

Fig. 2. The comparison between the FRGNG and GNG algorithm.

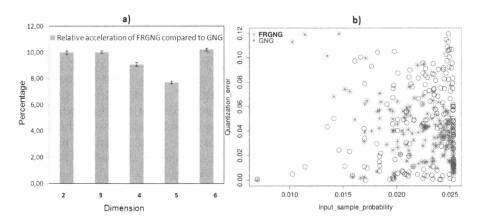

Fig. 3. The comparison between the GNG and FRGNG algorithm with respect to a) execution time and b) quantization error relative to the input sample probability.

new nodes the algorithm execution accelerates. In order to estimate the execution time of FRGNG for a given multidimensional mobile data input we performed additional experiments using two datasets coming from different cells of the same mobile operator and averaging the results over 10 runs of GNG and FRGNG algorithms. Figure 3 a) illustrates obtained results. In comparison with the execution time of the GNG algorithm the FRGNG shows slightly better results reaching the gain of around 10% compared to GNG.

Additionally, we evaluate and compare the GNG and FRGNG approaches with respect to the achieved quantization error as a function of input sample probability defined per input observation. Figure 3 b) shows the results obtained while feeding the FRGNG and GNG with 4-dimensional input (first 4 KPIs listed above). As the probability of the input sample increases the GNG tends to decrease the quantization error with which the samples are represented, whereas the FRGNG does not show such bias behavior - the quantization error of input samples is not dependent on the sample probability, the only target is to keep the quantization error of all samples below certain threshold.

6   Conclusions

In this work we proposed the novel unsupervised neural network model Fixed Resolution GNG. The FRGNG builds upon the Growing Neural Gas algorithm for incremental learning. However, the FRGNG solves the major problem of the original GNG algorithm, i.e. unbalanced representation of input data space in favor of frequent data samples, as well as the problem of the critical stopping criteria. Due to its accent on the input dataset quality measure our algorithm is able to gradually discard the input samples that are well represented and consider only the input data samples that still do not have a good representation. This potentially decreases the execution time of the algorithm, especially when large datasets are considered. Furthermore, the stopping criteria of the algorithm is natively imposed. The algorithm stops once the input dataset is empty.

Our analysis on the data taken from the mobile network performance measurements illustrate the advantages of the FRGNG over the GNG in terms of uniform representation of the input samples regardless of the probability distribution of the input dataset. In such a way all relevant network states are captured by the FRGNG which is a prerequisite for efficient network management and anomaly detection. Furthermore, the execution time of the FRGNG algorithm is lower than for the GNG applied over same data sets. In order to get a better insight on the FRGNG performance larger data sets taken over longer periods of measurements and considering higher number of input KPIs need to be performed. This is the part of our ongoing work.

References

1. Kohonen, T.: Self-organized formation of topologically correct feature maps. Biological Cybernetics **69**, 43–59 (1982)
2. Fritzke, B.: A growing neural gas network learns topologies. In: Advances in Neural Information Processing Systems, vol. 6, no. 3, pp. 625–632. MIT Press, Cambridge (1995)
3. GNG R package. https://github.com/kudkudak/Growing-Neural-Gas
4. Satizábal, H.F., Pérez-Uribe, A., Tomassini, M.: Prototype proliferation in the growing neural gas algorithm. In: Kůrková, V., Neruda, R., Koutník, J. (eds.) ICANN 2008, Part II. LNCS, vol. 5164, pp. 793–802. Springer, Heidelberg (2008)
5. Cselenyi, Z.: Mapping the dimensionality, density and topology of data: The growing adaptive neural gas. Computer Methods and Programs in Biomedicine, 141–156 (2005)
6. Villmann, T., Claussen, J.C.: Magnification Control in Self-Organizing Maps and Neural Gas. Neural Computation, 446–449 (2006)

7. Marsland, S., Shapiro, J., Nehmzow, U.: A self-organizing network that grows when required. Neural Networks, 1041–1058 (2002)
8. Quintana-Pacheco, Y., Ruiz-Fernández, D., Magrans-Rico, A.: Growing Neural Gas approach for obtaining homogeneous maps by restricting the insertion of new nodes. Neural Networks (2014). http://dx.doi.org/10.1016/j.neunet.2014.01.005
9. Canales, F., Chacón, M.: Modification of the growing neural gas algorithm for cluster analysis. In: Rueda, L., Mery, D., Kittler, J. (eds.) CIARP 2007. LNCS, vol. 4756, pp. 684–693. Springer, Heidelberg (2007)

Fuzzy Modeling

A Neural-Fuzzy Network Based on Hermite Polynomials to Predict the Coastal Erosion

George E. Tsekouras[1], Anastasios Rigos[1], Antonios Chatzipavlis[2], and Adonis Velegrakis[2]

[1] Department of Cultural Technology and Communication, University of the Aegean, Mytilene, Greece
gtsek@ct.aegean.gr, a.rigos@aegean.gr
[2] Department of Marine Sciences, University of the Aegean, Mytilene, Greece
a.chatzipavlis@marine.aegean.gr, afv@aegean.gr

Abstract. In this study, we investigate the potential of using a novel neural-fuzzy network to predict the coastal erosion from bathymetry field data taken from the Eresos beach located at the SW coastline of Lesvos island, Greece. The bathymetry data were collected using specialized experimental devices deployed in the study area. To elaborate the data and predict the coastal erosion, we have developed a neural-fuzzy network implemented in three phases. The first phase defines the rule antecedent parts and includes three layers of hidden nodes. The second phase employs truncated Hermite polynomial series to form the rule consequent parts. Finally, the third phase intertwines the information coming from the above phases and infers the network's output. The performance of the network is compared to other two relative approaches. The simulation study shows that the network achieves an accurate behavior, while outperforming the other methods.

Keywords: Coastal erosion · Bathymetry profile · Neural-fuzzy network · Hermite polynomials

1 Introduction

Coastal erosion is one of the most important environmental problems in coastal zones worldwide having devastating impacts on the economy and livelihood, as most of the world's population is concentrated in coastal territories [1]. Coastal erosion is the process of wearing away material from the coast's profile due to imbalance in the supply and export of material from certain sections. It takes place during strong winds, high waves, high tides, storm surge conditions, and results to coastline retreat due to accelerate nearshore sediment transport processes [2]. The phenomenon is expected to exacerbate in the future due to global warming and sea level rise [3]. Given the aforementioned context, predicting the coastal erosion in terms of the morphological response of a coast under increased storminess has become an urgent issue. Towards this direction, there have been developed several numerical models able to simulate the hydrodynamic/morphodynamic processes occurring on coasts while

© Springer International Publishing Switzerland 2015
L. Iliadis and C. Jayne (Eds.): EANN 2015, CCIS 517, pp. 195–205, 2015.
DOI: 10.1007/978-3-319-23983-5_19

storms [4-7]. However, their implementation is subjected to certain limitations such as: (a) the nonlinearities due to the uncertainty level of the pre- and post-storm hydro-dynamical, sedimentological and morphological field data, and (b) the high computational cost.

To alleviate the above problems, this paper proposes a novel neural-fuzzy network (NFN) specially designed to predict the coastal erosion. NFNs attempt to offset the approximate reasoning of fuzzy systems with the learning mechanisms and connectionist structures of neural networks [8-10]. To cope with the nonlinearities involved in the problem, we embed into the network's structure high order Hermite polynomials. The implementation of the network is carried out through a number of comparative simulation experiments.

The material is organized as follows. Section 2 describes the coastal erosion problem and the experimental setup to obtain the data. Section 3 provides a detailed analysis of the neural-fuzzy network. Section 4 illustrates the simulation experiments. Finally, the paper concludes in section 5.

2   Experimental Methodology and Data Acquisition

The study area is the beach of Eresos, which is a typical micro-tidal sediment starved pocket beach of the Aegean archipelago located along the SW coast of the Lesvos island, Greece (see Fig. 1). Fig. 2 states that during the last years the beach is under increasing erosion. Our experimental methodology consists of topographic and bathymetric surveys that have been extracted on an annual basis by a meteorological station operating since March 2009 providing hourly wind speed and directional data. In addition, a sophisticated camera system deployed during the last 5 years, has been recording daily shoreline changes. Repeated topographic and bathymetric measurements (totally 7 surveys, 1 every 1.5 months) have been collected from the 8 stations (S1-S8), depicted in Fig. 1, between Sept. 2010-Sept. 2014. These measurements have been extracted by a RTK-DGPS attached to a small boat equipped with a depth-meter.

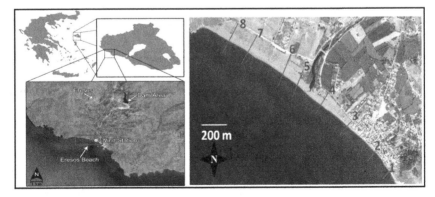

Fig. 1. Location of the Eresos beach (left) and beach topographic stations (right).

The basic concepts involved in the coastal erosion are the cross-shore coast profile also called bathymetry profile and the longshore sediment bar. The former is a cross-section taken perpendicular to a given coast contour. The latter is a submerged ridge of sediment built on the nearshore by waves and currents, generally parallel to the shore and in depths less than 5-10 m. Fig. 3(a) depicts a bathymetry profile, where a longshore bar and the shoreline position (corresponding to zero elevation) are shown. During winter the hydrodynamic conditions are more intense and the bar moves from the onshore (at the berm area) to the nearshore. The opposite occurs during summer period.

Fig. 2. Eresos beach in March 2010 (left) and in March 2013 (right).

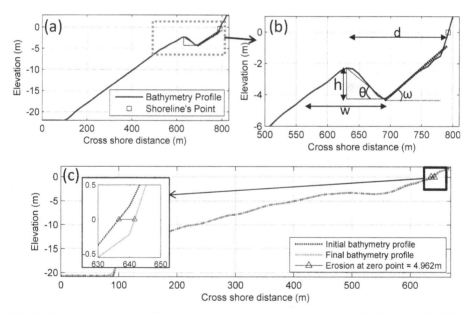

Fig. 3. (a) A bathymetry profile showing a longshore bar and the shoreline's point. (b) The characteristics of a longshore bar. (c) The coastal erosion is the distance between the initial profile and the final profile at the shoreline's point (i.e. zero position) and is represented by the length of the line segment connecting the triangles.

The way the longshore bar affects the coastal erosion depends on the bar's characteristics such as: w = "width", h = "height", d = "the distance from the bar's top point to the zero elevation point", θ = "the angle of the bar's interior opposite to the shoreline", and ω = "the mean seabed's slope from the zero elevation point to the point that the longshore bar begins". These characteristics are illustrated in Fig. 3(b). Waves are breaking in the nearshore when depth becomes small enough and the height of the longshore bar reduces the depth at this certain point where the bar is located. As a consequence, the higher waves generated during increased storm conditions break at this nearshore point and not onshore, as it would happen if the bar didn't exist, and the amount of the wave energy reaching the coastline is reduced. Therefore, a longshore bar can be viewed as the "natural defense" of the coast against erosion. In this paper we shall use the concepts "initial bathymetry profile" and "final bathymetry profile" to denote the bathymetry profiles before and after one or more storm events. The distance between the shoreline's points in these two cases defines the coastal erosion. Fig 3(c) shows such an example, where the corresponding erosion is equal to 4.962 meters. To this end, the variables used in this paper to quantify the bathymetric profile are: $x_1 = h$, $x_2 = w/2$, $x_3 = d$, $x_4 = \tan\theta$, and $x_5 = \tan\omega$. In addition, three more variables taken from local meteorological data during storm events are also considered: x_6 = "the storm duration (in hours)", x_7 = "the offshore significant wave height", and x_8 = "the wave peak spectral period". Finally, the set of the variables is completed by the: x_9 = "the mean sediment grain size", which takes certain values depending the grain size of the beach. The variables $x_1 - x_9$ are used as the input variables of the neural network, while the output variable is: y = "coastal erosion (in meters)". Our prime concern is to generate the data for the variables $x_1 - x_5$, because the rest of them are taken from other sources. To generate these data we used the experimental methodology described previously in order to quantify the bathymetry profiles of the stations S1-S8 (see Fig.1). The final result is a set of $N = 708$ input data, each of which is represented as $\mathbf{x}_k = \begin{bmatrix} x_1 & x_2 & ... & x_9 \end{bmatrix}$ with $k = 1, 2, ..., N$. Thus, the dimensionality of the neural-fuzzy network's input space is $p = 9$. Finally, the output data that concern the coastal erosion were generated by the XBeach model [4, 7] when subjected to the aforementioned input data. Among others, the implementation of this model returns the bathymetry of the beach after a storm event [7].

3   The Proposed Neural-Fuzzy Network

The basic configuration of the proposed NFN is given in Fig. 4. The network consists of three phases: (a) the antecedent phase, (b) the consequent phase, and (c) the inference mechanism. The first two phases synthesize the structure of the rule base, while the third one infers the estimated output. In what follows, we assume that we are given N input-output data pairs $\left\{ \mathbf{x}_k; y_k \right\}\big|_{k=1}^{N}$ with $\mathbf{x}_k \in \mathfrak{R}^p$ and $y_k \in \mathfrak{R}$.

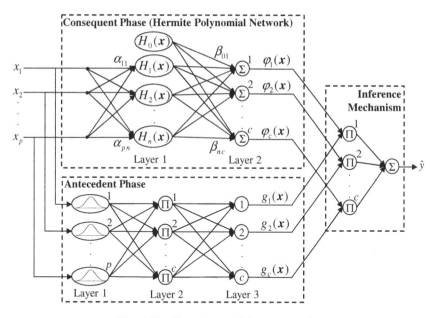

Fig. 4. The Hermite neural-fuzzy network.

3.1   Antecedent Phase

The antecedent phase consists of three layers (see Fig. 4). In Layer 1, each universe of discourse is partitioned into c fuzzy sets of the form

$$A_{ij}\left(x_{kj}\right) = \exp\left(-\left(\left(x_{kj} - v_{ij}\right)/\sigma_{ij}\right)^2\right) \tag{1}$$

where v_{ij} is the fuzzy set center, σ_{ij} the width ($1 \le k \le N$, $1 \le i \le c$, and $1 \le j \le p$).

We emphasize that there are c fuzzy rules and each fuzzy set A_{ij} is linked to only one rule. Layer 2 elaborates on the information provided by the Layer 1 in terms of fuzzy nodes that use the product T-norm operator. To do so, we define as many fuzzy nodes as the number of fuzzy rules, and the output of each fuzzy node is,

$$f_i(\boldsymbol{x}_k) = \prod_{j=1}^{p} A_{ij}\left(x_{kj}\right) \tag{2}$$

with $1 \le k \le N$, and $1 \le i \le c$. Note that $f_i(\boldsymbol{x}_k)$ is a multidimensional fuzzy set and its contribution is to calculate the firing strengths of the corresponding rules.

Layer 3 consists of c nodes, the outputs of which are calculated as,

$$g_i\left(\boldsymbol{x}_k\right) = f_i\left(\boldsymbol{x}_k\right) \bigg/ \sum_{s=1}^{c} f_s\left(\boldsymbol{x}_k\right) \tag{3}$$

3.2   Consequent Phase

This phase utilizes truncated Hermite polynomial series to construct the rule consequent parts. The Hermite polynomials are defined as [11, 12],

$$H_n(x) = (-1)^n e^{x^2} \frac{d^n}{dx^n}\left(e^{-x^2}\right) \tag{4}$$

with n being the polynomial order. They are orthogonal in $(-\infty, \infty)$ with respect to the normal distribution e^{-x^2} [11],

$$\langle H_n(x), H_q(x) \rangle = \int_{-\infty}^{+\infty} e^{-x^2} H_n(x) H_q(x) dx = \begin{cases} 0 & , \text{ if } n \neq q \\ 2^n n! \sqrt{\pi}, & \text{ if } n = q \end{cases} \tag{5}$$

where $\langle \cdot, \cdot \rangle$ stands for the inner product. The individual polynomials can be derived via the next recurrence relations,

$$H_0(x) = 1, \quad H_1(x) = 2x, \quad \text{and} \quad H_n(x) = 2x H_{n-1}(x) - 2(n-1) H_{n-2}(x) \tag{6}$$

In view of Fig. 4, the network's consequent phase includes two layers. The Layer 1 includes n nodes with activation functions the Hermite polynomials $H_0(x)$, $H_1(x)$,..., $H_n(x)$. Since the output of $H_0(x)$ is always unity, the corresponding node is not linked to the input variables and therefore, it plays the role of the bias. The rest of the nodes are linked to the input variables by using the coefficients $\alpha_{j\ell}$ with $1 \leq j \leq p$ and $1 \leq \ell \leq n$. The input of the ℓth node is the linear combination $\sum_{j=1}^{p} \alpha_{j\ell} x_{kj}$ and the activation function is:

$$H_\ell(\mathbf{x}_k) = H_\ell\left(\sum_{j=1}^{p} \alpha_{jl} x_{kj}\right) \tag{7}$$

The Layer 2 utilizes only summation nodes. The number of these nodes equals the number of fuzzy rules used by the network (i.e. c nodes). The output of the ith node is

$$\varphi_i(\mathbf{x}_k) = \sum_{\ell=1}^{n} \beta_{\ell i} H_\ell(\mathbf{x}_k) \tag{8}$$

with $1 \leq i \leq c$ and $1 \leq k \leq N$. The above functions are truncated series of order n.

3.3   Inference Mechanism

Based on the above analysis, the ith fuzzy rule of the network is,

$$R^i : IF \ x_1 \ is \ A_{i1} \ and \ x_2 \ is \ A_{i2} \ and \ ... and \ x_p \ is \ A_{ip} \ THEN \ y \ is \ \varphi_i(\mathbf{x}) = \sum_{\ell=1}^{n} \beta_{\ell i} H_\ell(x) \tag{9}$$

Using the relations (3), (8), and (9) we can establish an one-to-one correspondence between the functions $g_i(x_k)$ and $\varphi_i(x)$, and taking into account the form of the rule in (10), the inferred network's output reads as follows,

$$\tilde{y}_k = \sum_{i=1}^{c} g_i(x_k)\varphi_i(x_k) = \sum_{i=1}^{c}\left[g_i(x_k)\left(\sum_{\ell=1}^{n}\beta_{\ell i}H_\ell(x_k)\right)\right] \tag{10}$$

3.4   Network's Learning Process

There are three kinds of parameters involved in the network's structure namely, the fuzzy set centers, the fuzzy set widths and the coefficients $a_{j\ell}$ and $\beta_{\ell i}$ of the Hermite polynomial phase. To obtain the fuzzy set centers we use the well-known fuzzy c-means algorithm, and to calculate the fuzzy set widths we employ the corresponding fuzzy covariance matrix [13, 14]. In order to estimate the coefficients of the consequent phase, we use the Armijo rule [15] to minimize the network's square error:

$$J_{SE} = \sum_{k=1}^{N}|y_k - \tilde{y}_k|^2 = \sum_{k=1}^{N}\left|y_k - \sum_{i=1}^{c}\left[g_i(x)\left(\sum_{\ell=1}^{n}\beta_{\ell i}H_\ell(x)\right)\right]\right|^2 \tag{11}$$

To do so we define the vector,

$$z = \left[z_1,....,z_{pn},z_{pn+1},....,z_{(p+c)n+c}\right]^T = \left[\alpha_{11},....,\alpha_{pn},\beta_{01},...,\beta_{nc}\right]^T \tag{12}$$

and for the $t+1$ iteration, the learning rule is,

$$z(t+1) = z(t) - \eta(t)\nabla J_{SE}(z(t)) \tag{13}$$

where $\eta(t) = \gamma^\tau$ with $\gamma \in (0,1)$. The parameter τ is the smallest positive integer such that,

$$J_{SE}\left(z(t) - \eta(t)\nabla J_{SE}(z(t))\right) - J_{SE}\left(z(t)\right) < -\varepsilon\eta(t)\left\|\nabla J_{SE}(z(t))\right\|^2 \tag{14}$$

with $\varepsilon \in (0,1)$. Finally, the partial derivatives in (13)-(14) can easily be derived.

4   Simulation Study

Based on the analysis of section 2, the data set includes $N = 708$ input-output data pairs of the form $\{x_k; y_k\}|_{k=1}^{N}$ with $x_k = \begin{bmatrix} x_1 & x_2 & ... & x_9 \end{bmatrix}$ and $y_k \in \Re$. The data set was divided into a training set consisting of the 60% of the original data, and a testing set consisting of the rest 40%. In order to perform a rigorous comparative study, we designed a Takagi-Sugeno-Kang fuzzy system called TSKFS, and a Gaussian type radial basis function neural network called RBFNN. The training procedure of the TSKFS is as follows. The fuzzy set centers were obtained by applying the fuzzy c-means algorithm over the input training data set. The respective fuzzy set widths were estimated in terms of the corresponding fuzzy covariance matrix [13, 14]. Finally the consequent coefficients were determined using the gradient descent algorithm. The training procedure for the RBFNN encompasses the subsequent steps. The centers of the basis functions are extracted using the input-output clustering method developed in [16]. The respective widths were estimated using the maximum distance between all pairs of cluster centers [16]. Finally, the connection weights were determined using the least squares method.

Table 1. Mean *RMSE* values and the standard deviations obtained by the proposed network for polynomial orders $n=2$ and $n=3$ and various numbers (c) of nodes in the antecedent phase

	$n=2$		$n=3$	
	Training Data	Testing Data	Training Data	Testing Data
$c=3$	2.3642 ± 0.1141	2.5390 ± 0.1222	2.4055 ± 0.1498	2.6464 ± 0.2272
$c=5$	2.3569 ± 0.0952	2.5450 ± 0.1473	2.2110 ± 0.0771	2.2937 ± 0.1196
$c=6$	2.2560 ± 0.0614	2.3940 ± 0.1049	2.1906 ± 0.0948	2.2595 ± 0.1450
$c=8$	2.2088 ± 0.0380	2.3886 ± 0.0634	2.2188 ± 0.0516	2.3490 ± 0.0779
$c=9$	2.1992 ± 0.1241	2.3876 ± 0.1986	2.1676 ± 0.0602	2.2895 ± 0.0774
$c=10$	2.1999 ± 0.0406	2.3633 ± 0.0606	2.1681 ± 0.0705	2.2812 ± 0.1300

Table 2. Mean RMSE values and the standard deviations obtained by the proposed network for polynomial orders $n=4$ and $n=5$ and various numbers (c) of nodes in the antecedent phase

	$n=4$		$n=5$	
	Training Data	Testing Data	Training Data	Testing Data
$c=3$	2.5407 ± 0.2871	2.8181 ± 0.4150	2.4977 ± 0.2786	2.6769 ± 0.5780
$c=5$	2.3784 ± 0.1141	2.5045 ± 0.1638	2.4321 ± 0.3322	2.6002 ± 0.4865
$c=6$	2.4066 ± 0.0755	2.5631 ± 0.1721	2.4623 ± 0.3098	2.6967 ± 0.3808
$c=8$	2.3924 ± 0.1704	2.6002 ± 0.2645	2.4619 ± 0.2987	2.5734 ± 0.4234
$c=9$	2.3379 ± 0.1781	2.5132 ± 0.2734	2.5218 ± 0.2865	2.5552 ± 0.3234
$c=10$	2.3429 ± 0.1362	2.5372 ± 0.2066	2.5955 ± 0.4252	2.5624 ± 0.5066

Table 3. Mean RMSE values and the standard deviations obtained by the TSKFS and the RBFNN for various numbers (c) of rules/nodes

	TSKFS		RBFNN	
	Training Data	Testing Data	Training Data	Testing Data
$c=3$	3.2111 ± 0.0197	3.7829 ± 0.0255	3.2670 ± 0.0211	3.8192 ± 0.0290
$c=5$	3.1011 ± 0.0153	3.4837 ± 0.0155	3.1877 ± 0.0108	3.6159 ± 0.0250
$c=6$	3.0307 ± 0.0064	3.2549 ± 0.0095	3.1389 ± 0.0093	3.5862 ± 0.0240
$c=8$	3.0584 ± 0.0121	3.1859 ± 0.0295	3.1249 ± 0.0203	3.4186 ± 0.0164
$c=9$	2.9023 ± 0.0064	3.1030 ± 0.0184	3.1154 ± 0.0184	3.3878 ± 0.0186
$c=10$	2.9546 ± 0.0129	3.1200 ± 0.0284	3.1057 ± 0.0175	3.3083 ± 0.0174

The performance index to conduct the simulations was the root mean square error,

$$RMSE = \sqrt{\frac{1}{N}\sum_{k=1}^{N}\left|y_k - \tilde{y}_k\right|^2}$$ (15)

For the three networks we considered various number of rules/nodes, while for each number of rules/nodes we run 20 different initializations. In addition, for the proposed network we considered various polynomial orders, as well. The results are depicted in Tables 1-3, where we can easily verify the superior performance of the Hermite polynomial network. The best *RMSE* for the training data is reported for $c = 9$ and $n = 3$, while for the testing data for $c = 6$ and $n = 3$. In general, we can safely conclude that the polynomial orders $n = 2$ and $n = 3$ efficiently model the system.

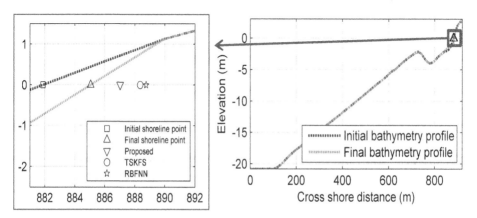

Fig. 5. A bathymetry profile (right) and the corresponding coastal erosion predictions (left) obtained by the proposed network, the TSKFS, and the RBFNN.

It is very interesting to see how the above results are translated into meaningful observations as far as the coastal erosion is concerned. Fig. 5 refers to a specific testing data and shows the predictions obtained by the three methods using $c = 8$, and for the Hermite polynomial network $n = 3$. Based on this figure, the proposed network clearly achieves the best performance because its prediction is closer to the shoreline's point of the final bathymetry, meaning that the corresponding erosion (in meters) is closer to original value. The *RMSE*s for the simulation in Fig. 5 were 2.3867 for the proposed network, 3.1678 for the TSKFS, and 3.4007 for the RBFNN.

5   Conclusions

In this paper we have presented a novel neural-fuzzy network to predict the coastal erosion from bathymetry data. The network consisted of three phases namely, the antecedent phase, the consequent phase and the inference mechanism. The novelty of this contribution lied mainly on the second phase that used Hermite polynomials to define the local regression models involved in the rule consequent parts.

The study area was the beach of Eresos located at Lesvos island, Greece. A set of significant variables were identified as important to carry out the prediction problem. Based on these variables, we applied a specialized experimental methodology to capture all the information related to the problem and to generate the network's input-output training data. Finally, the comparative simulation study showed the network's capability to predict the coastal erosion.

Acknowledgments. This research has been co-financed by the European Union (European Social Fund – ESF) and Greek national funds through the Operational Program "Education and Lifelong Learning" of the National Strategic Reference Framework (NSRF) - Research Funding Program: THALES. Investing in knowledge society through the European Social Fund.

References

1. Prigent, C., Papa, F., Aires, F., Jimenez, W., Rossow, W.B., Matthews, E.: Changes in land surface water dynamics since the 1990s and relation to population pressure. Geophysical Research Letters **39**, L08403 (2012). doi:10.1029/2012GL051276
2. Marcos, M., Jorda, G., Gomis, D., Perez, B.: Changes in storm surges in southern Europe from a regional model under climate change scenarios. Global and Planetary Change **77**, 116–128 (2011)
3. Vousdoukas, M.I., Ferreira, O., Almeida, L.P., Pacheco, A.: Toward reliable storm-hazard forecasts: XBeach calibration and its potential application in an operational early-warning system. Ocean Dynamics **62**, 1001–1015 (2012)
4. Roelvink, D., Reniers, A., van Dongeren, A., de Vries, J.V.T., Lescinsky, J., McCall, R.: Modeling storm impacts on beaches, dunes and barrier islands. Coastal Engineering **56**, 1133–1152 (2009)
5. Karambas, T.V., Karathanassi, E.K.: Boussinesq modeling of Longshore currents and sediment transport. Journal of Waterway, Port, Coastal and Ocean Engineering, American Society of Civil Engineers (ASCE) **130**(6), 277–286 (2004)
6. Leont'yev, I.O.: Numerical modelling of beach erosion during storm event. Coastal Engineering **29**, 187–200 (1996)
7. Harley, M., Armaroli, C., Ciavola, P.: Evaluation of XBeach predictions for a real-time warning system in Emilia-Romagna, Northern Italy. Coastal Research **64**, 1861–1865 (2011)
8. Banakar, A., Fazle Azeemb, M.: Parameter identification of TSK neuro-fuzzy models. Fuzzy Sets and Systems **179**, 62–82 (2011)
9. Chen, C., Wang, F.W.: A self-organizing neuro-fuzzy network based on first order effects sensitivity analysis. Neurocomputing **118**, 21–32 (2013)
10. Zheng, S.-J., Li, Z.-Q., Wang, H.-T.: A genetic fuzzy radial basis function neural network for structural health monitoring of composite laminated beams. Expert Systems with Applications **38**, 11837–11842 (2011)
11. Andrews, G.E., Askey, R., Roy, R.: Special functions. Cambridge University Press, UK (2000)
12. Ma, L., Khorasani, K.: Constructive feedforward neural networks using Hermite polynomial activation functions. IEEE Transactions on Neural Networks **16**(4), 821–833 (2005)

13. Tsekouras, G., Sarimveis, H., Kavakli, E., Bafas, G.: A hierarchical fuzzy-clustering approach to fuzzy modeling. Fuzzy Sets and Systems **150**, 245–266 (2005)
14. Tsekouras, G., Sarimveis, H., Bafas, G.: A simple algorithm for training fuzzy systems using input–output data. Advances in Engineering Software **34**, 247–259 (2003)
15. Armijo, L.: Minimization of functions having Lipschitz continuous first partial derivatives. Pacific Journal of Mathematics **16**(1), 1–3 (1966)
16. Tsekouras, G.E., Tsimikas, J.: On training RBF neural networks using input–output fuzzy clustering and particle swarm optimization. Fuzzy Sets and Systems **221**, 65–89 (2013)

RBF Neural Networks
and Radial Fuzzy Systems

David Coufal$^{(\boxtimes)}$

Institute of Computer Science AS CR, Pod Vodárenskou věží 2,
Praha 8, Czech Republic
david.coufal@cs.cas.cz

Abstract. RBF neural networks are an efficient tool for acquisition and representation of functional relations reflected in empirical data. The interpretation of acquired knowledge is, however, generally difficult because the knowledge is encoded into values of the parameters of the network. Contrary to neural networks, fuzzy systems allow a more convenient interpretation of the stored knowledge in the form of IF-THEN rules. This paper contributes to the fusion of these two concepts. Namely, we show that a RBF neural network can be interpreted as the radial fuzzy system. The proposed approach is based on the study of conjunctive and implicative representations of the rule base in radial fuzzy systems. We present conditions under which both representations are computationally close and, as the consequence, a reasonable syntactic interpretation of RBF neural networks can be introduced.

Keywords: RBF neural networks · Radial fuzzy systems · Conjunctive and implicative rule bases

1 Introduction

In the world of nature-inspired models of computation, neural and fuzzy computing possess a solid position. It is well known that neural networks primarily focus on acquisition of functional relations from empirical data in the form of a regression model. In contrast, fuzzy modeling aims on a syntactic description of functional relations. This description is provided in terms of IF-THEN rules that have the form of fuzzy logic formulas and incorporate fuzzy sets to model linguistic terms.

Both computational paradigms are backed by the well-developed theories and their own techniques for grasping information provided by a real-world environment. For a long time there have been attempts for creative combination of both (or even more) modeling approaches. This movement has become known under the term of soft-computing [10]. Our contribution follows the soft-computing line.

The neural networks are extremely good devices for processing data in terms of non-linear regression. They develop (learn) the regression function that for given inputs computes such the outputs that minimize the selected distance

© Springer International Publishing Switzerland 2015
L. Iliadis and C. Jayne (Eds.): EANN 2015, CCIS 517, pp. 206–215, 2015.
DOI: 10.1007/978-3-319-23983-5_20

to the desired outputs (a supervised learning task on a set of training data). In RBF neural networks, the regression function is implemented in the form of a weighted sum of certain selected points from the range of the regression function. The weights correspond to activation levels of radial computational units that represent the neurons of the network.

Fuzzy systems deal with a linguistic description of functional relations and mathematize it. Mathematization is provided by means of translating linguistic terms into fuzzy sets and combining them into IF-THEN rules. A group of IF-THEN rules then determines the rule base of the fuzzy system.

The rule base is the carrier of the knowledge stored in the fuzzy system. There are two basic mathematical representations of the rule base - the *conjunctive* and the *implicative* one. Under the conjunctive representation, the knowledge representation is data-driven. In this case, the rule base can be seen as the list of prototypical examples taken from the relation the fuzzy system accommodates. In contrast, under the implicative representation the rule base is seen as the set of conditions expressed syntactically in terms of fuzzy logic formulas.

Conjunctively represented radial fuzzy systems can be shown to be computationally equivalent to RBF neural networks and vice versa. Hence both devices can be transformed to each other. A conjunctive rule base can be formally represented in the implicative way. Under certain assumptions, it can be shown that this representation is computationally close to the original conjunctive one. Hence one can propose to interpreted the given RBF neural network as the conjunctive radial fuzzy system and then interpret its rule base implicatively. As a result, one gets the syntactic representation of the original RBF neural network in terms of the implicative radial fuzzy system.

The goal of this paper is to mathematically develop the above transformation idea and to study conditions under which conjunctive and implicative representation are computationally equivalent so that the proposed formal translation is reasonable.

The rest of the paper is organized as follows. The next section is the review section. The third section deals with the computational aspects of conjunctive and implicative representations of the rule base in the radial fuzzy systems. The fourth section concludes the paper.

2   RBF Neural Networks and Radial Fuzzy Systems

In this section we review the very basics of RBF neural networks, radial fuzzy systems and fuzzy logic in order to the paper be self-contained. The review is based on the classical textbooks [5], [7], [9], [4] and [6].

2.1   RBF Networks

The concept of the radial basis function neural network is very well known [1], [5]. In this paper, we consider the RBF network in the standard three-layered MISO (multiple-input single-output) configuration. That is, the

RBF network represents a function from $\mathbb{R}^n$ to $\mathbb{R}$ space, $n \in \mathbb{N}$. Let the hidden layer consist of $m \in \mathbb{N}$ processing units - neurons. The neurons are mathematically represented by radial functions of the form

$$\phi_j(\boldsymbol{x}) = act(||\boldsymbol{x} - \boldsymbol{a}_j||), \quad j = 1, \ldots, m. \tag{1}$$

In the formula, $act : \mathbb{R} \to \mathbb{R}$ is the so-called activation function that is considered to be be continuous, non-increasing with $act(0) = 1$ and $\lim_{z \to \infty} act(z) = 0$. The point $\boldsymbol{a}_j \in \mathbb{R}^n$ is the center of the radial function ϕ_j. Clearly, ϕ_j corresponds to a non-increasing function of the distance of the argument $\boldsymbol{x}$ from the central point $\boldsymbol{a}_j$. The distance is measured by a norm $|| \cdot ||$ in the $\mathbb{R}^n$ space.

The function RBF $: \mathbb{R}^n \to \mathbb{R}$ implemented by the RBF network has the form

$$\mathrm{RBF}(\boldsymbol{x}) = \sum_{j=1}^{m} w_j \cdot \phi_j(\boldsymbol{x}) = \sum_{j=1}^{m} w_j \cdot act(||\boldsymbol{x} - \boldsymbol{a}_j||) \tag{2}$$

where $w_j, j = 1, \ldots, m$ is the set of network's weights.

Concerning setting of parameters of the RBF network, namely, the number m of neurons, their central points $\boldsymbol{a}_j$ and weights $w_j, j = 1, \ldots, m$, this task is split into the structure and parameter learning subtasks. The first one is typically carried out by using a sort of a clustering algorithm. The second one corresponds to an optimization task. In our case, the latter is the task of linear regression. In a multi-layered version, the celebrated back propagation algorithm is usually employed here.

2.2   Basics of Fuzzy Systems

A fuzzy system in the MISO configuration computes a function from some input space $X \subseteq \mathbb{R}^n$ to some output space $Y \subseteq \mathbb{R}$. The special about fuzzy systems is how this function is implemented. Canonically, this is made by the combination of the four building blocks called the *fuzzifier*, the *rule base*, the *inference engine* and the *defuziffier* [7], [9].

The fuzzifier transforms points of the input space into fuzzy sets specified on this space. In this paper, we will consider the singleton fuzzifier that is the most common in applications. The singleton fuzzifier associates a given input $\boldsymbol{x}^* \in X$ with the fuzzy set A_{fuzz} such that $A_{fuzz}(\boldsymbol{x}) = 1$ for $\boldsymbol{x} = \boldsymbol{x}^*$, and $A_{fuzz}(\boldsymbol{x}) = 0$ otherwise.

The rule base of the fuzzy system is given by the set of $m \in \mathbb{N}$ IF-THEN rules. The j-th rule represents the fuzzy relation $R_j(\boldsymbol{x}, y)$ on $X \times Y$ space specified as

$$R_j(\boldsymbol{x}, y) = A_{j1}(x_1) \star \cdots \star A_{jn}(x_n) \triangleright B_j(y) = A_j(\boldsymbol{x}) \triangleright B_j(y). \tag{3}$$

In the formula, A_j and B_j fuzzy sets correspond to the antecedent and consequent fuzzy sets of the rule, respectively. The antecedent A_j is composed from the individual fuzzy sets $A_{ji}, i = 1, \ldots, n$ using a fuzzy conjunction which is implemented by the t-norm $\star$, i.e., we have $A_j(\boldsymbol{x}) = A_{j1}(x_1) \star \cdots \star A_{jn}(x_n)$ for the antecedent. Clearly, $x = (x_1, \ldots, x_n)$. The interpretation of the $\triangleright$ symbol depends on the representation of the whole rule base.

Under the conjunctive representation, $\triangleright$ corresponds to the fuzzy conjunction and the relation that implements the j-th rule reads as $R_j(x, y) = A_j(x) \star B_j(y)$. The whole rule base is then determined by a disjunctive combination of individual fuzzy relations

$$RB_{conj}(x, y) = \bigvee_{j=1}^{m} A_j(x) \star B_j(y). \tag{4}$$

In (4), maximum is the typical choice for implementing the fuzzy disjunction $\bigvee$. In the conjunctive case, the rule base RB_{conj} can be seen as the list of prototypic points from the relation on the input-output space that the fuzzy system implements.

Under the implicative representation, $\triangleright$ corresponds to the residuated fuzzy implication $\rightarrow$ devised from the t-norm $\star$ by the process of residuation [4], [6]: $u \rightarrow v = \sup_z \{z \mid z \star u \leq v ; u, v, z \in [0, 1]\}$. The j-th rule is then implemented as the fuzzy relation $R_j(x, y) = A_j(x) \rightarrow B_j(y)$; and the whole rule base is combined conjunctively

$$RB_{impl}(x, y) = \bigwedge_{j=1}^{m} A_j(x) \rightarrow B_j(y). \tag{5}$$

In (5), the minimum operation is typically used to represent the fuzzy conjunction $\bigwedge$. In the implicative case, the rule base can be seen as the set of the conditions that are simultaneously imposed on the points of the relation the fuzzy system implements.

2.3   Radial Fuzzy Systems

Radial fuzzy systems (RFSs) employ radial functions for representing membership functions of fuzzy sets in their rules [2], [3]. Moreover, the RFSs exhibit the so-called *radial property*. This is the shape preservation property in antecedents of IF-THEN rules. The property makes the computational model of radial systems more tractable and the antecedents of rules correspond to multivariate radial functions. Let us be more specific.

Definition 1. *A fuzzy system is called radial if:*

(i) *There exists a continuous function* $act : [0, +\infty) \rightarrow [0, 1]$, $act(0) = 1$ *such that: (a) either there exists* $z_0 \in (0, +\infty)$ *such that* act *is strictly decreasing on* $[0, z_0]$ *and* $act(z) = 0$ *for* $z \in [z_0, +\infty)$ *or (b)* act *is strictly decreasing on* $[0, +\infty)$ *and* $\lim_{z \rightarrow +\infty} act(z) = 0$.

(ii) *Fuzzy sets in the antecedent and consequent of the j-th rule are specified as*

$$A_{ji}(x_i) = act\left(\left|\frac{x_i - a_{ji}}{b_i}\right|\right), \quad B_j(y) = act\left(\frac{\max\{0, |y - c_j| - s\}}{d}\right) \tag{6}$$

where $n, m \in \mathbb{N}$; $i = 1, \ldots, n$; $j = 1, \ldots, m$; $\boldsymbol{x} \in \mathbb{R}^n$, $\boldsymbol{x} = (x_1, \ldots, x_n)$; $\boldsymbol{a}_j \in \mathbb{R}^n$,
$\boldsymbol{a}_j = (a_{j1}, \ldots, a_{jn})$; $\boldsymbol{b} = (b_1, \ldots, b_n)$, $b_i > 0$, $c_j \in \mathbb{R}$; $d_j > 0$, $s > 0$.
(iii) *For each* $\boldsymbol{x} \in \mathbb{R}^n$ *the radial property holds, i.e.,*

$$A_j(\boldsymbol{x}) = A_{j1}(x_1) \star \cdots \star A_{jn}(x_n) = act(\,||\boldsymbol{x} - \boldsymbol{a}_j||_{p,\boldsymbol{b}}), \qquad (7)$$

where $||\cdot||_{\boldsymbol{b}}$ *is the scaled* ℓ_p *norm for some* $p \geq 1$, *i.e.,* $||\boldsymbol{u}||_{p,\boldsymbol{b}} = (\sum_i |u_i/b_i|^p)^{1/p}$.

Let us shortly comment on the definition. The first two conditions refer to the specification of the individual fuzzy sets. In fact, they require that one-dimensional fuzzy sets are specified as radial functions with the central points a_{ji}, c_j and scaling parameters b_i and d. The parameter s is the shifting parameter that makes the shape of the consequent fuzzy sets trapezoid-like. In Fig. 1 there are presented examples of radial fuzzy sets.

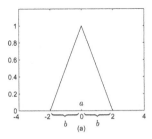 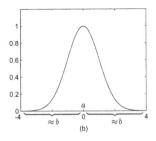

Fig. 1. Examples of radial fuzzy sets; (a) triangular $A(x) = \max\{0, 1 - |x - a|/b\}$ and (b) Gaussian $A(x) = \exp[-((x - a)/b)^2]$ fuzzy set.

The third condition is the radial property. The property requires the shape of individual fuzzy sets, which is determined by the *act* function, to be retained under the combination by the *t*-norm $\star$. From (7) we see that individual fuzzy sets are transformed into the multi-dimensional one, and this fuzzy set has the same shape as the individual fuzzy sets.

Table 1. Examples of *t*-norms and corresponding *act* functions.

t-norm	$act(z)$	$A_j(\boldsymbol{x})$
$x \star y = \max\{0, x + y - 1\}$	$\max\{0, 1 - z\}$	$\max(0, 1 - ||\boldsymbol{x} - \boldsymbol{a}_j||_{1,\boldsymbol{b}})$
$x \star y = x \cdot y$	$\exp(-z^2)$	$\exp(-||\boldsymbol{x} - \boldsymbol{a}_j||_{2,\boldsymbol{b}}^2)$
$x \star y = \exp\left(-[(-\ln(x))^{1/2} + (-\ln(y))^{1/2})]\right)^2$	$\exp(-z^2)$	$\exp(-||\boldsymbol{x} - \boldsymbol{a}_j||_{1,\boldsymbol{b}}^2)$

Remark that the radial property is not trivial. That is, the *t*-norm and the shape has to be matched somehow in order to the radial property holds. In [2], there is shown that in the case of continuous Archimedean *t*-norms the matching is provided by setting $act(z) = t^{(-1)}(qz^p)$ where $t^{(-1)}$ is the pseudo-inverse of the additive generator of the *t*-norm and $q > 0, p > 1$ are parameters. Table 1 presents some allowed pairs connected with the Łukasiewicz, product and Aczél-Alsina *t*-norm with $\lambda = 1/2$, respectively [6].

2.4   Computational Model of Radial Fuzzy Systems

The specification of the computational model of a fuzzy system depends on the implementation of the inference engine and the chosen deffuzification method.

The inference engine performs a sort of projection of the fuzzy set yielded by the fuzzifier through the fuzzy relation represented by the rule base. The projection can be seen as the composition of the fuzzified input with the knowledge stored in the fuzzy system. The most standard inference engine employed in fuzzy computing is the CRI inference engine referring to the compositional rule of inference [7], [9]. Mathematically, when the singleton fuzzifier is employed, it yields for an input $\boldsymbol{x}^*$ the fuzzy set B specified as

$$B(y) = \sup_{\boldsymbol{x}} \{ \boldsymbol{x} \in X \,|\, A_{fuzz}(\boldsymbol{x}) \star RB(\boldsymbol{x}, y) \} = RB(\boldsymbol{x}^*, y). \tag{8}$$

The defuzzier transforms the fuzzy set B into the point y^* from the output space Y. There are plenty of methods of defuzzification present [7], [9]; and their use depends on the representation of the rule base.

In the case of the conjunctive representation when $B(y) = \bigvee_j A(\boldsymbol{x}^*) \star B_j(y)$, the method of centroids is popular: $y^* = \left(\sum_{j=1}^m A_j(\boldsymbol{x}^*) \cdot y^*_{B_j} \right) / \left(\sum_{j=1}^m A_j(\boldsymbol{x}^*) \right)$. In radial fuzzy systems the centroids $y^*_{B_j}$ of B_j sets correspond to their central points, i.e., $y^*_{B_j} = c_j$, $j = 1 \ldots, m$.

The problem with this method is that the output is not defined for denominator being 0 nor is generally continuous when we set $y^* = 0$ for the zero denominator. That is why the formula of centroids method is further simplified into the form (in fact, this corresponds to a variant of the Takagi-Sugeno fuzzy system [8])

$$y^* = \sum_{j=1}^m A_j(\boldsymbol{x}^*) \cdot c_j. \tag{9}$$

In the case of the implicative representation, the fuzzy set B yielded by the CRI inference engine reads as $B(y) = \bigwedge_j A_j(\boldsymbol{x}^*) \to B_j(y)$. Since for any residuated implication one has $u \to v = 1$ iff $u \le v$, we see that $A_j(\boldsymbol{x}^*) \to B_j(y)$ are modified B_j sets such that $B'_j(y) = A_j(\boldsymbol{x}^*) \to B_j(y) = 1$ iff $A_j(\boldsymbol{x}^*) \le B_j(y)$. That is, B'_j sets are normal with the kernels $I_j(\boldsymbol{x}^*) = \{y \,|\, A_j(\boldsymbol{x}^*) \le B_j(y)\}$.

Any fuzzy conjunction of B'_j sets then yields a fuzzy set that is also normal with the kernel $I(\boldsymbol{x}^*) = \bigcap_j I_j(\boldsymbol{x}^*)$, under the assumption that $\bigcap_j I_j(\boldsymbol{x}^*) \ne \emptyset$. If this is true for any input $\boldsymbol{x}^* \in X$, then the implicative fuzzy system is called *coherent*. If the system under study is coherent, then it is natural to take as the defuzzified point the middle point of $I(\boldsymbol{x}^*)$. In the case of the radial systems it can be shown that $I(\boldsymbol{x}^*)$ corresponds to a closed interval. Hence

$$y^* = \frac{L(I(\boldsymbol{x}^*)) + R(I(\boldsymbol{x}^*))}{2} \tag{10}$$

where $L(I(\boldsymbol{x}^*)$ and $R(\boldsymbol{x}^*))$ are the left and right limit points of $I(\boldsymbol{x}^*)$, respectively.

3 Combination of RBF Networks and Radial Fuzzy Systems

In this section we propose an approach how to interpret the RBF neural network in terms of the implicative radial fuzzy system.

We start with the computational equivalence of RBF networks and radial conjunctive fuzzy systems. The equivalence is based on the comparison of computational models (2) and (9). These models coincide if the number of neurons equals the number of IF-THEN rules and if $\phi_j = A_j$ and $w_j = c_j$ for all $j = 1, \ldots, m$.

Concerning the first equality $\phi_j(\boldsymbol{x}) = A_j(\boldsymbol{x})$, $\boldsymbol{x} \in X = \mathbb{R}^n$ it rests on the radial property of the radial fuzzy systems. It is clear that the representation of neurons and the employed t-norm in the fuzzy system must be matched in order to the radial property holds. In the case of the second equality $w_j = c_j$, we see that it requires the central points of consequents to coincide with the network's weights. The width and shift parameters d and s, respectively, do not affect the computational equivalence between (2) and (9). Hence they can be set freely, however, in the case of the d parameter we will require that the fuzzy system is *strictly coherent*.

Definition 2. *A radial fuzzy system is said to be strictly coherent if for any pair of rules $j, k \in \{1, \ldots, m\}$ the following holds*

$$d \cdot \|\boldsymbol{a}_j - \boldsymbol{a}_k\|_{p,b} \geq |c_j - c_k|. \tag{11}$$

Based on the above definition, we set up d in such a way that the built conjunctive fuzzy system is strictly coherent. Hence we search for such the minimal $d = d_{coh}$ that (11) holds. Formally, we have

$$d_{coh} = \min_d \{d \cdot \|\boldsymbol{a}_j - \boldsymbol{a}_k\|_{p,b} \geq |c_j - c_k|, \ \forall j, k \in \{1, \ldots, m\}\}. \tag{12}$$

This is a straightforward optimization task. Solving this problem we set $d = d_{coh}$ which finalizes the specification of the parameters in the conjunctive radial fuzzy system. The specification of d does not impact the computational equivalence between (2) and (9), however, this specification is important from the implicative representation point of view, because if the implicative system is strictly coherent then it is also coherent in the sense of Section 2.4.

Let us consider conjunctive radial system with the set of its parameters. Let $\rightarrow$ be the residuated implication derived from the t-norm $\star$ which is employed in the conjunctive fuzzy system. We ask what will happen if we interpret the rule base implicatively with the values of the parameters retained.

Clearly, if the conjunctive and implicative representations are computationally close then the interpretation chain: the RBF network $\Rightarrow$ the conjunctive RFS $\Rightarrow$ the implicative RFS makes sense. In the following two lemmas we will show that both representations are computationally close to each other under certain assumptions.

Lemma 1. *Let $\star$ be a t-norm and $\rightarrow$ its residuated implication. Let the rule base of a radial fuzzy system consists of $m \in \mathbb{N}$ IF-THEN rules built up on the basis of antecedents A_j and consequents B_j, $j = 1, \ldots, m$. Let the radial fuzzy system be strictly coherent and the norm of the radial property be the scaled ℓ_1 norm, i.e., $p = 1$, then for the conjunctive and implicative representations of the rule base one has*

$$RB_{conj}(\boldsymbol{x}, y) \leq RB_{impl}(\boldsymbol{x}, y) \tag{13}$$

for any $\boldsymbol{x} \in \mathbb{R}^n$ and $y \in \mathbb{R}$.

Proof. We start by proving the inequality

$$A_j(\boldsymbol{x}) \star A_k(\boldsymbol{x}) \star B_j(y) \leq B_k(y) \tag{14}$$

for $j, k \in \{1, \ldots, m\}$ and $\boldsymbol{x} \in \mathbb{R}^n$, $y \in \mathbb{R}$. Indeed, by the triangle inequality, non-increasing character of the *act* function and the radial property for the ℓ_1 scaled norm $\|\cdot\|_{1,b}$, we have the following chain:

$$\|\boldsymbol{x} - \boldsymbol{a}_j\|_{1,b} + \|\boldsymbol{x} - \boldsymbol{a}_k\|_{1,b} \geq \|\boldsymbol{a}_j - \boldsymbol{a}_k\|_{1,b},$$
$$act(\|\boldsymbol{x} - \boldsymbol{a}_j\|_{1,b} + \|\boldsymbol{x} - \boldsymbol{a}_k\|_{1,b}) \leq act(\|\boldsymbol{a}_j - \boldsymbol{a}_k\|_{1,b}),$$
$$act(\|\boldsymbol{x} - \boldsymbol{a}_j\|_{1,b}) \star act(\|\boldsymbol{x} - \boldsymbol{a}_k\|_{1,b}) \leq act(\|\boldsymbol{a}_j - \boldsymbol{a}_k\|_{1,b}),$$
$$A_j(\boldsymbol{x}) \star A_k(\boldsymbol{x}) \leq act(\|\boldsymbol{a}_j - \boldsymbol{a}_k\|_{1,b}). \tag{15}$$

The triangle inequality of the second kind reads $|c_j - c_k| \geq |\,|y - c_j| - |y - c_k|\,|$. Therefore due to the strict coherence we have

$$d \cdot \|\boldsymbol{a}_j - \boldsymbol{a}_k\|_{1,b} \geq |\,|y - c_j| - |y - c_k|\,|,$$
$$d \cdot \|\boldsymbol{a}_j - \boldsymbol{a}_k\|_{1,b} + |y - c_j| - s \geq |y - c_k| - s,$$
$$d \cdot \|\boldsymbol{a}_j - \boldsymbol{a}_k\|_{1,b} + \max\{0, |y - c_j| - s\} \geq \max\{0, |y - c_k| - s\},$$
$$act(\|\boldsymbol{a}_j - \boldsymbol{a}_k\|_{1,b} + \max\{0, |y - c_j| - s\}/d) \leq act(\max\{0, |y - c_k| - s\}/d).$$

Due to the radial property for the the ℓ_1 scaled norm and (15), the left-hand side of the last inequality is greater than

$$act(\|\boldsymbol{x} - \boldsymbol{a}_j\|_{1,b}) \star act(\|\boldsymbol{x} - \boldsymbol{a}_k\|_{1,b}) \star act(\max\{0, |y - c_j| - s\}/d)$$

and implies the inequality (14).

For any t-norm $\star$ and its residuated implication $\rightarrow$ one has $u \star v \rightarrow w = u \rightarrow (v \rightarrow w)$ and $u \rightarrow v = 1$ iff $u \leq v$ for $u, v, w \in [0, 1]$, see [4]. Hence we can update (14) as follows

$$A_j(\boldsymbol{x}) \star A_k(\boldsymbol{x}) \star B_j(y) \leq B_k(y),$$
$$A_j(\boldsymbol{x}) \star B_j(y) \leq A_k(\boldsymbol{x}) \rightarrow B_k(y),$$
$$\max_{j=1}^{m}\{A_j(\boldsymbol{x}) \star B_j(y)\} \leq A_k(\boldsymbol{x}) \rightarrow B_k(y),$$
$$\max_{j=1}^{m}\{A_j(\boldsymbol{x}) \star B_j(y)\} \leq \min_{k=1}^{m}\{A_k(\boldsymbol{x}) \rightarrow B_k(y)\},$$
$$RB_{conj}(\boldsymbol{x}, y) \leq RB_{impl}(\boldsymbol{x}, y).$$

This concludes the proof.   □

There is the natural question of how computationally close are both representations. In terms of the fuzzy logic, the closeness relation is interpreted as the fuzzy equivalence relation [4]. The fuzzy equivalence is introduced by the formula $x \equiv y = (x \to y) \star (y \to x)$ for $x, y \in [0, 1]$. For any fuzzy equivalence one has $x \equiv y = 1$ if $x = y$. If $x \neq y$, then the measure of difference is given either as $x \to y$, if $x \geq y$ or by $y \to x$ if $x \leq y$. For example for Łukasiewicz t-norm it reads as $x \equiv y = 1 - |x - y|$. The next lemma is due to Hájek [4].

Lemma 2. *Let $\star$ be a t-norm and $\to$ its residuated implication. Let the rule base of a radial fuzzy system consists of $m \in \mathbb{N}$ IF-THEN rules built up on the basis of antecedents A_j and consequents B_j, $j = 1, \ldots, m$. Then for the conjunctive and implicative representations of the rule base it holds*

$$\min_{j=1}^{m}\{A_j(\boldsymbol{x}) \star A_j(\boldsymbol{x})\} \to (RB_{impl}(\boldsymbol{x}, y) \leq RB_{conj}(\boldsymbol{x}, y))$$

for any $\boldsymbol{x} \in \mathbb{R}$ and $y \in \mathbb{R}$.

Proof. Remind once again that $u \star v \to w = u \to (v \to w)$. Further, it can be shown that $(u \to v) \to [(u \star w) \to (v \star w)] = 1$ for any $u, v, w \in [0, 1]$, see [4]. Hence one has the following chain:

$$(A_j(\boldsymbol{x}) \to B_j(y)) \to [(A_j(\boldsymbol{x}) \star A_j(\boldsymbol{x})) \to (A_j(\boldsymbol{x}) \star B_j(y))],$$
$$(A_j(\boldsymbol{x}) \star A_j(\boldsymbol{x})) \to [(A_j(\boldsymbol{x}) \to B_j(y)) \to (A_j(\boldsymbol{x}) \star B_j(y))],$$
$$(A_j(\boldsymbol{x}) \star A_j(\boldsymbol{x})) \to [\min_{j=1}^{m}\{A_j(\boldsymbol{x}) \to B_j(y)\} \to (A_j(\boldsymbol{x}) \star B_j(y))],$$
$$(A_j(\boldsymbol{x}) \star A_j(\boldsymbol{x})) \to [\min_{j=1}^{m}\{A_j(\boldsymbol{x}) \to B_j(y)\} \to \max_{j=1}^{m}\{A_j(\boldsymbol{x}) \to B_j(y)\}],$$
$$\min_{j}\{A_j(\boldsymbol{x}) \star A_j(\boldsymbol{x})\} \to [RB_{impl}(\boldsymbol{x}, y) \to RB_{conj}(\boldsymbol{x}, y)].$$

This concludes the proof. ☐

The combination of both lemmas gives us the final answer on how computationally close are both representation of the rule base in the radial fuzzy systems. In order to state the theorem, we will denote as the *degree of equivalence* the number

$$\text{DOE} = \inf_{\boldsymbol{x} \in X}\{\min_{j}\{A_j(\boldsymbol{x}) \star A_j(\boldsymbol{x})\}\}.$$

Theorem 1. *Let $\star$ be a t-norm and $\to$ its residuated implication. Let the rule base of a radial fuzzy system consists of $m \in \mathbb{N}$ IF-THEN rules built up on the basis of antecedents A_j and consequents B_j, $j = 1, \ldots, m$. Let the radial fuzzy system be strictly coherent and the norm of the radial property be the scaled ℓ_1 norm, i.e., $p = 1$, then for the conjunctive and implicative representations of the rule base one has*

$$\text{DOE} \leq (RB_{impl}(\boldsymbol{x}, y) \equiv RB_{conj}(\boldsymbol{x}, y)).$$

Proof. By the above lemmas and the specification of $\equiv$ relation we get

$$\min_{j}\{A_j(\boldsymbol{x}) \star A_j(\boldsymbol{x})\} \to [RB_{impl}(\boldsymbol{x}, y) \equiv RB_{conj}(\boldsymbol{x}, y)].$$

Therefore also $\text{DOE} \leq [RB_{impl}(\boldsymbol{x}, y) \equiv RB_{conj}(\boldsymbol{x}, y)]$ because $u \to v = 1$ iff $u \leq v$, $u, v \in [0, 1]$. ☐

Theorem 1 tells us that if the scaled ℓ_p norm occurring in the radial property is the scaled ℓ_1 norm and d is set in such a way that the fuzzy system is strictly coherent, then $\text{DOE} \leq RB_{conj}(\boldsymbol{x}, y) \equiv RB_{impl}(\boldsymbol{x}, y)$ for any $\boldsymbol{x} \in \mathbb{R}^n$ and $y \in \mathbb{R}$. Therefore for any input $\boldsymbol{x}^*$ one has also $\text{DOE} \leq RB_{conj}(\boldsymbol{x}^*, y) \equiv RB_{impl}(\boldsymbol{x}^*, y)$.

4 Conclusions

The main result presented in the paper is the characterization of the mutual relation between conjunctive and implicative representations of the rule base in the radial fuzzy systems. We have shown that we can measure or control the closeness of computational equivalence by means of the degree of equivalence.

The proposed use of the result follows the line of fusion of concepts of the RBF neural networks and radial fuzzy systems. In fact, we have shown how the RBF neural network can be interpreted in syntactic way in terms of the radial implicative fuzzy system. The process of transforming the RBF neural networks can be also understood as the method for establishing the radial fuzzy systems on the basis of empirical data. In this case, the well-developed machinery for learning the RBF neural networks can be exploited.

Acknowledgments. This work was supported by COST grant LD13002 provided by the Ministry of Education, Youth and Sports of the Czech Republic.

References

1. Broomhead, D.S., Lowe, D.: Multivariate functional interpolation and adaptive networks. Complex Systems **2**, 321–355 (1988)
2. Coufal, D.: Radial implicative fuzzy systems. In: The IEEE International Conference on Fuzzy Systems, FUZZ-IEEE 2005, Reno, Nevada, USA, pp. 963–968 (2005)
3. Coufal, D.: Coherence of radial implicative fuzzy systems. In: IEEE International Conference on Fuzzy Systems, FUZZ-IEEE 2006, Vancouver, Canada, pp. 903–970 (2006)
4. Hájek, P.: Metamathematics of Fuzzy Logic. Kluwer Academic Publishers (1998)
5. Haykin, S.S.: Neural Networks and Learning Machines, 3rd edn. Prentice Hall, Upper Saddle River (2009)
6. Klement, E.P., Mesiar, R., Pap, A.: Triangular Norms. Kluwer Academic Publishers, Dordrecht (2000)
7. Klir, G.J., Yuan, B.: Fuzzy sets and Fuzzy logic - Theory and Applications. Prentice Hall, Upper Saddle River (1995)
8. Nguyen, H.T., Sugeno, M. (eds.): Fuzzy systems Modeling and Control. The Handbooks of Fuzzy Sets Series, vol. 2. Kluwer Academic Publishers, Boston (1998)
9. Wang, L.X.: A Course in Fuzzy Systems and Control. Prentice Hall (1997)
10. Zadeh, L.A.: What is Soft Computing? Soft Computing **1**(1) (1997)

Evolving Fuzzy-Neural Method for Multimodal Speech Recognition

Mario Malcangi and Philip Grew[✉]

Department of Computer Science, Università degli Studi di Milano, Milan, Italy
{malcangi,grew}@di.unimi.it

Abstract. Improving automatic speech recognition systems is one of the hottest topics in speech-signal processing, especially if such systems are to operate in noisy environments. This paper proposes a multimodal evolutionary neuro-fuzzy approach to developing an automatic speech-recognition system. To make inferences at the decision stage about audiovisual information for speech-to-text conversion, the EFuNN paradigm was applied. Two independent feature extractors were developed, one for the speech phonetics (speech listening) and the other for the speech visemics (lip reading). The EFuNN network has been trained to fuse decisions on audio and decisions on video. This soft computing approach proved robust in harsh conditions and, at the same time, less complex than hard computing, pattern-matching methods. Preliminary experiments confirm the reliability of the proposed method for developing a robust, automatic, speech-recognition system.

Keywords: Audiovisual Speech Recognition (AVSR) · Evolutionary Fuzzy Neural Network (EFuNN) · Speech-To-Text (STT) · Decision fusion · Multimodal speech recognition

1 Introduction

Personal-communication and information systems are fast evolving toward wearables. Consequently, the human-machine interface and the interaction paradigms need to feel natural. Multimodal, speech-recognition systems will be one of the most important enabling technologies for deeply embedded and wearable systems [1, 2].

Audiovisual speech recognition (AVSR) [3, 4, 5, 6, 7] is a technique that combines speech recognition and image recognition by integrating utterance hearing and lip reading to improve speech recognition, especially in harsh audio conditions. The common approach to implementing AVSR systems is to merge audio features and video features into a single pattern-matching framework. This approach leads to a high complexity, making it very difficult to tune the system when it is running in real conditions [8, 9, 10, 11]. An alternative approach is to develop the AVSR system so that the utterance-recognition system and the lip-reading system work independently of each other at the feature-matching stage and then fuse the decision at a later stage. Several advantages emerge from this two-stage AVSR framework. The two most important concern the system's reliability and flexibility and its capacity to adopt soft

© Springer International Publishing Switzerland 2015
L. Iliadis and C. Jayne (Eds.): EANN 2015, CCIS 517, pp. 216–227, 2015.
DOI: 10.1007/978-3-319-23983-5_21

computing paradigms, such as fuzzy logic and neural networks, thus lowering complexity and enhancing performance under harsh operating conditions.

A great deal of attention has been paid to integrating visual speech reading with audio speech recognition because of its potential in applications such as human-computer interaction (HCI), speaker recognition and identification, talking avatars, sign-language recognition, and video surveillance. Multimodal speech recognition (mainly audiovisual) is founded on the natural ability of human beings to communicate by integrating various sensory signals at the decision layer. Speech recognition in human beings is context-sensitive. For example, the McGurk and MacDonald experiment [12] demonstrated the existence of a human capacity for perception that creates interaction between auditory and visual information [13]. As a result, decisions about the meaning of a speech sound may differ according to specific audiovisual context. This experiment shows that audiovisual speech processing at recognition time enables two embedded capabilities, one for merger and one for combining. Merging and combining at the phoneme-recognition stage are powerful abilities that enable the AVSR to fuzzily remedy ambiguities in deciding on the most probable phoneme-to-grapheme transcription of the uttered speech. Soft computing-based implementation of such fusion and combination proved effective, especially when the AVSR system has two layers, a lower layer acting as feature extractor and an upper layer acting as decision fuser [14, 15, 16, 17].

Fuzzy logic implementation of the AVSR decision layer was effective, though some drawbacks related to its modeling need to be resolved. Two main drawbacks involve knowledge development (the rule set and membership functions) and exploding the rules. These problems were addressed with different approaches [18], but the evolutionary paradigm proved ideal for optimizing both [19] of them. The evolving fuzzy logic paradigm is also optimal for predicting, because it adjusts the prediction proprieties embedded in the process to match the right phoneme sequence and transcribe speech to text.

The evolving fuzzy neural-network (EFuNN) paradigm [20] was applied to implement the decision layer for a previously developed AVSR [15]. The purpose was to develop an inner layer between the phoneme-viseme-matching layer and the speech-to-text-transcription layer, so fusion and combination could be executed before matching errors due to noise were propagated to the phoneme-to-grapheme-transcription layer.

1.1   AVSR Motivation

In the movie "2001: A Space Odyssey," HAL, the computer that controls the spacecraft, says to Dave, a crew member: "Dave, although you took very thorough precautions in the pod against my hearing you, **I could see your lips move.**"

It is known that speech intelligibility improves when the listener is able to see the face of the speaker, mainly due to lip movements. Some phonemes might be confused in the audio domain (e.g. /m/ and /n/) but not in the visual domain because they correspond to distinct visemes. Phoneme-to-viseme mapping highlights this peculiarity in human understanding of speech. There are fewer visemes than phonemes, so there is a many-to-one mapping between phoneme subsets and a given viseme. This means that, in order to be effective, reading speech visually requires understanding speech acoustically. However, merely understanding audio speech proves incomplete because its lacks the ability to discriminate between similar sounds with different meanings.

This is a strong motivation to adopt the AVSR approach for automatic speech recognition (ASR), especially considering the different contexts in which it might be applied:

- Speech-to-text recognition in harsh, noisy conditions
- In-car natural interfaces
- Impaired access to systems
- Automatic captioning of audiovisual streams
- Audiovisual data mining
- Natural user interfaces in handheld and wearable devices.

Speech understanding is fundamentally multimodal, that is, each perceptual function, alone, can recognize the semantics of the utterance. When the auditory and visual perceptual functions are active jointly, they fuse separate recognition results at a higher level to yield optimal speech understanding. Emotion also contributes to this fusion process, because muscles move during utterance according to mood and depending on prosody.

2   Framework

2.1   System Architecture

The architecture of the proposed AVSR system has three layers (Fig. 1). The first layer consists of three parallel operating units devoted to extracting and classifying features. It is a hard computing layer based on (audio and video) digital signal-processing algorithms (DSP). The second layer is a fusion and combination unit. This soft computing layer is based on a fuzzy logic engine (FLE). The third layer is a speech-to-text-transcription unit. It is based on artificial intelligence (AI).

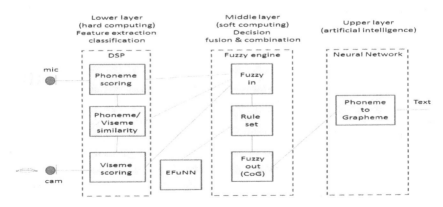

Fig. 1. Audiovisual speech-recognition system with decision layer based on fuzzy logic engine fed with rules tuned using EFuNN paradigm

System input consists of audio and video streams. Audio information is captured by a microphone, conditioned, and digitalized at 16 kHz/16 bit. Video information is captured by a camera recording at 24 frames per second.

2.2   The Feature-Extraction Unit

Phonemes and visemes (Table 1) are matched, scored and encoded at lower the system. The matching is based on signal processing methods for classification and scoring, frame by frame, the utterance and the related visual sequence.

Table 1. Viseme-class coding and the corresponding phoneme symbols.

Viseme-class code	Phoneme(s) symbol
0	Pause
1	ae, ax, ah
2	aa
3	ao
4	ey, eh, uh
5	er
6	y, iy, ih, ix
7	w, uw
8	ow
9	aw
10	oy
11	ay
12	h
13	r
14	l
15	s, z
16	sh, ch, jh, zh
17	th, dh
18	f, v
19	d,t,n
20	k, g, ng
21	p, b, m

The feature-extraction unit consists of three distinct subsystems, one processing the utterance (Fig. 3a) to match the phonemes, the other processing the video frames (Fig. 3b) to match the visemes, the third measuring the similarity of the matched phoneme and viseme. Phoneme and viseme matching are independent systems. The similarity-scoring subsystem depends on the matching and scoring subsystems for phonemes and visemes.

The phoneme-extraction unit segments the audio stream into short intervals (20.85 milliseconds), measures the features (pitch, formants, and intensity), and executes the classification (phoneme: score, duration).

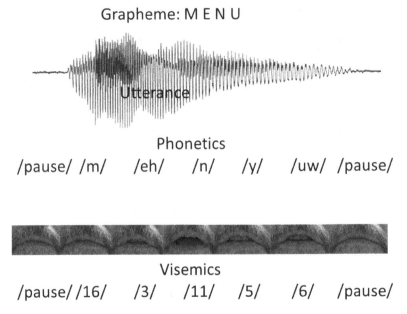

Fig. 2. Utterance of the word "menu", its phonemic transcription, and the corresponding visemes

The following features were used:
   Root mean square (RMS):

$$RMS_j = \sqrt{\frac{1}{N}\sum_{m=0}^{N-1} s_j^{\,2}(m)}$$

(1)

Zero-crossing rate (ZCR):

$$ZCR_j = \sum_{m=0}^{N-1} 0.5\left|sign(s_j(m)) - sign(s_j(m-1))\right|$$

(2)

Auto correlation (AC):

$$AC_j = \sum_{i=1}^{N} \sum_{j=1}^{N+1-i} s_j(j)s_j(i+j-1) \tag{3}$$

Cepstral linear prediction coefficients (CLPC):

$$CLPC_j = a_m + \sum_{k=1}^{m-1}\left(\frac{k}{m}\right)c_k a_{m-k} \tag{4}$$

Frame-by-frame (j) uttered speech is encoded in feature vectors that are matched with a set of phoneme templates to classify and score the j-st. Euclidean distance metrics are applied to match each frame. Phoneme duration is measured as the number of contiguous audio frames (windows) the current phoneme matches.

The viseme-extraction unit measures the lip features (height, width, duration) on each video frame (1/24 s) and executes the classification (viseme: score, duration). Visemes are identified by the height-to-width ratio of lip contour. To measure lip features, mouth contour was located after the face was detected and lip position was then delimited. Four key lip points (Fig. 3) were computed, two vertical and two horizontal, delimiting effective lip contour. Height and width were measured to create a relative index from the ratio of height to width. The viseme is then identified and scored, applying a matching method based on a set of templates and its Euclidean distance metrics. Viseme duration is then measured as the number of contiguous visual frames the viseme matches.

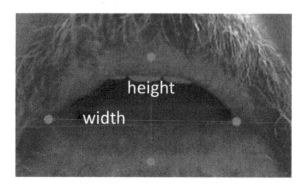

Fig. 3. After four key lip points were computed, height and width were measured to create a relative index from the ratio of height to width

The phoneme-viseme-comparison unit is a lookup table that scores how much the current phoneme is phonetically similar to the current viseme [23, 24].

The fusion-and-combination unit predicts the phoneme on a window-by-window basis, running a fuzzy logic engine tuned according to the EFuNN paradigm. The output of this unit is a stream of phonemes (one phoneme per window) ready for phoneme-to-grapheme transcription.

The phoneme-to-text-transcription unit applies a ruleset to transform each phoneme into the corresponding alphabetic representation, yielding the final text transcription of the uttered word (or a homophone).

3   The Method

The fuzzy logic-based inference paradigm was applied to draw inferences about phonemes from a set of audio and visual features, handling uncertainty due to noise and high variability in both audio and visual information. The main issue in designing the fuzzy logic engine was setting the rules. This was accomplished by applying the evolving neuro-fuzzy paradigm EFuNN, a neuro-fuzzy structure that evolves by creating and modifying its nodes and connections by learning from input.

The EFuNN paradigm connects using a feed-forward architecture of five layers of neurons and can be trained with neural-network methods [19, 20]. Its evolving architecture does not oblige it to adapt an *a priori* architecture, because it starts with a minimal set of initial nodes and then grows or shrinks at training and learning time, depending on the data input. This strategy avoids the problem of catastrophic forgetting and enables the network to be trained further with new data, retaining the effects of previous learning because new nodes are created without removing the old ones, thus preserving previous knowledge. Pruning and aggregation at training-time avoid overtraining during learning by removing weak connections and their nodes.

The EFuNN is a five-layered, feed-forward, artificial neural network, in which each layer performs one specialized function of a fuzzy logic engine: input, condition, rule, action, and output (Fig. 4). The input layer (layer 1) represents (crisp) input variables that are presented to the nodes on the condition layer. The nodes on the condition layer (layer 2) are fuzzy membership functions that perform fuzzification on crisp input. The rule layer is the evolving layer (layer 3), which can create and aggregate the nodes, adapting them to changes in fuzzified input data. The nodes in this layer model the rules that embed the correspondence of input to output. The action layer (layer 4) consists of fixed-shape, fuzzy membership functions that fuzzily quantify output values. This layer computes the degree to which an output vector belongs to an output membership function (MF). The output layer (layer 5) defuzzifies the action output.

The layers perform their functions as follows:

- Layer 1: input (crisp values)
- Layer 2: condition (input membership functions)
- Layer 3: association (rules)
- Layer 4: action (output membership functions)
- Layer 5: output (crisp values)

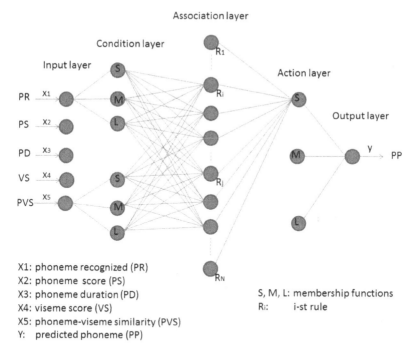

Fig. 4. EFuNN evolving architecture applied to fuse phoneme-viseme classification and predict phoneme occurrence

The learning algorithm consists mainly of certain key actions, such as updating connections, aggregating nodes, pruning nodes, and extracting rules. At layer 3, the rule nodes cluster the input-output data associations. Two connection weights, W1 and W2, are adjusted so that w1 is related to the fuzzified input vector and W2 is related to the corresponding output vector. To adjust W1, supervised learning based on output error is applied. To adjust W2, similarity using the cluster method is applied.

To train and test the fuzzy engine, a sequence of pattern data was recorded from the output of the phoneme-extractor and viseme-extractor units. Data vectors $x(t)$ of the input and the corresponding output were assembled to train and test the EFuNN. The data vector consists of five input measurements: phoneme recognized, phoneme score (PS), phoneme duration (PD), viseme score (VS), and phoneme-viseme similarity (PVS); as well as one output: the predicted phoneme (PP); i.e.:

$$x(n) = [PR, PS, PD, VS, PVS, PP] \tag{5}$$
$$n=t/Tw$$

The vector, indexed by time-window number (Tw), is compiled throughout the utterance. The size of Tw is compatible with the quasi-stationary characterization of speech (from 20 to 40 ms). However, it is also compatible with the duration of a visual frame (1/24 s, that is 41.7 ms). So, the time window was set to 20.85 ms, half the duration of a visual frame.

The dataset was generated from basic uttered phoneme sequences (syllables), with and without added background noise. It consists of 520 patterns x(n), one per frame, each fully describing the association between input and output. The dataset was split randomly to yield two data subsets, one with 80% of the vectors for training, the other with 20% for testing.

The NeuCom [26] environment was chosen for modeling and simulating the EFuNN by applying the following setup:

- Sensitivity threshold: 0.9
- Error threshold: 01
- Number of membership functions: 3
- Learning rate for W1: 0.1
- Learning rate for W1: 0.1
- Pruning: On
- Node age: 60
- Aggregation: On

The sensitivity and error threshold affect the generation of new rule nodes. If the sensitivity among inputs increases, then the network is more likely to create new rule nodes. If the error threshold among the actual outputs and the desired output decreases, then the network is more likely to increase rule nodes. As the threshold increases, the network tends to retain its learning over a long time. If pruning is on, i.e. the network's ability to remove connections between the layers while maintaining its original training performance, the network is less likely to reproduce a rule node that was pruned previously. The learning rate influences the training process. As the learning rate increases, the node will saturate faster, reducing its generalization capabilities. As the age threshold increases, the network's ability to retain what it has learned over the time increases. If aggregation is on, the network tries to aggregate the rules to form global behavior descriptions, avoiding an increase in size that would unwieldy. The number and the shape of the membership functions relate directly to the dynamics of the input and output data and to how they are measuring data in the crisp domain. The higher the number, the higher the interconnections at the input and output layers.

4   Performance Evaluation

The EFuNN successfully learned from its input and proved able to fuse audio and visual information with minimal errors. The performance of the EFuNN was tested in its ability to recover the confusion of similar phonemes. To do this, the word MENU with the right phoneme sequence has been first inputted, than the same word with the /m/ and /n/ phonemes exchanged has been inputted. Each phoneme has been imputed forty times with additive noise. The results (Fig. 5) showed that the EFuNN's self-teaching ability is adequate to learn from data, recovering the confusion of similar phonemes (e. g. /m/ and /n/ in the uttered word "MENU"). The fuzzy engine developed how to fuse and combine phonemes and visemes independently recognized by independent audio and visual units.

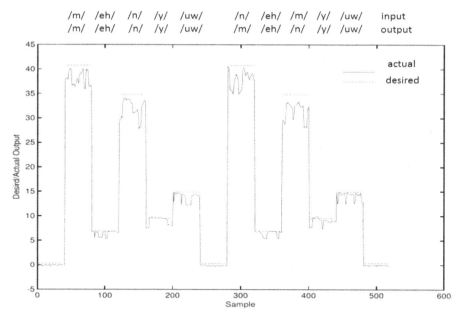

Fig. 5. Desired and actual output values of the trained EFuNN engine at test time

5   Conclusion and Future Development

Preliminary results showed that the evolving fuzzy neural network (EFuNN) paradigm can be successfully applied to develop a fuzzy logic-based inference engine for
fusing and combining phonemes and visemes at the decision level. Several advantages were obtained, mostly concerning performance. These included increased reliability due to reduced system complexity.

Future development will focus on two issues. One is to apply the dynamic evolving
fuzzy neural-network paradigm [25] to the fuzzy logic engine for fusing and combining phonemes with visemes. This would allow evolving capabilities to be embedded
in the system. The second issue is to apply the EFuNN paradigm to the lower layer of
the system, whose task is recognizing phonemes and visemes. This layer is currently
based on hard computing methods.

Artificial intelligence methods need to be applied to the upper layer of the system,
where robust speech-to-text-transcription processing is required. An ANN will first
be applied to convert phonemes to text. Other AI methods then need to be used to
produce text, including prosody information.

References

1. Kölsch, M., Bane, R., Höllerer, T., Turk, M.: Touching the Visualized Invisible: Wearable AR with a Multimodal Interface. IEEE Computer Graphics and Applications, May/June 2006
2. Marshall, J., Tennent, P.: Mobile interaction does not exist. In: CHI 2013 Extended Abstracts on Human Factors in Computing Systems (CHI EA 2013), pp. 2069–2078. ACM, New York (2013)
3. Basu, S., Neti, C., Senior, A., Rajput, N., Subramaniam, A., Verma, A.: Audio-visual large vocabulary continuous speech recognition in the broadcast domain. In: IEEE Workshop on Multimedia Signal Processing, pp. 475–481 (1999)
4. Massaro, D.W.: Bimodal speech perception: a progress report. In: Stork, D.G., Hennecke, M.E. (eds.) Speechreading by Humans and Machines: Models, Systems, and Applications, pp. 79–101. Springer, New York (1996)
5. Benoît, C., Guiard-Marigny, T., Le Goff, B., Adjoudani, A.: Which components of the face do humans and machines best speechread? In: Stork, D.G., Hennecke, M.E. (eds.) Speechreading by Humans and Machines: Models, Systems, and Applications, pp. 315–328. Springer, New York (1996)
6. Salama, E.S., El-khoribi, R.A., Shoman, M.E.: Audio-Visual Speech Recognition for People with Speech Disorders. International Journal of Computer Applications **96**(2), 51–56 (2014)
7. Bernstein, L.E., Auer Jr, E.T.: Word Recognition in Speechreading. In: Stork, D.G., Hennecke, M.E. (eds.) Speechreading by Humans and Machines: Models, Systems, and Applications, pp. 17–26. Springer, New York (1996)
8. Kaucic, R., Dalton, R., Blake, A.: Real-time lip tracking for audio-visual speech recognition applications. In: Buxton, B.F., Cipolla, R. (eds.) ECCV 1996. LNCS, vol. 1065, pp. 376–387. Springer, Heidelberg (1996)
9. Yang, J., Waibel, A.: A real-time face tracker. In: Proc. WACV, pp. 142–147 (1996)
10. Steifelhagen, R., Yang, J., Meier, U.: Real time lip tracking for lipreading. In: Proceedings of Eurospeech (1997)
11. Malcangi, M., de Tintis, R.: Audio based real-time speech animation of embodied conversational agents. In: Camurri, A., Volpe, G. (eds.) GW 2003. LNCS (LNAI), vol. 2915, pp. 350–360. Springer, Heidelberg (2004)
12. McGurk, H., MacDonald, J.: Hearing lips and seeing voices. Nature **264**, 746–748 (1976)
13. Wright, D., Wareham, G.: Mixing sound and vision: The interaction of auditory and visual information for earwitnesses of a crime scene. Legal and Criminological Psychology **10**(1), 103–108 (2005)
14. Noda, K., Yamaguchi, Y., Hiroshi, K., Okuno, G., Ogata, T.: Audio-visual speech recognition using deep learning. Applied Intelligence **42**(4), 722–737 (2015). Springer
15. Malcangi, M., Ouazzane, K., Patel, P.: Audio-visual fuzzy fusion for robust speech recognition. In: The 2013 International Joint Conference on Neural Networks (IJCNN), pp. 582–589 (2013)
16. Patel, P., Ouazzane, K., Whitrow, R.: Automated visual feature extraction for bimodal speech recognition. In: Proceedings of IADAT-micv 2005, pp. 118–122 (2005)
17. Stork, D.G., Wolff, G.J., Levine, E.P.: Neural network lipreading system for improved speech recognition. In: Proceedings International Joint Conf. on Neural Networks, vol. 2, pp. 289–295 (1992)

18. Joo, M.G.: A method of converting conventional fuzzy logic system to 2 layered hierarchical fuzzy system. In: Proceedings of the IEEE International Conference on Fuzzy Systems, pp. 1357–1362 (2003)
19. Kasabov, N.: Evolving fuzzy neural networks – algorithms, applications and biological motivation. In: Yamakawa, T. and Matsumoto, G. (eds.) Methodologies for the conception, design and application of the soft computing, World Computing, pp. 271–274 (1998)
20. Kasabov, N.: EFuNN, IEEE Tr SMC (2001)
21. Patel, P., Ouazzane, K.: Validation and performance evaluation of an automated visual feature extraction algorithm. In: proceedings of IMVIP 2006, pp. 68–73 (2006)
22. Malcangi, M.: Softcomputing Approach to segmentation of Speech in Phonetic Units. International Journal of Computer and Communications 3(3), 41–48 (2009)
23. Cappelletta, L., Harte, H.: Phoneme-to-viseme mapping for visual speech recognition. In: Proceedings of the International Conference on Pattern Recognition Applications and Methods, pp. 322–329 (2012)
24. Kohonen, T.: The 'Neural' Phonetic Typewriter. Computer 21(3), 11–22 (1988)
25. Kasabov, N., Song, Q.: DENFIS: Dynamic Evolving Neural-Fuzzy Inference System and Its Application for Time-Series Prediction. IEEE Trans. on Fuzzy Systems 10, 144–154 (2002)
26. http://www.kedri.aut.ac.nz/areas-of-expertise/data-mining-and-decision-support-systems/neucom

Robotics and Control

Azimuthal Sound Localisation with Electronic Lateral Superior Olive

Anu Aggarwal$^{(\boxtimes)}$

University of Maryland, College Park, MD, USA
aaagganu@gmail.com

Abstract. Since the lateral superior olive (LSO) is the first nucleus in the auditory pathway where binaural inputs converge, it is thought to be involved in azimuthal localization of sounds by calculating the interaural level difference (ILD). The electronic LSO can be used for azimuthal localization in robotics. Thus, in this paper we demonstrate the design, fabrication and test results from a silicon chip which performs azimuthal localization based on the Reed and Blum's model of the population response of the LSO in brain.

Keywords: Azimuthal localization · ILDci · LSO · Reed & Blum's model

1   Introduction

Azimuthal localization [1] of sound means detecting the direction along the azimuth from which the sound is incident. It is of survival value to a species as it helps the animal in spatial navigation, localizing the prey, protecting from predators etc. The LSO is thought to perform this function in biology. Thus, implementing the functionality of the LSO in silicon can provide a mechanism for azimuthal localization in autonomous robotic spatial navigation.

Incoming sounds provide information about both the range and the azimuth. With change in range, the intensity of incoming sound at both ears changes by the same amount so that both ears receive the same intensity. But with change in azimuth, the intensity of sound received at the two ears is different. Thus, the interaural intensity or level difference (ILD[1]), in the former case is zero while in the latter case is greater than zero. Thus, the ILD is a function of the direction along the azimuth from which the sound is received. The ILD has been empirically observed to affect the output of the LSO neurons whose firing as a function of direction is a sigmoid shaped response curve. On this curve, there is an ILD below which the LSO neuron does not fire, called its ILD of complete inhibition, or ILDci. At this ILD, the inputs from left and right ear reaching the LSO neuron are equal. ILD is more important in azimuthal localization than interaural time difference in animals with a small head size and response to frequencies in the ultrasonic range. Moreover, several sounds are received by both ears at close time intervals. So, to estimate the range and azimuth of each sound cor-

---

[1] ILD is the difference between levels (intensity) of sound inputs received at the two ears.

© Springer International Publishing Switzerland 2015
L. Iliadis and C. Jayne (Eds.): EANN 2015, CCIS 517, pp. 231–240, 2015.
DOI: 10.1007/978-3-319-23983-5_22

rectly, the response to a subsequent sound should not be influenced by response to that received before it.

The ears receive sounds of several frequencies which are segregated by the inner ear cochlea. The fibers from each ear relay in the ipsilateral cochlear nucleus (AVCN) and the contralateral medial nucleus of the trapezoid body (MNTB). Outputs from these nuclei relay in the superior olivary complex which contains the lateral superior olive (LSO). The LSO neurons are arranged in columns. Each LSO neuron receives excitatory inputs from sounds of a narrow range of frequencies from the ipsilateral ear through the ipsilateral AVCN and inhibitory inputs from sounds of same frequencies from the contralateral ear through the contralateral AVCN and the ipsilateral MNTB. The azimuthal sound localization is based on the population response of the LSO according to the Reed and Blum's model [2].

Thus, in this work we demonstrate the population response of the LSO (the basis for azimuthal localization) in contrast to works done before this which demonstrate the sigmoid shaped response curve of single LSO neurons. Further, most of the earlier work [3, 4, 5, 6] was done using the difference computation [7, 8, 9] rather than the conductance model [10] of neuron response, which is biologically more plausible and is also used in the current work. In all the prior implementations, the sounds of different intensities received from the environment were coded as spike trains of different length or frequencies which were used to excite a single LSO neuron. However, to implement the LSO population response in silicon using similar principles, several synapse circuits will be required which shall make the system very big and unwieldy. Thus, in this paper we routed the inputs from the two ears to the LSO neuron (based on the conductance model [10]) through two second order synapse circuits [11] each of which produces conductance proportional to the number or frequency of input spikes. These input spikes in turn represented the intensity of inputs received at the two ears. This reduced the number of synapses needed to provide input to each LSO neuron from several to only two and hence, enabled the system to be fit onto a small area of a tiny chip unit (1.275X1.275 mm).

2 Neuro-Computational Models

The system design is inspired by the mathematical models proposed for the system (the Reed and Blum's model) and that for individual neuron (conductance model) and synapses (alpha function or more generally second order function). The latter are more general models for the synapse and neuron while the former is specific to the LSO.

2.1 The Reed and Blum's Model of the LSO

According to this model [2], the LSO neurons are arranged in iso-frequency columns which receive excitatory inputs from the ipsilateral and inhibitory inputs from the contralateral ear. These inputs to the LSO are arranged such that the strength of excitatory inputs increases and that of inhibitory inputs decreases from top to bottom of the column. Wherever the two inputs are equal, ILDci is achieved and hence, the LSO neuron does not exhibit any response. As mentioned earlier, the ILD depends on the azimuthal direction from which the sound is incident (primarily). And according to this model, the ILDci depends on the neuron number along the LSO column. Thus, the ILD=ILDci for a unique combination of azimuthal direction and neuron number. Thus, the neuron number which stops firing first encodes for the direction along the azimuth from which the sound is incident. Also, the intensity of sound, which depends on the range from which it is incident, has been seen empirically to alter the latency of neuronal response.

2.2 Conductance Model of the Neuron

According to the model [10], the neuron responds based on weighted average of inhibitory and excitatory conductances incident on it (from the two ears in case of the LSO), its membrane capacitance merely slows down the response as a result of which response to the current input does not depend on that to the past inputs (as required-incoming sounds should not interfere), and the time constant in this model is inversely proportional to the intensity of sound so that the sounds with higher intensity shall have shorter latency as compared to those with lower intensity but same ILD (as required for range localization).

3   System Design and Circuits

An LSO chip was designed, fabricated and tested [13], [15] to realize the function of the LSO in silicon. On this chip, an array of 32 synapse- neuron circuits (Fig. 1) was designed to mimic the arrangement of synapse-neurons in the LSO. Each neuron in the array receives input from an inhibitory synapse representing input from the contralateral ear and an excitatory synapse representing input from the ipsilateral ear. The strength of the excitatory inputs was increased and that of the inhibitory inputs was decreased along the length of the column from top to bottom by using two resistor ladders to vary the parameter V_{syn} on the synapse circuit. Each synapse circuit was that of the second order synapse and each neuron circuit was that of the conductance neuron (Fig. 2). The array of synapse-neuron circuits, was used for azimuthal localization based on the Reed and Blum's model of the LSO.

The synapse circuit has already been described in [11]. The circuit for the neuron gets two inputs- excitatory and inhibitory (V_{exc} and V_{inh}) from the two synapses, which are integrated on C_{mem} to V_x. The V_e node of the excitatory synapse was connected to the gates of M1, M5 and that of the inhibitory synapse to the gate of M6. The V_e from the excitatory synapse pulls the V_x down to V_{we} and V_e from the inhibitory synapse pulls the V_x up to Vdd. The resulting V_x is the weighted average of the inhibitory and the excitatory inputs and is compared to V_{thr} and if it is comparable to V_{thr}, then the neuron fires. *Rout* is the spike generated which is the request sent to the address event representation (AER) [12] and acknowledge (*ack*) is the signal coming back from the AER which resets the neuron and prepares it to fire again. The time for resetting is adjusted by V_{refr}. The excitatory conductance was controlled by the input current I'_e. Note that the excitatory inputs were provided at two different points in the circuit (defined by V_{exc} on transistors M1 and M5) one of which produces I_e and the other I'_e. The inhibitory conductance is controlled by the input current I_i (defined by V_{inh} on M6).

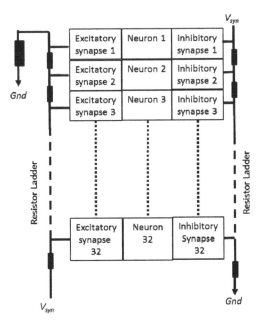

Fig. 1. Schematic of the arrangement of the synapse neuron circuits in the array representing the column of LSO neurons on the LSO chip.

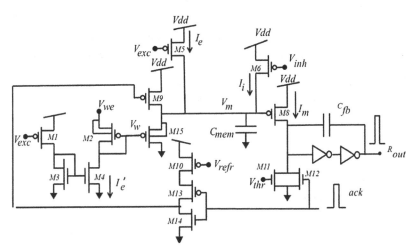

Fig. 2. Circuit schematic of the conductance neuron used in this work.

Starting from the drain equations (subthreshold mode) for M15, M2, and M8,

$$I_e + I_i = I_0 e^{\frac{\kappa(V_m - V_w)}{V_T}} \tag{1}$$

$$I_e' = I_0 e^{\frac{\kappa(V_{we} - V_w)}{V_T}} \tag{2}$$

$$I_m = I_0 e^{\frac{\kappa(Vdd - V_m)}{V_T}} \tag{3}$$

The value of the "membrane potential" current (I_m), when the two inputs, V_{exc} and V_{inh} are constant, can be described by the following equation,

$$I_m = \left(\frac{I_e'}{I_e + I_i}\right).e^{\kappa\left(\frac{Vdd - V_{we}}{V_T}\right)} \tag{4}$$

In the case where I'_e is much larger than $I_e + I_i$, Im becomes large (near the excitatory synaptic reversal potential defined by V_{we}) and when I_i is much larger than I_e, the membrane current will be near zero. The I_m in steady-state becomes a function of the weighted average of the two conductances which is what is required for implementing the ILD response of the LSO neuron.

4   Tests and Results

The LSO chip (Fig. 3) was fabricated using 0.5 µm ON[2] semiconductor technology. With no critical sizing issues as part of our design, all the transistors in the circuits were designed to have W/L=2.4µm/2.4µm and values of all the capacitors were 0.1pF. Test PCB was designed to test the chip. The test results were obtained using oscilloscope or digital logic analyzer and then plotted using MATLAB [14]. Spike inputs (V_{spk}) to the synapse circuit were always digital pulses, 4V in amplitude and 0.5 µs in duration.   In all cases, synapse circuit parameters were V_{t1}=3.50V and V_{t2}=4.50V.

4.1   Neuron and Synapse Circuits

Firstly, *Rout* was recorded at different values of threshold voltage, V_{thr}=0.55, 0.60, 0.64V using an oscilloscope. The results in Fig. 3 (left column) indicate that the neuron is working properly and that the spike frequency is inversely proportional to the threshold voltage. In this experiment, excitatory and inhibitory inputs were applied directly to the neuron circuit. Another feature of the LSO neuron is that the response to an input does not depend upon that to prior inputs so that the response to consecutives sounds is independent. 3 consecutive bursts were repeated at: 10, 3.33, and 2ms.

---

[2] ON is a semiconductor technology which was used for fabrication of the chip.

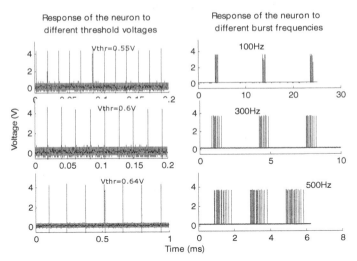

Fig. 3. (left) Spiking outputs from the neuron circuit during initial testing. (V_{refr} =4.1V, V_{we}=4.1V, V_{inh}=4.3V, V_{exc}=4.1V.) (right) The neuron response to subsequent bursts is not affected by that to priors. (V_{thr}=0.65V, V_{refr}=4.1V, V_{we}=4.1V, V_{inh}=4.2V, V_{syn}=2.57V, V_f=2.88V, V_s=3.76V.)

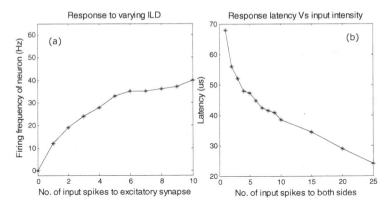

Fig. 4. (a) Sigmoid shaped response curve of the LSO neuron. (V_{thr}= 0.6, V_{refr}=4.1V, V_{we}=4.1V, V_{syn}=2.57V, V_f=2.88V, V_s=3.76V.) (b) As the input strength increases, the response latency decreases. (V_{thr}=0.65V, V_{refr}=4.1V, V_{we}=4.1V, V_{syn}=2.57 V, V_f=2.88V, V_s=3.76V. NB: burst inputs were provided.)

As shown in Fig. 3 (right), the response of the neuron to consecutive pulses did not change due to the priors. The next experiment (Fig. 4 (a)) shows the response of the neuron to varying ILDs. For the measurements, different number of input spikes were applied to the excitatory synapse while none was provided to the inhibitory synapse. The results demonstrate that the ILD curve has shape similar to the response curve of an LSO neuron as seen empirically. In the next experiment, we tried to ascertain if this neuron's response latency decreases with increasing intensity of inputs (for range localization) as observed for the LSO neuron in biology. To test this, similar inputs were applied to both synapses and their intensity, depicted by the number of inputs was changed. The results in Fig. 4(b) show that as the intensity of incoming sound increases, the response latency of the neuron decreases, indicating that these circuits can correctly mimic this function of the biological LSO.

4.2   Population Response

To test the population response of the LSO and azimuthal localization based on the Reed and Blum's model, initially V_{thr} was changed from 2V to 2.8V in steps of 5mV each and corresponding neuron firing was plotted as rasters (Fig. 5). As per the chip design, the strength of inputs changes linearly along the length of the array of the LSO neurons, so, the response of the LSO neurons also should vary linearly in that order (thermometer response) as seen in the figure. However, some of the neurons had similar response to that of their neighbors because of small ladder step size. Thus, instead of 32, only 22 different ILDci were seen. Thus, the resolution of azimuthal localization is ~8 degrees (180/22).

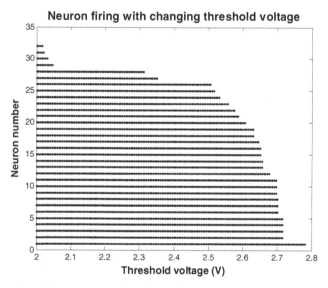

Fig. 5. Rasters from 32 neurons on the chip. (V_{refr}=4.3V, V_{we}=4.15V, V_{syn}=2V, V_f=2.85V, V_s=3.16V.)

4.3   Azimuthal Localization

In this test, V_{spk} to the excitatory synapse was changed from 100 Hz to 2200 Hz while that to inhibitory synapse was kept at 100 Hz. All the pulses were 0.5 μs, 4V magnitude. Since the ILD received depends on the azimuthal direction from which the sound is received, we assigned each ILD used in testing the population response to an azimuthal direction. According to the Reed and Blum's model, the ILDci of each neuron along the LSO column is different and thus, the neuron number that stops firing for an ILD or an azimuthal direction is unique to that direction. Tabulating this mapping from azimuthal direction to ILD to ILDci and the neuron number which stops firing, provides a look up table (Table 1) for the azimuthal localization using the population response of the LSO. As shown in the table, if the neuron number 12 along the column stops firing, then it can be inferred that the sound is coming from 33 degrees along the azimuth as seen from the animal's center. Other parameters used for the test were: V_{thr}=2.57V, V_{refr}= 4.3V, V_{bias}=4.15V, V_{syn}=2V, V_s=3.15V, V_f=2.85V, $V_{\tau l}$ =3.5V, $V_{\tau 2}$ =4.5V.

Table 1. Look up table for the neuron number encoding azimuthal direction on either LSO.

Neuron number experiencing ILDci	ILD	Direction (degrees)
1	100/100	0
2, 3, 4, 5	200/100	8
6, 7, 8	300/100	16
9, 10, 11	400/100	25
12	500/100	33
13-14	600/100	41
15-16	700/100	49
17	800/100	57
18-19	900/100	66
20	1000/100	74
21	1100/100	82
22	1200/100	90
23	1300/100	98
24	1400/100	107
25	1500/100	115
26	1600/100	123
27	1700/100	131
28	1800/100	139
29	1900/100	148
30	2000/100	156
31	2100/100	164
32	2200/100	172

5   Conclusion

This work describes the analog VLSI design, fabrication, measurements and data analysis from a chip implementing the population response of the LSO and azimuthal localization based on that. All the above was implemented using second-order synapse circuits; and the conductance neurons. Ability to linearly add inputs and to change the strengths of inputs by changing the synapse circuit parameters enabled implementation of the Reed and Blum's model of population response of the LSO on a part of a single tiny chip unit. Processing the ILDci data from all the 32 neurons showed that it could help in azimuthal localization with a resolution of ~8 degrees. This chip has potential for use as the back end system in robotic spatial navigation to perform azimuthal localization. Actual demonstration of its use in a robotic buggy could not be done because of lack of access to a front end system which could convert input sound into spike trains proportional to the ILD, which could form inputs to the chip documented in this work.

Acknowledgement. The work was performed under the supervision of Prof R. W. Newcomb, University of Maryland College Park, MD, USA. An earlier version of the neuron circuit was developed jointly with Prof T.K. Horiuchi.

References

1. Purves, D., Augustine, G.J., Fitzpatrick, D., Hall, W.C., LaMantia, A.S., White, L.E.: Neuroscience. Sinauer Associates, Sunderland (2012)
2. Reed, M.C.: Blum., J. J: A model for the computation and encoding of azimuthal information by the lateral superior olive. J. Acoust. Soc. Am **88**(3), 1442–1453 (1990)
3. Horiuchi, T., Hynna, K.: Spike based VLSI modelling of the ILD system in echolocating bat. Neural Networks **14**, 755–762 (2001)
4. Shi, R.Z., Horiuchi, T.K.: A neuromorphic VLSI model of bat interaural level difference processing for azimuthal echolocation. IEEE Transactions on Circuits and Systems I **54**(1), 74–88 (2007)
5. Horiuchi, T.K., Cheely, M.: A systems view of a neuromorphic VLSI echolocation system. In: IEEE International Symposium on Circuits and Systems, pp. 605–608 (2007)
6. Abdalla, H.A.N.: Estimation of elevation and Azimuth in a neuromorphic VLSI bat echolocation system. (unpublished) Ph.D. Thesis (2009)
7. Park, T.J., Grothe, B., Pollak, G.D., Koch, U.: Neural delays shape selectivity to inter aural intensity differences in the Lateral Superior Olive. J. Neurosci. **16**(20), 6554–6556 (1996)
8. Schnupp, J.W.H., Carr, C.: On hearing with more than one ear: lessons from evolution. Nature Neuroscience **12**(6), 692–697 (2009)
9. Zacksenhouse, M., Johnson, D.H., Williams, J., Tsuchitani, C.: Single neuron modelling of LSO unit responses. J. Neurophysiol. **79**, 3098–3110 (1998)
10. Horiuchi, T.K., Bansal, C., Massoud, T.M.: Binaural intensity comparison in the echolocating bat using synaptic conductance. In: IEEE International Symposium on Circuits and Systems, TWN Taipei, pp. 2153–2156 (2009)
11. Aggarwal, A., Horiuchi, T.K.: Neuromorphic VLSI second order synapse. The Electronics Letters **51**(4), 319–321 (2015)

12. Boahen, K.A.: Communicating Neuronal Ensembles between Neuromorphic Chips. Neuromorphic Systems Engineering. The Springer International Series in Engineering and Computer. Science **447**, 229–259 (1998)
13. Cadence PSPICE simulator, SPB_16.3. https://www.ema-eda.com
14. MATLAB R2011b. http://www.mathworks.com
15. NEURON simulator http://www.neuron.yale.edu/neuron

Smart Cameras

Tampering Detection in Low-Power Smart Cameras

Adriano Gaibotti[1,2], Claudio Marchisio[1], Alexandro Sentinelli[1],
and Giacomo Boracchi[2(✉)]

[1] STMicroelectronics, Advanced System Technology, Via Camillo Olivetti 2,
20864 Agrate Brianza, MB, Italy
{adriano.gaibotti,claudio.marchisio,alexandro.sentinelli}@st.com
[2] Dipartimento di Elettronica, Informazione e Bioingegneria (DEIB),
Politecnico di Milano, Via Ponzio 34/5, 20133 Milano, MI, Italy
giacomo.boracchi@polimi.it

Abstract. A desirable feature in smart cameras is the ability to autonomously detect any tampering event/attack that would prevent a clear view over the monitored scene. No matter whether tampering is due to atmospheric phenomena (e.g., few rain drops over the camera lens) or to malicious attacks (e.g., occlusions or device displacements), these have to be promptly detected to possibly activate countermeasures. Tampering detection is particularly challenging in battery-powered cameras, where it is not possible to acquire images at full-speed frame-rates, nor use sophisticated image-analysis algorithms.

We here introduce a tampering-detection algorithm specifically designed for low-power smart cameras. The algorithm leverages very simple indicators that are then monitored by an outlier-detection scheme: any frame yielding an outlier is detected as tampered. Core of the algorithm is the partitioning of the scene into adaptively defined regions, that are preliminarily defined by segmenting the image during the algorithm-configuration phase, and which shows to improve the detection of camera displacements. Experiments show that the proposed algorithm can successfully operate on sequences acquired at very low-frame rate, such as one frame every minute, with a very small computational complexity.

Keywords: Tampering detection · Smart cameras · Displacement detection · Blurring detection · Low-power cameras · Low-frame rate

1 Introduction

Cameras operating outdoor and in harsh environments are exposed to atmospheric phenomena and intentional attacks that might prevent the correct image acquisition. Rain, snow, dust lying on the camera lens cause blurry (Figure 1(a)) or partially occluded (Figure 1(b)) pictures, while wind might displace the camera (Figure 1(c) - 1(d)). Similarly, an attacker can intentionally change the camera focus, spray some opaque or glossy liquid, displace or occlude the camera. We refer to these events/attacks as tampering. In many situations, tampering is

© Springer International Publishing Switzerland 2015
L. Iliadis and C. Jayne (Eds.): EANN 2015, CCIS 517, pp. 243–252, 2015.
DOI: 10.1007/978-3-319-23983-5_23

(a)   (b)   (c)   (d)

Fig. 1. Examples of tampering events due to atmospheric phenomena. (a) rain drops resulting in a blurry picture, (b) snow partially occluding the camera view, (c)-(d) wind displacing the camera.

not straightforward to detect (e.g. when the camera is not physically damaged and does not go out-of-order), and image analysis techniques are the only viable option.

Tampering detection is an essential feature in surveillance systems [1], which are expected to autonomously detect any tampering, and promptly report alerts. Tampering, in fact, might result in images that are useless for monitoring purposes: even a mild blur might hinder the identification of important details such as licence plates. Surveillance cameras most often operate at normal frame-rates (e.g. around few frames per second) and are connected to the power supply. Tampering detection for surveillance cameras have been quite investigated in the literature; in particular, camera displacements and occlusions are typically detected by comparing the current frame against an estimate of the scene background, while blurring is detected by monitoring the high-frequency components of images, as these are expected to drop. In [2] and [3] tampering is detected by comparing the histograms of background and current frame, and by monitoring the energy in wavelet or Fourier domain. Two background models, estimated over different time intervals, are used in [3]. In [4], tampering detection is performed by combining background subtraction together with edge detection and normalized cross-correlation. Background estimates and block matching in [5] enable the displacement detection, while the number of SURF [6] keypoints is used to detect blurring. A buffer of recent frames rather than an explicit background can be used as in [7]. All the above solutions, including [8,9], are computationally demanding and are not viable options for low-power cameras.

In this work, we expressly target low-power and ultra-low-power smart cameras, like *SecSoC* (Security System on Chip), an innovative prototype based on a cluster of ReISC (Reduced Energy Instruction Set Computer) cores, designed and produced by STMicroelectronics. SecSoC is a battery-powered device, characterized by a constrained computational power (clock rates 82.5 MHz at 1.2V, and sub-1MHz at 0.6V) and reduced memory (1.25 MB). While low-power cameras are not meant for critical surveillance applications, they might be easily employed for monitoring wide environments thanks to their low cost and maintenance requirements.

Tampering detection in low-power cameras (eventually organized in wireless sensor network) is more challenging than in conventional surveillance systems, and this problem has not been much investigated so far [10,11]. Beside computational aspects – such as the number of operations per pixels allowed – the big issue is that low-power smart cameras typically operate at very low-frame rates (e.g., less than one frame per minute), thus the acquired sequence does not evolve smoothly. These aspects prevent the use of learned background models and the analysis of foreground variations. In fact, when dynamic environments are monitored at low frame-rates, changes in the scene and in the light conditions might produce consecutive frames that are very different (see Figure 2). Smart cameras have to promptly distinguish between *normal* changes (due to illumination or movements in the scene), and changes due to camera tampering, to raise an alert and eventually avoid the transmission of tampered frames.

We address the detection of camera blurring and displacement (Section 2), and propose an algorithm (Section 3) that relies on indicators that can be easily computed in low-power smart cameras. We monitor the average image intensity or frame difference (to detect camera displacement), and the average gradient norm (to detect blurring), and detect tampered frames by analyzing outliers in these indicators. In particular, we show that separately monitoring these indicators over different regions of the image can substantially improve the displacement-detection performance. Image regions are defined during an initial configuration phase (Section 3.1), thus the algorithm operates at a negligible computational overhead with respect to monitoring the whole image (Section 3.2). Our algorithm thus represents a prompt trigger, to be possibly combined with other sequential monitoring techniques. Experiments (Section 4) show that leveraging image regions can substantially improve the displacement-detection performance.

2   Problem Formulation

Let z_t be frame acquired at time t

$$z_t(x) = \mathcal{D}_t[y_t](x), \quad \forall x \in X \tag{1}$$

where $\mathcal{D}_t$ denotes an operator transforming the original image y_t, and $x \in \mathbb{Z}^2$ indicates the pixel coordinates belonging to the regular pixel grid $X \subset \mathbb{Z}^2$. As far as there are no tampering attacks/events,

$$\mathcal{D}_t[y_t](x) = y_t(x) + \eta_t(x), \quad \forall x \in X \tag{2}$$

where η_t is a random variable accounting for image noise and y_t are acquired from the same viewpoint and camera orientation, even though typically $y_t \neq y_{t-1}$ because the depicted scene changes.

When, at time τ^*, an external disturbance introduces *blurring*, the image y_t is degraded by an unknown blur operator, and z_t becomes

$$\mathcal{D}_t[y_t](x) = \int_X y(s)h_t(x,s)ds + \eta_t(x) \quad \forall x \in X, t \geq \tau^* \tag{3}$$

(a)

(b)

Fig. 2. Examples of synthetically generated blurring (3-rd frame in (a)) and camera displacement (3-rd frame in (b)).

where $h_t(x, \cdot) > 0$ is the point-spread function at pixel $x \in X$.
A *camera displacement* at frame τ^* is instead modeled as

$$z_t(x) = \begin{cases} y_t(x) + \eta(x) \text{ per } t < T^* \\ w_t(x) + \eta(x) \text{ per } t \geqslant T^* \end{cases}, \quad (4)$$

where w_t and y_t refer to different viewpoints and/or camera orientations.

The proposed tampering-detection algorithm analyzes a sequence of frames $\{z_t, t = 1, \dots\}$ to detect the time instant τ^* when tampering like (3) or (4) occurs. We assume that T_0 tampering-free frames are provided for training. For simplicity, we consider grayscale frames: extensions to color images are straightforward.

3   Tampering Detection

Algorithm 1 presents the proposed tampering detection, which relies on simple indicators such as the average intensity (the *luma*, denoted by l), the average frame difference (*FD*, denoted by d) and the average norm of the gradient (denoted by g). The first two are meant to detect camera displacements, which would substantially change the image content, while the latter the blurring, which would attenuate the high frequencies of the image. As anticipated in Section 1, during the initial configuration, the scene is segmented in K disjoint regions $\{R_k, k = 1, \dots, K\}$, namely $R_k \subset \mathcal{X}, R_i \cap R_j = \emptyset, \forall i \neq j$. The employed segmentation is detailed in Section 3.1.

For each z_t, we compute the luma and frame difference separately on each of the K regions,

$$l_k(t) = \frac{1}{\#R_k} \sum_{x \in R_k} z_t(x) \quad (5)$$

$$d_k(t) = \frac{1}{\#R_k} \sum_{x \in R_k} (z_t(x) - z_{t-1}(x))^2, \quad k = 1, \dots, K, \quad (6)$$

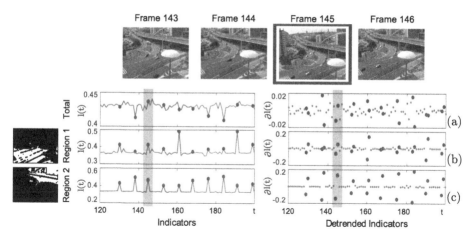

Fig. 3. Example of camera displacements. Left plots report the luma values, while right plots depict the detrended luma: we show both indicators computed over the full image (a) and on two reported regions (b and c). Red dots indicate a single displaced frame. Values in the highlighted areas refer to the frames depicted on top of the figure. The highlighted displacement yields a small peak in $l(t)$ and two outliers in the sequence of detrended indicators $\partial l(t)$, which can be clearly detected.

where $\#(\cdot)$ denotes the cardinality of a set. We can simultaneously monitor both luma and frame difference, even though computing d_k requires to store the previous frame and this also depends on memory availability of the device. In what follows, including Algorithm 1, we consider only the luma l, but the same procedures apply to the frame difference d as well. The average gradient norm is instead computed over the whole image

$$g(t) = \sum_{x \in X} \left(\sqrt{(z_t \circledast f_h)^2 (x) + (z_t \circledast f_v)^2 (x)} \right), \tag{7}$$

where f_h and f_v are the horizontal and vertical derivative filters, respectively, and $\circledast$ denotes the 2d convolution.

Figure 3 shows how a camera displacement affects the luma. First of all, we observe that a displaced frame changes $l(t)$ (introducing a peak in Figure 3(a)) and that this change is also visible in the indicators $l_k(t)$ for the regions (Figure 3(b) and Figure 3(c)). Second, we observe that outlier-detection methods [12] based on density estimates or on confidence intervals –that are very efficient to run– cannot be straightforwardly applied here. In fact, these methods are meant for independent and identically distributed (i.i.d.) random variables, while here indicators follow an unpredictable trend because of changes in the scene or in the illumination. Therefore, we perform a detrending [13] of the indicator sequence by a temporal derivative

$$\partial l_k(t) = l_k(t) - l_k(t-1), \quad k = 1, \ldots, K, \tag{8}$$

and we similarly define $\partial d_k(t)$ and $\partial g(t)$.

Algorithm 1. The Proposed Tampering-Detection Algorithm

Input: $\gamma_l, \gamma_g, \Gamma_l$, training frames $\{z_t, t = 1, \ldots, T_0\}$, regions $\{R_k, \ k = 1, \ldots, K\}$

Training phase:

1. Compute $\partial l_k(t)$ and $\partial g(t), t = 1, \cdots, T_o$
2. Compute $\overline{\partial l}_k, \overline{\partial g}, \sigma_{l_k}$ and σ_g.

Operational phase:

3. **for** $t = T_o + 1, \ldots, \infty$ **do**
4.   Get frame z_t, set $n_l = 0$;
5.   Compute $\partial g(t)$
6.   **if** $\partial g(t) < -\gamma_g \sigma_g \vee \partial l(t) > \gamma_g \sigma_g$ **then**
7. | raise a blurring alert in z_t
     end
8.   **for** $k = 1, \ldots, K$ **do**
9.   Compute $\partial l_k(t)$
10.   **if** $\partial l_k(t) < -\gamma_l \sigma_{l_k} \vee \partial l_k(t) > \gamma_l \sigma_{l_k}$ **then**
11. | $n_l = n_l + 1$
       end
     end
12.   **if** $n_l \geq \Gamma_l$ **then**
13. | raise a camera displacement alert in z_t
     end
end

The sequences of detrended indicators (reported in Figure 3) can be suitably monitored by the following confidence intervals:

$$[\overline{\partial l}_k(t) - \gamma_l \sigma_{l_k}, \overline{\partial l}_k(t) + \gamma_l \sigma_{l_k}], \quad k = 1, \ldots, K, \tag{9}$$

where $\overline{\partial l}_k(t)$ denotes the mean and σ_{l_k} the standard deviation of ∂l over R_k (Algorithm 1, line 1), computed from tampering-free frames provided for training (i.e. $z_t, t = 1, \ldots, T_o$) and $\gamma_l > 0$ is a tuning parameter. Similar intervals are built for $\partial d_k(t)$ and ∂g (line 2).

During operations, indicators are computed (lines 5 and 9), and any indicator falling outside its confidence region is considered an outlier. In particular, any outlier in ∂g yields a blurring alert (line 6), while camera-displacement alerts are raised when at least Γ_l indicators ∂l_k simultaneously yield outliers (line 12). The threshold Γ_l together with γ_l determine the displacement-detection promptness. The extreme configurations correspond to $\Gamma_l = 1$, where it is sufficient that a single region fires an outlier to raise a camera-displacement alert, and $\Gamma_l = K - 1$ where all but one region[1] have to simultaneously fire an outlier to raise an alert.

It is important to remark that tampering yields outliers in the transient of the detrended indicators (8): namely, a single tampered frame yields two outliers (as shown in Figure 3) and only the first and the last of a sequence of consecutive

---

[1] It is better to exclude a region since, for instance, the sky-region typically does not change when the camera is displaced, either horizontally or vertically.

tampered frames yield outliers in the detrended indicators. This is the reason why detrended indicators are monitored in a *one-shot* manner, targeting the detection of the first tampered-frame. On the one hand, this one-shot monitoring can provide prompt detections, which is particularly important at the low frame-rates we consider, on the other hand, the persistence of tampering is disregarded. To take this valuable information into account, some form of sequential monitoring should be applied to the indicator sequence as in [11], possibly combined with Algorithm 1.

3.1 Scene Segmentation

Scene has to be preliminarily segmented to define regions that Algorithm 1 takes as input. To this purpose, we use part of the training frames before T_0 to compute the feature vector $\mathbf{f}(x) \in \mathbb{R}^5$ for each pixel $x \in \mathcal{X}$

$$\mathbf{f}(x) = \left[r(x); c(x); \bar{l}(x); \sigma_l(x); \bar{g}(x); \sigma_g(x) \right] , \ \forall x \in X . \tag{10}$$

In (10), $r(x)$ and $c(x)$ denotes the row and column of x, respectively, $\bar{l}(x)$ and $\sigma_l(x)$ the mean and standard deviation of the intensity at x, computed over time, and $\bar{g}(x)$ and $\sigma_g(x)$ the mean of gradient norm and its standard deviation, computed over time. These feature vectors are meant to cluster pixels in regions having, over training frames, similar spatial appearance and temporal behavior.

As in superpixel methods [14], segmentation is performed by k-means clustering. Feature vectors (10) over the whole image are clustered by a weighted k-means [15] that scales each component of the Euclidean distance between a feature vector and a cluster centroid of a weight that is inversely proportional to the standard deviation over the cluster. This scaling compensates the fact that the components of the feature vector might span very different ranges. The number of clusters is defined by testing several values and then choosing the best solution according to the Calinski-Harabasz criterion [16].

Finally, morphological image-processing operations are executed to remove boundaries between different regions, and eventually regions that are too small. This defines the regions $\{R_k , \ k = 1, \ldots, K\}$, and two examples are reported in Figure 3.

3.2 Computational Complexity

The most computationally demanding operations of Algorithm 1 consist in computing the indicators, in particular g. Computing g requires, when using the Sobel filters in (7), 34 operations per pixels[2], while computing l and d requires 1 and 3 operations per pixel, respectively. Detecting outliers in the indicators requires two comparisons per frame: thus, monitoring regions instead of the whole image has a negligible impact on the overall computational complexity.

---

[2] Execution can be accelerated whether hardware implementations of the FFT transform are available.

Algorithm 1 has very low memory requirements: l and g indicators can be computed online, while d requires to store a single frame in memory. Segmentation is executed only during the initial configuration and can be performed on an external device connected to the smart camera. As such, Algorithm 1 can be properly executed on low-power and ultra low-power cameras.

4   Experiments

Experiments are meant to assess the advantages of separately monitoring regions in the considered tampering-detection framework. To this purpose, we show that Algorithm 1 detects camera displacements better than monitoring indicators on the whole image (*Whole Image* in Figure 4). We also show that it is important to adapt regions to the image content, since operating on K regions obtained by clustering $[r(x), c(x)]$ leads to performance loss (*Voronoi* in Figure 4). For both Voronoi and regions defined as in Section 3.1, we consider the two extreme configurations where alerts are raised at the first region firing an outlier or when $K - 1$ regions simultaneously fire outliers.

We recorded 8 sequences from webcams monitoring different urban areas, yielding overall 12200 frames. From each frame, we have cropped the central area, removing the 50 top-most and bottom-most rows, and the 50 left-most and right-most columns[3]. We have synthetically introduced 10% of tampered frames: displacements have been simulated by moving the central cropping area of a random shift having magnitude between 20 and 50 rows and columns (as in Figure 2(b)), while blurring (as in Figure 2(a)) was simulated by convolution against a Gaussian kernel, with standard deviation randomly defined in the range $[1, 5]$.

A typical figure of merit for assessing detection performance is the Receiver Operating Characteristic (ROC) curve, where each point corresponds to a pair $(\text{FPR}_\gamma, \text{TPR}_\gamma)$ defined as

$$\text{FPR}_\gamma = \frac{\text{FP}_\gamma}{\text{TN}_\gamma + \text{FP}_\gamma}, \quad \text{and} \quad \text{TPR}_\gamma = \frac{\text{TP}_\gamma}{\text{TP}_\gamma + \text{FN}_\gamma}.$$

Here, TP_γ represents the number of true positives (tampered frames correctly detected), FP_γ of false positives (tampering-free frames detected), TN_γ of true negatives (tampering-free frames not detected) and FN_γ of false negatives (missed tampered frames), for a given value of γ. ROC curves have been computed by varying γ_l, γ_d and γ_g in a range of values between 0.1 and 50.

Discussion. ROC curves in Figure 4(a)-(b) and the Area Under the Curve (AUC) values reported in the caption confirm that camera displacements can be better detected by monitoring FD than luma, and that it is convenient to define regions by segmenting the scene. As an example, if we set $\gamma_d = 15.6$, Algorithm 1 operates at FPR = 1% and detects nearly all tampering since TPR = 99.92%. In contrast, to operate at FPR = 1% when monitoring the whole image or Voronoi regions we obtain TPR = 91.67% ($\gamma_d = 6.5$) and TPR = 89.05% ($\gamma_d = 18.64$),

---

[3] Sequences can be provided upon request.

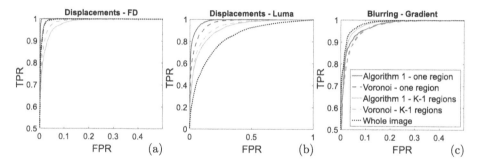

Fig. 4. (a) ROC curves relative to displacement detection based on FD: the AUC obtained by "Algorithm 1 - One region" is 99.89%, while the AUC obtained by on the whole image is 99.65%. (b) ROC curves relative to displacement detection based on luma: the AUC obtained by "Algorithm 1 - One region" is 98.44%, while the AUC obtained on the whole image is 84.07%. (c) ROC curves relative to blurring detection based on gradient: as opposed to (a) and (b), "Algorithm 1 - One region" (AUC 98.4%) is outperformed by monitoring the whole image, AUC = 99.13%. To highlight differences, the ROC curves in (a) and (c) are drawn over a smaller FPR, TPR range.eps

respectively. When monitoring l, Algorithm 1 achieves TPR = 73.98% at FPR = 1% ($\gamma_d = 5.8$), while monitoring the whole image is quite ineffective (TPR = 17.85%, $\gamma_d = 3.7$), and similarly perform Voronoi's regions (TPR = 53.02%, $\gamma_d = 6$). We remark that, in the considered low-power scenario, it is important to operate at low FPR to prevent useless data transmission (that would reduce the battery time-life) and the activation of eventual countermeasures. As far as the values of Γ_d and Γ_l are concerned, the most effective configuration consists in raising an alert as soon as a single region fires a very clear outlier, which implies setting quite wide intervals (9). This option is also preferable because it enables the detection of partial occlusions of the scene. We observe that displacements typically change all the Voronoi regions, and this is why Voronoi outperforms Algorithm 1 in the configuration $K - 1$ in both luma and FD.

Figure 4(c) confirms that blurring is more effectively detected by monitoring the whole image at once. This is probably due to the fact that, typically, regions do not include the most prominent edge of the scene, which are indeed the most informative parts to detect blurring.

5 Conclusions

We presented a tampering-detection algorithm that leverages an image segmentation to improve the detection of camera displacements, and that at the same time can detect blurring. The algorithm has low-computational complexity and has to be considered a prompt trigger for detecting tampering in low-power and ultra low-power cameras operating at low frame rates.

Ongoing work concerns approaching other types of tampering, such as degradations of the imaging sensor, and the integration of sequential monitoring schemes [11] to detect subtle tampering that persists over time. Moreover, we will investigate the use of superpixels methods [14] to segment the scene, by including suitable temporal information to the feature vectors.

References

1. Hampapur, A., Brown, L., Connell, J., Ekin, A., Haas, N., Lu, M., Merkl, H., Pankanti, S.: Smart video surveillance: exploring the concept of multiscale spatiotemporal tracking. IEEE Signal Processing Magazine **22**(2), 38–51 (2005)
2. Aksay, A., Temizel, A., Cetin, A.E.: Camera tamper detection using wavelet analysis for video surveillance. In: IEEE Int. Conf. on Advanced Video and Signal Based Surveillance, AVVS 2007, pp. 558–562. IEEE (2007)
3. Saglam, A., Temizel, A.: Real-time adaptive camera tamper detection for video surveillance. In: Sixth IEEE International Conference on Advanced Video and Signal Based Surveillance, AVSS 2009, pp. 430–435. IEEE (2009)
4. Gil-Jiménez, P., López-Sastre, R.J., Siegmann, P., Acevedo-Rodríguez, J., Maldonado-Bascón, S.: Automatic control of video surveillance camera sabotage. In: Mira, J., Álvarez, J.R. (eds.) IWINAC 2007. LNCS, vol. 4528, pp. 222–231. Springer, Heidelberg (2007)
5. Tsesmelis, T., Christensen, L., Fihl, P., Moeslund, T.B.: Tamper detection for active surveillance systems. In: IEEE Int. Conf. on Advanced Video and Signal Based Surveillance, AVVS 2013, pp. 57–62 (2013)
6. Bay, H., Tuytelaars, T., Van Gool, L.: SURF: speeded up robust features. In: Leonardis, A., Bischof, H., Pinz, A. (eds.) ECCV 2006, Part I. LNCS, vol. 3951, pp. 404–417. Springer, Heidelberg (2006)
7. Ribnick, E., Atev, S., Masoud, O., Papanikolopoulos, N., Voyles, R.: Real-time detection of camera tampering. In: IEEE Int. Conf. on Video and Signal Based Surveillance, AVSS 2006, pp. 10–10 (2006)
8. Harasse, S., Bonnaud, L., Caplier, A., Desvignes, M.: Automated camera dysfunctions detection. In: IEEE Southwest Symp. on Image Analysis and Interpretation, pp. 36–40 (2004)
9. Komorkiewicz, T.K.M., Gorgon, M.: FPGA implementation of camera tamper detection in real-time. In: Int. Conf. on Design and Architectures for Signal and Image Processing DASIP, pp. 1–8 (2012)
10. Perrig, A., Stankovic, J., Wagner, D.: Security in wireless sensor networks. Communications of the ACM **47**(6), 53–57 (2004)
11. Alippi, C., Boracchi, G., Camplani, R., Roveri, M.: Detecting external disturbances on the camera lens in wireless multimedia sensor networks. IEEE Trans. on Instr. and Meas. **59**(11), 2982–2990 (2010)
12. Chandola, V., Banerjee, A., Kumar, V.: Anomaly detection: A survey. ACM Computing Surveys (CSUR) **41**(3), 15 (2009)
13. Gustafsson, F.: Adaptive Filtering and Change Detection. Wiley, October 2000
14. Achanta, R., Shaji, A., Smith, K., Lucchi, A., Fua, P., Susstrunk, S.: Slic superpixels compared to state-of-the-art superpixel methods. IEEE Trans. Pattern Anal. Mach. Intell. **34**(11), November
15. Kottke, D.P., Sun, Y.: Motion estimation via cluster matching. IEEE Transactions on Pattern Analysis and Machine Intelligence **16**(11), 1128–1132 (1994)
16. Caliński, T., Harabasz, J.: A dendrite method for cluster analysis. Communications in Statistics-theory and Methods **3**(1), 1–27 (1974)

Pattern Recognition-Facial Mapping

Recognizing Handwritten Characters with Local Descriptors and Bags of Visual Words

Olarik Surinta[1]([✉]), Mahir F. Karaaba[1], Tusar K. Mishra[2],
Lambert R.B. Schomaker[1], and Marco A. Wiering[1]

[1] Institute of Artificial Intelligence and Cognitive Engineering (ALICE),
University of Groningen, Nijenborgh 9, Groningen, The Netherlands
{o.surinta,m.f.karaaba,l.r.b.schomaker,m.a.wiering}@rug.nl
[2] Department of Computer Science and Engineering,
National Institute of Technology, Rourkela, Odisha, India
tusar.k.mishra@gmail.com

Abstract. In this paper we propose the use of several feature extraction methods, which have been shown before to perform well for object recognition, for recognizing handwritten characters. These methods are the histogram of oriented gradients (HOG), a bag of visual words using pixel intensity information (BOW), and a bag of visual words using extracted HOG features (HOG-BOW). These feature extraction algorithms are compared to other well-known techniques: principal component analysis, the discrete cosine transform, and the direct use of pixel intensities. The extracted features are given to three different types of support vector machines for classification, namely a linear SVM, an SVM with the RBF kernel, and a linear SVM using L2-regularization. We have evaluated the six different feature descriptors and three SVM classifiers on three different handwritten character datasets: Bangla, Odia and MNIST. The results show that the HOG-BOW, BOW and HOG method significantly outperform the other methods. The HOG-BOW method performs best with the L2-regularized SVM and obtains very high recognition accuracies on all three datasets.

Keywords: Handwritten character recognition · Feature extraction · Bag of visual words · Histogram of oriented gradients · Support vector machines

1 Introduction

In this paper we propose the use of several feature descriptors for handwritten character recognition. Obtaining high accuracies on handwritten character datasets can be difficult due to several factors such as background noise, many different types of handwriting, and an insufficient amount of training examples. Our motivation for this study is to obtain high recognition accuracies for different datasets even when there are not many examples in these datasets. There are currently many character recognition systems which have been tested on the

© Springer International Publishing Switzerland 2015
L. Iliadis and C. Jayne (Eds.): EANN 2015, CCIS 517, pp. 255–264, 2015.
DOI: 10.1007/978-3-319-23983-5_24

standard benchmark MNIST dataset [16]. MNIST consists of isolated handwritten digits with a size of 28 × 28 pixels and contains 60,000 training images and 10,000 test images. Compared to other handwritten datasets such as the Bangla and Odia character datasets, MNIST is simpler as it contains much more training examples, the diversity of handwritten digits is smaller in MNIST, and the number of digits is much smaller than the number of characters in the Odia and Bangla datasets. Therefore it is not surprising that since the construction of the MNIST dataset a lot of progress on the best test accuracy has been made.

Currently the best approaches for MNIST make use of deep neural network architectures [20]. The deep belief network (DBN) [11] has been investigated for MNIST in [11], where different restricted Boltzmann machines (RBMs) are stacked on top of each other to construct a DBN architecture. Three hidden layers are used where the sizes of each layer are 500, 500 and 2000 hidden units, respectively. The recognition performance with this method is 98.65% on the MNIST dataset.

In [17], a committee of simple neural networks is proposed for the MNIST dataset, where three different committee types comprising majority, average and median committees are combined. Furthermore, deslanted training images are created by using principal component analysis (PCA) and the elastic deformations are used to create even more training examples. The trained 9-net committees obtained 99.61% accuracy on MNIST. This work has been extended in [4] where 35 convolutional neural networks are trained and combined using a committee. This approach has obtained an accuracy of on average 99.77%, which is the best performance on MNIST so far. This technique, however, requires a lot of training data and also takes a huge amount of time for training for which the use of GPUs is mandatory.

Although currently many deep learning architectures are used in the computer vision and machine learning community, in [6] an older method from computer vision, namely the bag of visual words approach [7] was used on the CIFAR-10 dataset and obtained a high recognition accuracy of 79.6%. This simpler method requires much less parameter tuning and much less computational time for training the models compared to deep learning architectures. Also many other feature extraction techniques have been used for different image recognition problems, such as principal component analysis (PCA) [9], restricted Boltzmann machines [14], and autoencoders [12].

The cleanliness of MNIST and lack of variation may make MNIST a bad reference for selecting feature extractor techniques that are suitable for Asian scripts. Different feature extraction methods have been used for the Bangla and Odia handwritten character datasets. In [13], the celled projection method is proposed. The recognition performance obtained on the Bangla digit dataset was 94.12%. In [21], the image pixel-based method (IMG) which uses directly the intensities of pixels of the ink trace is used. The IMG is shown to be a quite powerful method [21] when the training set size is increased and obtained a recognition accuracy of 96.4% on the Bangla digit dataset.

Fig. 1. Illustration of the handwritten character datasets. (a) Some examples of Bangla and (b) Odia handwritten characters and (c) MNIST handwritten digits.

Contributions: In this paper, we propose the use of histograms of oriented gradients (HOG), bags of visual words using pixel intensities (BOW), and bags of visual words using HOG (HOG-BOW) for recognizing handwritten characters. These methods are compared to the direct use of pixel intensities, the discrete cosine transform (DCT), and PCA on three datasets, namely Bangla, Odia and MNIST, shown in Fig. 1. There are some challenges in the Bangla and Odia character datasets, such as the writing styles (e.g., heavy cursivity and arbitrary tail strokes), as shown in Fig. 2, similar structures of different characters, background noise, and a lack of a large amount of handwritten character samples. We have evaluated the six feature extraction techniques with three types of support vector machines [23] as a classifier, namely a linear SVM, an SVM with a radial basis function (RBF) kernel, and a linear SVM with L2-norm regularization (L2-SVM).

The results show that the HOG-BOW method obtains the highest accuracies on the three handwritten datasets. Also the HOG and BOW feature descriptors work much better than the more traditional techniques such as PCA, DCT and the direct use of pixel intensities. The results also show a very high performance of HOG-BOW with the L2-SVM on the MNIST dataset. Its recognition performance is 99.43% without the use of elastic distortions to increase the dataset, without the use of ensemble learning techniques, and without the need for a large amount of training time.

Paper outline: This paper is organized in the following way. Section 2 describes the feature extraction techniques. Section 3 describes the handwritten character datasets which are used in the experiments. The experimental results of the feature extraction techniques and the classifiers are presented in Section 4. The conclusion is given in the last section.

2   Feature Extraction Methods

We study different kinds of feature extraction techniques to deal with the handwritten character datasets as described below.

(a)   (b)

Fig. 2. Illustration of the Bangla and Odia handwritten characters, in the first and second row, respectively. Some examples of (a) heavy cursive and (b) arbitrary tail strokes writing styles.

2.1 Image Pixel-Based Method (IMG)

The IMG method directly uses the pixel intensities of the handwritten character image. It is a simple method to construct a feature vector. In this method, the character image is resized to a fixed size in pixels [21] and this resulting image is treated as a feature vector. In this study, a 36×36 pixel resolution is selected, so that the feature-vector size becomes 1,296 dimensions.

2.2 Principal Component Analysis (PCA)

PCA is a well-known dimensionality reduction method, which extracts an effective number of uncorrelated variables (called 'principal components') from high-dimensional correlated variables (input data). In fact, a small number of principal components is sufficient to represent the actual data. Here, eigenvectors are computed from the training data which are used as a model which is applied on an image to compute the feature vectors. After conducting preliminary experiments, we have selected 80 eigenvectors for this approach.

2.3 Discrete Cosine Transform (DCT)

The DCT technique transforms the data from a spatial domain (image) into frequency components using only a cosine function. We use 2D-DCT in our experiments since 2D-DCT is more suitable for 2D image data. Here, the highest coefficient values are stored in the upper left and the lowest valued coefficients are stored in the bottom right of the output array [15]. The highest coefficient values are extracted in a zigzag form [18] and then represented as feature vectors. In the experiment, 60 coefficient values were selected in the feature vectors.

2.4 Histograms of Oriented Gradients (HOG)

The HOG descriptor was proposed in [8] for the purpose of human detection from images. To compute the HOG descriptor, the handwritten character image is divided into small regions [22], called 'blocks', $\eta \times \eta$. To compute the horizontal G_x and vertical G_y gradient components at every coordinate x, y of the handwritten character image, we use a simple kernel $[-1, 0, +1]$ as the gradient detector (*i.e.* Sobel or Prewitt operators) [2].

The gradient detector can be calculated as $G_x = f(x+1, y) - f(x-1, y)$ and $G_y = f(x, y+1) - f(x, y-1)$, where $f(x, y)$ is the intensity value at coordinate x, y. The gradient magnitude M and the gradient orientation θ are calculated as:

$$M(x, y) = \sqrt{G_x^2 + G_y^2}$$
$$\theta(x, y) = \tan^{-1} \frac{G_y}{G_x} \tag{1}$$

Furthermore, the image gradient orientations within each block are weighted into a specific orientation bin β of the histogram. Then, the HOG descriptors from all blocks are combined and normalized by applying the L2-norm [8]. In this experiment, we employed rectangular HOG (R-HOG) with non-overlapping blocks. The best η and β parameters for recognizing our handwritten character datasets use 36 rectangular blocks ($\eta = 6$) and 9 orientation bins, yielding a 324-dimensional feature vector.

2.5 Bag of Visual Words with Pixel Intensities (BOW)

The bag of visual words [7] has been widely used in computer vision research. In this approach, local patches that contain local information of the image are extracted and used as a feature vector. Then, a codebook is constructed by using an unsupervised clustering algorithm. In [1], some novel visual keyword descriptors for image categorization are proposed and it was shown that the soft assignment schemes outperform the more traditional hard assignment methods. In [6], it was shown that the BOW method outperformed other feature learning methods such as RBMs and autoencoders. In [5], the method was applied to text detection and character recognition in scene images. We will now explain the BOW method consisting of patch extraction, codebook computation, and feature extraction.

Extracting Patches from the Training Data. The sub-image patches X are extracted randomly from the unlabeled training images, $X = \{x_1, x_2, ..., x_N\}$ where $x_k \in \mathbf{R}^p$ and N is the number of random patches. The size of each patch is defined as a square with ($p = w \times w$) pixels. In our experiments we used $w = 15$, meaning 15×15 pixel windows are used.

Construction of the Codebook. The codebook is constructed by clustering the vectors obtained by randomly selecting patches. Here, the codebook C is computed by using the K-means clustering method on pixel intensity information contained in each patch. Let $C = \{c_1, c_2, ..., c_K\}, c \in \mathbf{R}^p$ represent the codebook [24], where K is the number of centroids. In our experiments we used 400,000 randomly selected patches to compute the codebooks.

Feature Extraction. To create the feature vectors for training and testing images, the soft-assignment coding scheme from [6] is used. This approach uses the following equation to compute the activity of each cluster given a feature vector x from a patch:

$$i_k(x) = max \{0, \mu(s) - s_k\} \tag{2}$$

where $s_k = \|x - c_k\|_2$ and $\mu(s)$ is the mean of the elements of s [6].

The image is split into four quadrants and the activities of each cluster for each patch in a quadrant are summed up. We use a sliding window on the train and test images to extract the patches. Because the stride is 1 pixel and the window size is 15×15 pixels, the method extracts 484 patches from each image to compute the cluster activations. The feature vector size is $K \times 4$ and because we use $K = 600$ clusters, the feature vectors for the BOW method have 2,400 dimensions.

2.6 Bag of Visual Words with HOG Features (HOG-BOW)

In the previous BOW method, the intensity values in each patch are extracted and used for clustering. With HOG-BOW, however, feature vectors from patches are computed by using the state-of-the-art HOG descriptor [8], and then these feature vectors are used to compute the codebook and the cluster activities. The HOG descriptor captures the gradient structure of the local shape and may provide more robust features. In this experiment, the best HOG parameters used 36 rectangular blocks and 9 orientation bins to compute feature vectors from each patch. As in BOW, HOG-BOW uses 4 quadrants and 600 centroids, yielding a 2,400 dimensional feature vector.

3 Handwritten Character Datasets and Pre-processing

The handwritten character images in the Bangla and Odia datasets were scanned into digital images at different pixel resolutions. The details of the handwritten character datasets and pre-processing steps will now be described.

3.1 Bangla Character Dataset

The Bangla basic character consists of 11 vowels and 39 consonants [3]. In the experiment, the dataset includes 45 classes and contains 5,527 character images. The dataset is divided into training and test sets, containing 4,627 and 900 samples, respectively. Samples of the Bangla characters are shown in Fig 1(a).

3.2 Odia Character Dataset

The Odia handwritten dataset was collected from 50 writers using a take-note device. This dataset consists of 47 classes, 4,042 training and 987 test samples. Some examples of the Odia characters are shown in Fig. 1(b).

3.3 MNIST Dataset

The standard MNIST dataset [16] is a subset of the NIST dataset. The handwritten images were normalized to fit into 28×28 pixels. The anti-aliasing technique is used while normalizing the image. The handwritten images of the MNIST dataset contain gray levels. The dataset contains 60,000 handwritten training images and 10,000 handwritten test images, see Fig. 1(c) for some examples.

An overview of the handwritten datasets is given in Table 1.

Table 1. Overview of the handwritten character datasets

Dataset	Color Format	No. of Writers	No. of Classes	Train	Test
Bangla character	Grayscale	Multi	45	4,627	900
Odia character	Binary	50	47	4,042	987
MNIST	Grayscale	250	10	60,000	10,000

3.4 Dataset Pre-processing

In order to prepare the handwritten character images from the Bangla and Odia datasets, a few pre-processing steps which include background removal, basic image morphological operations and image normalization are employed. First, the Bangla handwritten dataset contains different kinds of backgrounds and is stored in gray-scale images. On the other hand, the Odia handwritten dataset is stored in binary image format as shown in Fig 1(b). Hence, the background removal is applied only to the Bangla handwritten dataset. In this study, due to its simplicity and yet robustness feature, we selected Otsu's algorithm [19] for removing background noise and making a binary image.

Next, a basic morphological dilation operation is applied to the binary handwritten images from the previous step. Finally, many researchers investigated the effect of scale differences for handwritten character recognition [17]. In this study, we normalize the handwritten image into 36×36 pixels with the aspect ratio preserved.

4 Experimental Results

We evaluated the feature extraction techniques on the Bangla, Odia and MNIST datasets by using three different SVM algorithms [23]. We used an SVM with a linear kernel, an SVM with an RBF kernel and a linear L2-regularized SVM (L2-SVM) [10]. The results are shown in Table 2, Table 3 and Table 4. In each table we show the results of the feature extraction techniques with a different SVM. In all the tables, the results show that the HOG-BOW, the HOG and the BOW method significantly outperform the other methods. Furthermore, on all 9 experiments the HOG-BOW method performs best (highly significant differences according to a student's t-test are indicated in boldface).

Table 2. Results of training (10-fold cross validation with the standard deviation) and testing recognition performances (%) of the feature descriptors when combined with the linear SVM.

Algorithms	Bangla dataset		Odia dataset		MNIST dataset	
	10-cv	Test	10-cv	Test	10-cv	Test
PCA	54.87 ± 0.20	53.67	56.57 ± 0.32	53.60	93.29 ± 0.02	92.69
DCT	59.33 ± 0.32	52.33	60.77 ± 0.40	54.81	92.51 ± 0.06	91.32
IMG	56.25 ± 0.22	54.33	56.12 ± 0.57	56.23	94.13 ± 0.05	94.58
BOW	77.96 ± 0.21	77.17	79.30 ± 0.34	78.01	98.71 ± 0.02	98.47
HOG	81.17 ± 0.30	80.11	79.86 ± 0.20	80.45	98.62 ± 0.01	99.11
HOG-BOW	**82.07 ± 0.24**	82.44	**81.74 ± 0.49**	82.43	**99.09 ± 0.03**	99.16

Table 3. Results of training (10-fold cross validation with the standard deviation) and testing recognition performances (%) of the feature descriptors when combined with the SVM with the RBF kernel.

Algorithms	Bangla dataset		Odia dataset		MNIST dataset	
	10-cv	Test	10-cv	Test	10-cv	Test
IMG	63.25 ± 0.28	60.00	57.95 ± 0.42	60.28	96.95 ± 0.02	97.27
PCA	64.08 ± 0.30	61.11	60.57 ± 0.57	59.87	96.86 ± 0.02	96.64
DCT	70.18 ± 0.27	61.33	69.91 ± 0.34	63.63	98.18 ± 0.09	97.51
BOW	78.76 ± 0.38	77.17	81.29 ± 0.42	80.65	98.98 ± 0.01	98.97
HOG	83.11 ± 0.25	83.00	82.16 ± 0.27	83.38	99.13 ± 0.01	99.12
HOG-BOW	83.14 ± 0.18	83.33	**83.62 ± 0.17**	83.56	**99.30 ± 0.02**	99.35

Table 4. Results of recognition performances (%) of the methods when used with the L2-SVM.

Algorithms	Feature dimensionality	Handwritten character dataset		
		Test Bangla	Test Odia	Test MNIST
DCT	60	51.67	56.94	90.84
PCA	80	50.33	53.90	91.02
IMG	1,296	31.33	42.65	91.53
HOG	324	74.89	74.27	98.53
BOW	2,400	86.56	84.60	99.10
HOG-BOW	2,400	87.22	85.61	99.43

In terms of SVM algorithms, we can see different results. Here, the linear SVM obtains a worse performance compared to the SVM with the RBF kernel. The linear SVM seems, however, to better handle low dimensional input vectors, compared to the L2-SVM. The L2-SVM yields significantly better results if high dimensional feature vectors [10] are used such as with the HOG-BOW and the BOW method. In fact, the best results have been achieved with the L2-SVM with the HOG-BOW method. It is followed by the BOW and the HOG method, respectively. The HOG method outperforms the BOW method when using the SVM with an RBF kernel (see Table 3). The feature vector size of each feature extraction technique is shown in Table 4.

5 Conclusion

In this paper, we have demonstrated the effectiveness of different feature extraction techniques from computer vision for handwritten character recognition. We have shown that the HOG-BOW method combined with an L2-regularized SVM outperforms all other methods. The obtained accuracies with this method can be considered very high. On the MNIST dataset for example, HOG-BOW combined with the L2-regularized SVM obtains a recognition accuracy on the test set of 99.43% which is a state-of-the-art performance. The best method for MNIST [4] uses an ensemble of 35 convolutional neural networks and elastic deformations to increase the dataset and obtains around 99.77% accuracy. The proposed HOG-BOW method, however, is much faster, needs less training data and we have not yet evaluated its performance in an ensemble of different classifiers.

In future work we want to research different ways to improve the HOG-BOW even more. We are interested in examining other soft assignment coding schemes to compute cluster activities and we also want to construct an ensemble method to obtain even higher accuracies.

References

1. Abdullah, A., Veltkamp, R., Wiering, M.: Ensembles of novel visual keywords descriptors for image categorization. In: 2010 11th International Conference on Control Automation Robotics Vision (ICARCV), pp. 1206–1211, December 2010
2. Arróspide, J., Salgado, L., Camplani, M.: Image-based on-road vehicle detection using cost-effective histograms of oriented gradients. Visual Communication and Image Representation 24(7), 1182–1190 (2013)
3. Bhowmik, T.K., Ghanty, P., Roy, A., Parui, S.: SVM-based hierarchical architectures for handwritten Bangla character recognition. Document Analysis and Recognition (IJDAR) 12(2), 97–108 (2009)
4. Cireşan, D.C., Meier, U., Schmidhuber, J.: Multi-column deep neural networks for image classification. In: 2012 IEEE Conference on Computer Vision and Pattern Recognition (CVPR), pp. 3642–3649, June 2012
5. Coates, A., Carpenter, B., Case, C., Satheesh, S., Suresh, B., Wang, T., Wu, D., Ng, A.: Text detection and character recognition in scene images with unsupervised feature learning. In: 2011 International Conference on Document Analysis and Recognition (ICDAR), pp. 440–445, September 2011
6. Coates, A., Lee, H., Ng, A.Y.: An analysis of single-layer networks in unsupervised feature learning. In: 2011 International Conference on Artificial Intelligence and Statistics (AISTATS), pp. 215–223, April 2011
7. Csurka, G., Dance, C.R., Fan, L., Willamowski, J., Bray, C.: Visual categorization with bags of keypoints. In: 2004 8th European Conference on Computer Vision (ECCV), pp. 1–22 (2004)

8. Dalal, N., Triggs, B.: Histograms of oriented gradients for human detection. In: 2005 The IEEE Computer Society Conference on Computer Vision and Pattern Recognition (CVPR), pp. 886–893, vol. 1, June 2005

9. Deepu, V., Madhvanath, S., Ramakrishnan, A.: Principal component analysis for online handwritten character recognition. In: 2004 The 17th International Conference on Pattern Recognition (ICPR), vol. 2, pp. 327–330, August 2004

10. Fan, R.-E., Chang, K.-W., Hsieh, C.-J., Wang, X.-R., Lin, C.-J.: LIBLINEAR: A library for large linear classification. Machine Learning Research 9, 1871–1874 (2008)

11. Hinton, G.E., Osindero, S., Teh, Y.-W.: A fast learning algorithm for deep belief nets. Neural Computation 18(7), 1527–1554 (2006)

12. Hinton, G.E., Zemel, R.S.: Autoencoders, minimum description length and Helmholtz free energy. In: Cowan, J., Tesauro, G., Alspector, J. (eds.) Advances in Neural Information Processing Systems, vol. 6, pp. 3–10. Morgan-Kaufmann (1994)

13. Hossain, M., Amin, M., Yan, H.: Rapid feature extraction for bangla handwritten digit recognition. In: 2011 International Conference on Machine Learning and Cybernetics (ICMLC), vol. 4, pp. 1832–1837, July 2011

14. Karaaba, M.F., Schomaker, L., Wiering, M.: Machine learning for multi-view eye-pair detection. Engineering Applications of Artificial Intelligence 33, 69–79 (2014)

15. Lawgali, A., Bouridane, A., Angelova, M., Ghassemlooy, Z.: Handwritten Arabic character recognition: Which feature extraction method? Advanced Science and Technology 34, 1–8 (2011)

16. LeCun, Y., Cortes, C.: The MNIST database of handwritten digits (1998)

17. Meier, U., Ciresan, D., Gambardella, L., Schmidhuber, J.: Better digit recognition with a committee of simple neural nets. In: 2011 International Conference on Document Analysis and Recognition (ICDAR), pp. 1250–1254, September 2011

18. Mishra, T., Majhi, B., Panda, S.: A comparative analysis of image transformations for handwritten odia numeral recognition. In: 2013 International Conference on Advances in Computing, Communications and Informatics (ICACCI), pp. 790–793, August 2013

19. Otsu, N.: A threshold selection method from gray-level histograms. IEEE Transactions on Systems, Man and Cybernetics 9(1), 62–66 (1979)

20. Schmidhuber, J.: Deep learning in neural networks: An overview. Neural Networks 61, 85–117 (2015)

21. Surinta, O., Schomaker, L., Wiering, M.: A comparison of feature and pixel-based methods for recognizing handwritten bangla digits. In: 2013 International Conference on Document Analysis and Recognition (ICDAR), pp. 165–169, August 2013

22. Takahashi, K., Takahashi, S., Cui, Y., Hashimoto, M.: Remarks on computational facial expression recognition from HOG features using quaternion multi-layer neural network. In: Mladenov, V., Jayne, C., Iliadis, L. (eds.) EANN 2014. CCIS, vol. 459, pp. 15–24. Springer, Heidelberg (2014)

23. Vapnik, V.N.: Statistical Learning Theory. Wiley, September 1998

24. Ye, P., Kumar, J., Kang, L., Doermann, D.: Unsupervised feature learning framework for no-reference image quality assessment. In: 2012 IEEE Conference on Computer Vision and Pattern Recognition (CVPR), pp. 1098–1105, June 2012

Recognize Emotions from Facial Expressions Using a SVM and Neural Network Schema

Isidoros Perikos[✉], Epaminondas Ziakopoulos, and Ioannis Hatzilygeroudis

School of Engineering, Department of Computer Engineering & Informatics,
University of Patras, 26504, Patras, Greece
{perikos,ziakopoulo,ihatz}@ceid.upatras.gr

Abstract. Emotions are important and meaningful aspects of human behaviour. Analyzing facial expressions and recognizing their emotional state is a challenging task with wide ranging applications. In this paper, we present an emotion recognition system, which recognizes basic emotional states in facial expressions. Initially, it detects human faces in images using the Viola-Jones algorithm. Then, it locates and measures characteristics of specific regions of the facial expression such as eyes, eyebrows and mouth, and extracts proper geometrical characteristics form each region. These extracted features represent the facial expression and based on them a classification schema, which consists of a Support Vector Machine (SVM) and a Multilayer Perceptron Neural Network (MLPNN), recognizes each expression's emotional content. The classification schema initially recognizes whether the expression is emotional and then recognizes the specific emotions conveyed. The evaluation conducted on JAFFE and Kohn Kanade databases, revealed very encouraging results.

Keywords: Emotion recognition · Facial gestures · Human-computer interaction · Support vector machines · Multilayer Perceptron Neural Network

1 Introduction

Facial expressions form a universal language of emotions, which can instantly express a wide range of human emotional states and feelings. The facial expressions assist in various cognitive tasks; so, reading and interpreting the emotional content of human expressions is essential to deeper understand the human condition. In an early work on human facial emotions [10] it has been indicated that during a face-to-face human communication only 7% of the information of a message is communicated by the linguistic part of the message, such as spoken words, 38% is communicated by paralanguage (vocal part) and 55% is communicated by the facial expressions. So, the facial expressions constitute the most important communication medium in face-to-face interaction.

Giving to computer applications the ability to recognize the emotional state of humans from their facial expressions is a very important and challenging task with wide ranging applications. Indeed, automated systems that can determine emotions of a human based on his/her facial expressions, can improve the human computer

© Springer International Publishing Switzerland 2015
L. Iliadis and C. Jayne (Eds.): EANN 2015, CCIS 517, pp. 265–274, 2015.
DOI: 10.1007/978-3-319-23983-5_25

interaction and give computer systems the opportunity to customize and adapt its response. In intelligent tutoring systems, emotions and learning are inextricably bound together; so, recognizing the learner's emotional states could significantly improve the efficiency of the learning procedures delivered to him/her [1] [14]. Moreover, surveillance applications such as driver monitoring and elderly monitoring systems could benefit from a facial emotion recognition system, gaining the ability to deeper understand and adapt to the person's cognitive and emotional condition. Also, facial emotion recognition could be applied to medical treatment to monitor patients and detect their status. However, the recognition of the facial emotional state is considered to be very challenging task, due to the non-uniform nature of the human face and various conditions such as, lightening, shadows, facial pose and orientation [7].

In this work, a facial emotion recognition system developed to determine the emotional state of human facial expressions is presented. The system analyzes facial expressions and recognizes the six basic emotions as defined by Ekman [4], following an analytical, local-based approach. Initially, given a new image, the system detects human faces in it using the Viola-Jones algorithm [18]. Then, it locates and measures facial deformations of specific regions such as eyes, eyebrows and mouth and extracts geometrical characteristics such as locations, length, width and shape. These extracted features represent the deformations of the facial expression and based on them a classification schema, which consists of a Support Vector Machine (SVM) and a Multi-layer Perceptron Neural Network (MLPNN) classifier, recognizes each expression's emotional content. Initially, the SVM classifier characterizes the facial expression as emotional or neutral and then the MLPNN recognizes and classifies the emotional facial expression into the proper emotion category. The evaluation study was conducted on JAFFE and Kohn Kanade databases and revealed very encouraging results regarding the performance of the system in recognizing emotional expressions and specifying their emotional content on Ekman's scale.

The structure of the reminder of the paper is as follows. In section 2, related work is presented. In Section 3, the methodology and the system developed are illustrated. In Section 4, the evaluation conducted and the experimental results gathered are presented. Finally, Section 5 concludes the paper and presents direction for future work.

2   Related Work

In the literature there are a lot of research efforts on facial image analysis and emotion recognition [3][17]. Although a human can detect and interpret faces and facial expressions naturally with little or no effort, accurate facial expression recognition by machines is still a challenge. Authors, in the work presented in [2], present a method for emotional classification of facial expressions, which is based on histogram sequence of feature vector. The system is able to recognize five human expression categories: happy, anger, sad, surprise and neutral, based on the geometrical characteristics of the human mouth with an average recognition accuracy of 81.6%. The work presented in [15] recognizes facial emotions based on a novel approach using Canny, principal component analysis (PCA) technique for local facial feature extraction and an artificial neural network for the classification process. The average facial expression classification accuracy of the method is reported to be 85.7%. The authors in the work presented in [13], recognize four basic emotions: happiness, anger, surprise and sadness, focusing in preprocessing techniques for feature extraction such as, Gabor filters, linear discrimination analysis and PCA. They achieve in their experiments 93.8% average accuracy for images of the Jaffe face database with little noise and with particularly exaggerated expressions and an average accuracy of 79% in recognition of just smiling/non smiling expressions in the ORL database. In [11], authors use a SVM classifier to recognize the 6 basis Ekman emotions in Cohn-Kanade database. In the study, authors extract 22 features form facial still images and report an average accuracy of 87.9%. In [8], authors recognize the six basic Ekman's emotion in elders' facial expressions using a SVM classifier. Features are extracted for expressions based on the active shape model, which fits in the shape parameters, using techniques such as gradient descent. The SVM is trained on a dataset such as, JAFFE in Cohn-Kanade and MII, and authors report promising results. In [19], authors developed a system for the emotion classification through lower facial expressions using the adaptive SVM. The system extracts eight feature points from the mouth, chin and nose and feed them into the A-SVMs classifier to categorize expression. The system was tested on Jaffe database and reports an average accuracy of 74.5%.

3 Emotion Recognition System

The system, given a new image, analyzes it and tries to detect human faces in it using the Viola-Jones method [18]. Then, it analyzes the facial expression, extracts appropriate features and classifies it to the proper emotion category, as defined by Ekman. The steps are depicted in Figure 1.

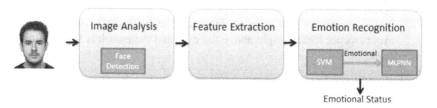

Fig. 1. The work flow chart of the system

When a face is detected in the image, the area of the face is located and is further analyzed. Initially, the facial area is normalized and the contrast is enhanced and after that, specific regions which contribute in recognizing the emotional content of the facial expression are specified. The specific regions, from which the proper features are extracted, are named Areas of Interests (AOIs) and are the region of the eyes, the area of the mouth and the eyebrows. The feature extraction process analyzes the AOIs and tries to extract proper facial features, which describe and model the characteristics of the facial region. The system follows an analytical local-based approach and so the feature extraction is implemented in the specific AOIs of the face. Finally, the emotion recognition is conducted by the classification schema, which specifies the appropriate emotional content of the expression on Ekman's scale. The classification schema consists of a Support Vector Machine (SVM) and a Multi-Layer Perceptron Neural Network (MLPNN) classifier, where the SVM specifies whether a facial expression is emotional or not and the MLPNN recognize the emotional content of the expressions recognized as emotional by the SVM.

3.1   Image Analysis and Face Detection

The system utilizes the Viola - Jones algorithm to locate human faces in an image. In general, the height and width of the face-box output by the Viola and Jones face detector are determined accurately and can vary by approximately 5% [16]. In case that more than one faces are detected, each face is analyzed separately. A human face is further analyzed in order to locate specific AOIs. The first step, aiming to enhance the contrast of the face, is to apply histogram equalization and then a sober filter to emphasize the edges of the face. Then, the exact location of the eyes is specified. The locations of the rest of the AOIs are detected based on the relative position of the eyes: the eyebrows are located right above the eyes and the mouth is located in the horizontal center of the face under the eyes. More specifically, the image is analyzed as follows:

1. Detect the face in the image using the Viola - Jones algorithm.
2. Apply histogram equalization to enhance contrast of the upper-half of the face, which contains the eyes.
3. Apply a Sobel filter to emphasize edges and transition of the face.
4. Cut the face into four horizontal zones.
5. Measure each zone's pixel density by averaging the number of the white (transitional) pixels in each individual zone.
6. Select the zone with the highest pixel density. This zone contains the eyes.
7. Locate eyebrows and mouth using facial features mask (new figure). The eyebrows are located right above the eyes and the mouth is located under the eyes.

The histogram equalization is very useful and is used to enhance the contrast of the image in order to obtain a uniform histogram and re-organizes the image's intensity distributions. After that, a Sobel filter is applied to emphasize edges and transition of the face and after its application, the image of the human face is represented as a matrix whose elements have only two possible values: zero and one. An element that has a value set to one represents an edge of a facial part in the image, such as an edge of an eye of the mouth etc. The pixel density of a zone is calculated to be the average sum of consecutive rows of the matrix of the zone. In this approach, the pixel density of a zone represents the complexity of the face in each particular zone. The zone with the highest complexity contains the eyes and the eyebrows. In Figure 2, the stages of the image analysis and the results of each process step are illustrated in an example facial image.

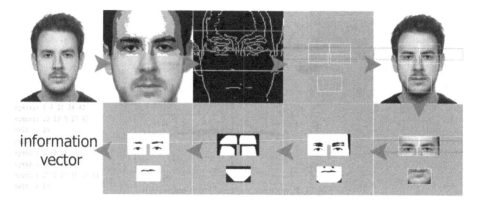

Fig. 2. The stages of the analysis of a facial image.

3.2   AOI Analysis and Feature Extraction

After the AOIs of the human face have been located, each AOI is further analyzed in order to extract proper features that describe its characteristics and convey meaningful information about the AOI deformation. The features extracted represent the geometrical characteristics of the AOI's deformation such as the facial part's height and width, and also try to model the element's shape. After the successful AOI isolation, the analysis of each AOI is performed in an effort to simplify the gathered information and extract the proper feature values. The steps of the AOIs analysis are the following:

1. Measure the average brightness of the face.
2. Set a brightness threshold. All the elements of the matrix with higher values than the threshold maximize their values and all the elements of the matrix with lower values than the threshold minimize their values
3. Relate the brightness threshold to the average brightness. Brighter faces require higher thresholds than darker faces, in order to minimize useful data loss from the application of the filter.
4. Set the brightness filter.
5. Apply the brightness filter to the AOI.
6. Cut of the excess AOI noise using proper individual masks.

Initially, a brightness filter is applied to simplify an AOI in order to remove unnecessary data. AOIs, after the filter application, are represented more smoothly and the features can be easily extracted. In order to measure the average brightness, a facial mask is applied to isolate the face area from the background image and also from the model's hair since these areas do not affect the brightness of the face. In Figure 3, the features extracted from each AOI in the example facial image presented above are illustrated.

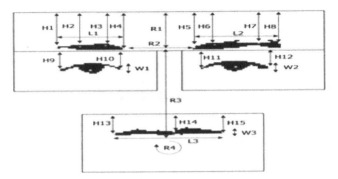

Fig. 3. Features extracted from each AOI

Left eyebrow

- H1: The height of the far left part.
- H2: The height of the part found on the 1/3 of the distance between the far left part and far right part.
- H3: The height of the part found on the 2/3 of the distance between the far left and far right part.
- H4: The height of the far right.
- L1: The length of the eyebrow.

Right eyebrow

- H5: The height of the far left part.
- H6: The height of the part found on the 1/3 of the distance between the far left and far right part.
- H7: The height of the part found on the 2/3 of the distance between the far left and far right part.
- H8: The height of the far right part.
- L2: The length of the eyebrow.

Left eye

- H9: The height of the far left part.
- H10: The height of the far right part.
- W1: The width of the eye.

Right eye

- H11: The height of the far left part.
- H12: The height of the far right part.
- W2: The width of the eye.

Mouth

- H13: The height of the far left part of the mouth.
- H14: The height of the part found on the 1/2 of the distance between the far left and far right part of the mouth.
- H15: The height of the far left part of the mouth.
- W3: The width of the mouth.
- L3: The length of the mouth.

Relative Values

- R1: The average height of the eyebrows.
- R2: The horizontal distance between the eyebrows.
- R3: The vertical distance between the eyebrows and the bottom of the mouth.
- R4: The ratio of the mouth length to the mouth width.

So, a facial expression is represented by 25 features. These features model the facial expression and contain the necessary data to assist the classifiers to classify an expression into the proper category.

3.3   Classification Schema

The classification schema consists of two classifiers, a SVM and a MLPNN. SVMs are very powerful binary classifiers. In general, a SVM classifier constructs a hyperplane or a set of hyperplanes in a high- or infinite-dimensional space, which can be used for the classification procedure. A good separation is achieved by the hyperplane that has the largest distance to the nearest training data point of any class (so-called functional margin), since in general the larger the margin the lower the generalization error of the SVM classifier. For the purposes of our work, the SVM is used to separate the facial expressions that convey emotion from the neutral ones. The second classification method is based on a MLPNN classifier which enables higher flexibility in training and also great adaptation to the problem of emotional classification [5]. In general, MLPNNs are feed-forward neural networks and can be used for solving of various multi-class, nonlinear classification problems. In our methodology, the MLPNN classifier used for the emotional classification of the facial expressions has three hidden layers, containing 14, 14 and 8 neurons respectively. Since there is no standard method to select the number of hidden layers and neurons, due to the nature of the Neural Networks, the selection is often, as in this case, the architecture with the best performance. The input layer of the MLPNN has 25 sensory neurons, in order to match the length of the information vectors. The MLPNN network is trained using the back propagation supervised learning technique, which provides a systematic way to update the synaptic weights of multi-layer perceptron networks. Both the MLPNN and the SVM classifiers were implemented in MATLAB toolkit.

4   Experimental Study

4.1   Data Collection

The methodology and the system developed were extensively evaluated on facial images from two popular and widely used databases, the Japanese Female Facial Expression Database (JAFFE) [9] and the Cohn-Kanade database [6]. The Jaffe database consists of 211 facial pictures of 10 different posers. Approximately the two thirds (140 images) of the images of the database were selected to be part of the training dataset and the remaining 71 images were part of the test dataset. The Cohn-Kanade is a popular database which includes 486 sequences from 97 posers. For the needs and purposes of this study, the total number of the selected images was 209, where 140 images were part of the training dataset and 69 images were part of the test dataset respectively.

4.2   System Evaluation

Initially, an evaluation of the system's performance in characterizing a facial expression either as emotional or as neutral was conducted. The performance of the mechanism is presented in Table 1.

Table 1. SVM Performance Results

Metric	Jaffe	Kohn Kanade	Total
Accuracy	92.3	100	96.4
Precision	98.3	100	99.2
Sensitivity	93.4	100	96.9
Specificity	90	100	95

The results indicate that the system has a very good performance in determining whether a facial expression conveys emotions or not. Indeed, the SVM achieves excellent performance in the Kohn Kanade database, mainly due to the fact that Kohn Kanade models present neutral expressions in a very consisted and inactive way. Also, high sensitivity indicates that the mechanism can accurately identify emotional gestures that indeed are emotional.

The second part of the evaluation study assessed the performance of the MLPNN mechanism of the system in recognizing the emotional content of an emotional facial expression. The facial expressions that were emotional and correctly recognized as emotional by the SVM mechanism, were used for this part of the experiment. Given that we have a multiple class output, we use the following metrics: average accuracy, precision and f-measure. The performance of the system is presented in Table 2.

Table 2. Performance Results of the MLPNN

Metric	Jaffe	Kohn Kanade	Total
Average Accuracy	89,7 %	81,4%	85,5%
Precision	87,3%	82,8%	86,7%
F-measure	88,9%	81.4%	86,7%

The MLPNN demonstrates quite satisfactory performance in recognizing the emotional status of emotional expressions. The results are better in the JAFFE database and we believe that the reason is the strength that the JAFFE models express the emotions. Indeed, in some cases the models of the JAFFE database express the different emotions very vividly and with strong facial deformations.

The third part of the study evaluated the overall performance of the system developed. The overall performahce of the system and the confusion matrix of the performance on both databases are presented in Table 3 and Table 4 respectively.

The overall evaluation results indicate that the system is performing very well in both databases. A noticeable point concerns the high performance of the system in recognizing facial gestures that express happiness. This is mainly due to the fact that joy is a very strong emotion that in most cases is expressed by vivid facial deformations.

Table 3. The Performance Results of the System

Metric	Jaffe	Kohn Kanade	Total
Average Accuracy	85.9 %	84,1%	85%
Precision	87,5%	85,6%	86%
F-measure	87.4%	85.4%	86.1%

Table 4. The Combined Confusion Matrix of System's Performance.

	Joy	Sadness	Surprise	Fear	Disgust	Anger	Neutral
Joy	22	0	0	0	0	1	0
Sadness	0	12	0	6	0	1	0
Surprise	0	0	14	3	0	1	1
Fear	0	3	0	16	0	0	1
Disgust	0	3	0	2	13	1	0
Anger	0	1	0	1	0	17	1
Neutral	0	0	2	2	1	0	15

5   Conclusion and Future Work

In this paper, we present an emotion recognition system, which recognizes the six basic emotions defined by Ekman in facial expressions. Initially, it detects human faces in images using the Viola-Jones algorithm. Then it locates and measures characteristics of specific regions and extracts proper features and geometrical aspects form each region. These extracted features represent the facial expression and based on them a classification schema recognizes each expression's emotional content. The classification schema initially recognizes whether the expression is emotional, based on a SVM classifier, and then recognizes the specific emotions conveyed by the expression, based on a MLPNN. The evaluation conducted on JAFFE and Kohn Kanade databases revealed very encouraging results.

As a future work, we plan to conduct a bigger scale evaluation and also examine the system's performance in additional databases. Furthermore, an extension could be made in order to assist the system to analyse expressions from different poses and face orientations. Finally, the classification schema could be extended and the exploration of a neuro-fuzzy approach could enhance the system's classification performance. Exploring this direction is a main aspect of our future work.

References

1. Akputu, K.O., Seng, K.P., Lee, Y.L.: Facial emotion recognition for intelligent tutoring environment. In: 2nd International Conference on Machine Learning and Computer Science (IMLCS 2013), pp. 9–13 (2013)
2. Aung, D.M., Aye, N.A.: Facial expression classification using histogram based method. In: International Conference on Signal Processing Systems (2012)
3. Bettadapura, V.: Face expression recognition and analysis: the state of the art. arXiv preprint arXiv:1203.6722 (2012)
4. Ekman, P.: Basic emotions. In: Handbook of Cognition and Emotion, pp. 45–60 (1999)
5. Filko, D., Goran, M.: Emotion recognition system by a neural network based facial expression analysis. Autom.- J. Control Meas. Electron Comput. Commun. **54**, 263–272 (2013)
6. Kanade, T., Cohn, J.F., Tian, Y: Comprehensive database for facial expression analysis. In: Proceedings of the Automatic Face and Gesture Recognition (2000)
7. Koutlas, A., Fotiadis, D.I.: An automatic region based methodology for facial expression recognition. In: IEEE International Conference on Systems, Man and Cybernetics, SMC 2008, pp. 662–666 (2008)
8. Lozano-Monasor, E., López, M.T., Fernández-Caballero, A., Vigo-Bustos, F.: Facial expression recognition from webcam based on active shape models and support vector machines. In: Pecchia, L., Chen, L.L., Nugent, C., Bravo, J. (eds.) IWAAL 2014. LNCS, vol. 8868, pp. 147–154. Springer, Heidelberg (2014)
9. Lyons, M.J., Akamatsu,S., Kamachi, M., Gyoba, J.: Coding facial expressions with Gabor wavelets. In: Proceedings of the Third IEEE International Conference on Automatic Face and Gesture Recognition, pp. 200–205. IEEE Computer Society, Los Alamitos (1998)
10. Mehrabian, A.: Communication without words. Psychology Today **2**(4), 53–56 (1968)
11. Michel, P., El Kaliouby, R.: Facial expression recognition using support vector machines. In: The 10th International Conference on Human-Computer Interaction, Greece (2005)
12. Perikos, I., Ziakopoulos, E., Hatzilygeroudis, I.: Recognizing emotions from facial expressions using neural network. In: Iliadis, L. (ed.) AIAI 2014. IFIP AICT, vol. 436, pp. 236–245. Springer, Heidelberg (2014)
13. Přinosil, J., Smékal, Z., Esposito, A.: Combining features for recognizing emotional facial expressions in static images. In: Esposito, A., Bourbakis, N.G., Avouris, N., Hatzilygeroudis, I. (eds.) HH and HM Interaction. LNCS (LNAI), vol. 5042, pp. 56–69. Springer, Heidelberg (2008)
14. Shen, L., Wang, M., Shen, R.: Affective e - learning: Using "emotional" data to improve learning in pervasive learning environment. Educational Technology & Society **12**(2), 176–189 (2009)
15. Thai, L.H., Nguyen, N.D.T., Hai, T.S.: A facial expression classification system integrating canny, principal component analysis and artificial neural network. arXiv preprint arXiv:1111.4052 (2011)
16. Valstar, M.F., Mehu, M., Jiang, B., Pantic, M., Scherer, K.: Meta-analysis of the first facial expression recognition challenge. IEEE Transactions on Systems, Man, and Cybernetics, Part B: Cybernetics **42**(4), 966–979 (2012)
17. Verma, A., Sharma, L.K.: A Comprehensive Survey on Human Facial Expression Detection. International Journal of Image Processing (IJIP) **7**(2), 171 (2013)
18. Viola, P., Jones, M.J.: Robust real-time face detection. International Journal of Computer Vision **57**(2), 137–154 (2004)
19. Visutsak, P.: Emotion Classification through Lower Facial Expressions using Adaptive Support Vector Machines. JMMT: Journal of Man, Machine and Technology **2**(1), 12–20 (2013)

Multimodal Data Fusion for Person-Independent, Continuous Estimation of Pain Intensity

Markus Kächele[1]([✉]), Patrick Thiam[1], Mohammadreza Amirian[1],
Philipp Werner[2], Steffen Walter[3], Friedhelm Schwenker[1], and Günther Palm[1]

[1] Institute of Neural Information Processing, Ulm University, Ulm, Germany
{markus.kaechele,patrick.thiam,mohammadreza.amirian,
friedhelm.schwenker,guenther.palm}@uni-ulm.de
[2] Institute of Information Technology, University of Magdeburg,
Magdeburg, Germany
philipp.werner@ovgu.de
[3] Department of Psychosomatic Medicine and Psychotherapy, Ulm University,
Ulm, Germany
steffen.walter@uni-ulm.de

Abstract. In this work, a method is presented for the continuous estimation of pain intensity based on fusion of bio-physiological and video features. The focus of the paper is to analyse which modalities and feature sets are suited best for the task of recognizing pain levels in a person-independent setting. A large set of features is extracted from the available bio-physiological channels (ECG, EMG and skin conductivity) and the video stream. Experimental validation demonstrates which modalities contribute the most to a robust prediction and the effects when combining them to improve the continuous estimation given unseen persons.

1 Introduction

Robust and accurate estimation of signals from the domain of Human computer interaction (HCI) is a goal that has been pursued to a high extent during the last decade. With current algorithms for the recognition of speech, actions and affective states, this goal has been achieved to a certain extent. Consequently, in recent years the focus of attention of many researchers has shifted from HCI to bio-medical applications such as the recognition of depression [7] or the severity of pain of persons in clinical and non-clinical settings.

Pain is a feedback mechanism that prevents harmful behaviour and can result from physical or mental injuries. The automated recognition of pain is of increased interest in situations in which the severity of pain cannot be communicated. Such situations can occur in clinical settings when persons are not addressable anymore (e.g. because of degenerative diseases). While recent advances in automated pain recognition have mainly focused on prediction from facial expressions [2,6,9,10], it is increasingly considered an attractive setting for research on

© Springer International Publishing Switzerland 2015
L. Iliadis and C. Jayne (Eds.): EANN 2015, CCIS 517, pp. 275–285, 2015.
DOI: 10.1007/978-3-319-23983-5_26

multimodal recognition systems. Such recognition systems incorporate a multitude of different modalities such as bio-physiological channels (e.g. heart rate, skin conductance) [14] or linguistic and paralinguistic measures into complex classification systems. Pain recognition based on multimodal signals has recently been proven to be very effective, most often beating unimodal systems significantly [8,17]. In this work, a systematic analysis of different bio-physiological and visual features is performed to determine their applicability for the recognition of pain. In a further step, the accustomed setting of pain recognition as a two class problem (i.e. pain vs. no pain) is exchanged for a continuous domain with different pain intensity levels and is posed as a regression problem. To this date, research in automatic pain estimation has mainly focused on fixed classes and besides [9] few other notable works exist that exchange the setting with a fully continuous one. To our knowledge, this work presents the first system for continuous, person-independent pain estimation from physiological and video channels.

The remainder of this work is organized as follows. In the next Section, the dataset is introduced along with its modalities. In Section 3 the utilized features and the steps for the preparation of the data are given, followed by experimental validation. The paper is concluded with a discussion and an outlook on future work.

2   Dataset and Modalities

In these experiments the BioVid Heat Pain database [15] is analysed. It comprises of 90 participants (45 male, 45 female) of the age groups 18-35 years, 36-50 years and 51-65 years, in equal proportions. Pain stimulation was applied 20 times for each of 4 calibrated intensities (level 1 to 4). Together with additional baseline measurements, a total number of 100 stimulations was available for each participant. During the experiments, high resolution video and a bio-physiological amplifier were recorded. The physiological channels included electromyography (EMG) (zygomaticus, corrugator and trapezius muscles), skin conductance level (SCL) and an electrocardiogram (ECG). The reader is referred to Figure 1 for details.

3   Feature Extraction

In this study, the feature extraction was carried out on windows of length 5.5 seconds (green area in Figure 1) starting 1 second after the pain level is reached. This measurement method is in accordance with the other works performed on this dataset [8,16,17] that use the same setting. This was done to (1) prevent to run into subsequent baseline measurements and (2) to consider particularities of the channels such as the reaction lag of about 2 seconds for the SCL channel.

3.1   Biophysiology

The feature extraction from the bio-physiological channels was preceded by bandpass filtering, detrending and artifact correction.

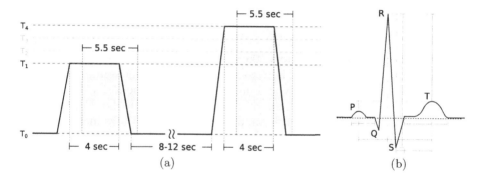

Fig. 1. Left: Pain stimulation. The stimuli levels are represented by temperatures T_0 to T_4. The features are extracted from the green window of length 5.5 seconds. **Right:** QRS complex with extracted distance features. A number of time and amplitude distances have been extracted between the P, Q, R, S and T points (red).

EMG: For the EMG channel, a butterworth bandpass filter of order 4 with cutoff frequencies $[20, 250]$ Hz was applied to isolate the bursts that carry the information about the muscle activity. Subsequently, the following features have been extracted: *mean* and *standard deviation* of the signal and the first and second derivatives for the filtered signal as well as for the filtered and normalized signal. Based on Welch's power spectrum density estimation the *ratio* of low frequency to very low frequency was computed.

The amplitude of the signal gave rise to the features *peak amplitude* (value of highest peak) and *range* (difference between highest and lowest value), as well as the *root of the squared mean* and the *mean of the absolute (MAV)* signal. Analogously to [11], the features *Integrated EMG (IEMG)*, *modified MAV (MMAV1+2)*, *slope of MAV (MAVSLP)*, *simple square integral (SSI)*, *waveform length (WL)*, *slope sign change (SSC)* and *the Willison amplitude (WAMP)* were implemented.

Additionally the ν-*Order*$=\sqrt[\nu]{E\{|x_k|^\nu\}}$ and *logDetect* $= \exp(\frac{1}{N}\sum_i \log(|x_i|))$ [13] have been implemented.

Furthermore, *histogram coefficients* that serve as a rough density estimate were also used [11] as well as *coefficients* obtained from fitting an autoregressive model using the Burg method.

The rate of vibrations of the facial muscles and the variability of this rate was evaluated in this study using the frequency domain analysis of the EMG signals. The *central, mode and mean frequencies* are measures of the rate of vibrations, while the *bandwidth* represents its variability. The amount of *zero crossings* was also considered as a measure of the vibration rate of the muscles [16].

As previously reported in the biomechanics and physiology literature, the EMG signals are not stationary [1]. Therefore some of the signal's features vary over time. The *stationary median, area, mean* and *variance* are the measures computed in order to evaluate and track the non-stationary properties of the EMG signals.

Because the muscles activity changes during an external stimulus by means of the sympathetic nervous system, the regularity of the EMG signal changes as well. *Approximation entropy* is introduced as the first measure used in this study to evaluate the irregularity and unpredictability of the EMG signals. The *sample entropy* was calculated similarly by excluding self-matches. Additionally by removing the baseline from the sample entropy and by introducing a fuzzy membership function, the *fuzzy entropy* [5] can be defined. All entropies together alongside the typical *Shannon entropy* were used to measure the regularity of the EMG signals in time domain. Similarly, the *spectral entropy* and *SEPFT* (Shannon entropy of the peak frequency shifting) are measuring the regularity of the EMG signals in the frequency domain [4].

Electrocardiogram: The processing of the ECG channel included piecewise linear detrending followed by detection of the so called QRS complexes (characteristic waveform of the heart beats). This was accomplished by detecting and pooling of local maxima over short time windows. Together with the physiological constraints (i.e. the refractory period of the heart), the R peaks could reliably be detected. Because the time windows were too short for common ECG features such as pNN50 or TiNN , the characteristic phases of the detected QRS complexes in the 5.5 second window were analysed. The features included the *amplitudes* of P, Q, R, S and T points, the *time delay* between points and the *angles* of the Q and S valleys. For details, the reader is referred to Fig. 1.

The wavelet transform decomposition coefficients were introduced as reliable features for heart rhythm detection in [19]. Necessary aligning was done based on the detected R peaks. The window length used for the segmentation was 650 ms, including 250 ms before and 400 ms after the R peak. After practical experiments a Daubechies wavelet of order 8 was chosen for the feature extraction. Four level wavelet decomposition was applied to obtain the desired features. The approximation coefficients of the wavelet decomposition a_1 were considered as the features of each heart beat since the detail coefficients d_1 are commonly noise and d_2, d_3 and d_4 contain the high frequency components of the signal. Ultimately the *mean* of the approximation coefficients of each heart beat within the entire window were considered as features for the recognition task.

Skin conductance level: The skin conductance level indicates the activity of the sweat glands in the skin and is directly controlled by the sympathetic nervous system. For the SCL channel, the same features as for the EMG channel were used. Additionally, statistical features such as the *skewness, kurtosis, ratio* between maximum and minimum, *number* of *SCR occurrences, mean amplitudes* of those occurrences, *temporal slope* of the signal, *spectral entropy* and *normalised length density* were computed.

The tonic and phasic decomposition of the SCL channel provides an unbiased quantification. Accordingly, in order to obtain complementary features of SCL data, it is decomposed into phasic and tonic components using continuous and discrete decomposition analysis. The data was downsampled by a factor of $1/16$ before decomposition to reduce the computational complexity. Furthermore, the data was smoothed using a Gaussian window and a moving average smoothing method to reduce the artifacts.

The continuous (discrete) data decomposition analysis calculates the phasic and tonic components of the SCL data by deconvolving the data by the general response shape (nonnegative deconvolution) [3]. Then the SCL features are extracted according to the decomposition. The features used for pain estimation were *number of SCL responses* within in a window, *response latency* of the first significant response, *average* and *area* of phasic driver, *maximum phasic activity* and *mean* of tonic activity.

Additionally, the *global mean, maximum positive deflection, number* of SCL responses and *sum of the SCL amplitudes* of significant SCL responses were used.

3.2 Video Feature Extraction

The following section describes the process undertaken to extract relevant features from the video sequences. Each video sequence has a length of 5.5 seconds and depicts the demeanour of a single participant during a specific pain stimulus. Two categories of features were extracted from the video sequences: geometric-based features and appearance-based features. A combination of both feature categories improves the robustness of the classification system. In both cases, the extraction process consisted of first detecting the face of the participant in each frame of a video sequence followed by locating a set of landmarks by the application of a constraint local model on the facial region. The computed landmarks consisted of 66 $2D$ facial points describing eyes, nose, and mouth regions as well as the shape of the detected facial region. Based on these landmarks, geometric-based features were extracted at frame level. Furthermore statistical measures were computed from the latter and combined to form the feature vector at the sequence level. Furthermore, the landmarks were used to compute specific facial regions that are subsequently used to extract appearance-based features describing the dynamics of those regions during the whole sequence.

Geometric-based Features. At frame level, the computed set of 66 facial points was used to compute specific features from the detected facial regions. The first features consisted of the euclidean distance between each single facial point and the center of gravity of the facial points set.

$$shape_{center} = \frac{1}{n} \sum_{i=1}^{n} p_i = \frac{1}{n} (\sum_{i=1}^{n} x_i, \sum_{i=1}^{n} y_i) \quad \text{with } p_i = (x_i, y_i) \quad (1)$$

$$\forall i \in [1, n], \quad f_i = \| p_i - shape_{center} \| \quad (2)$$

Here p_i denotes the facial landmarks. Subsequently 18 more euclidean distances (see Figure 2) were computed between specific facial points to capture the dynamic of the facial expression in each frame. Therefore each frame was represented by a 84 dimensional vector resulting from the concatenation of all computed distances. At the sequence level, 10 statistical measures (mean, standard deviation, min, max, median, range, inter-quartile range, median absolute deviation, mean absolute deviation, trimmed mean) were computed for each distance in the span of a video sequence. The resulting 840 dimensional distance vector encoded the dynamics of the facial region depicted in the video during the stimulus. The standardized distance vector was used as feature vector.

Appearance-based Features. Based on the landmarks computed at frame level, three regions of interest were defined and computed (see Figure 2). Subsequently the gradient direction of each region was computed after the latter has been converted to a grayscale image. Finally the local binary patterns on three orthogonal planes (LBP-TOP) operator [18] with uniform patterns was applied to the set of extracted facial regions to generate the feature vectors. Prior to the application of the LBP-TOP operator, each video sequence was divided into a set of 4 equally sized frame-blocks. The LBP histograms resulting from the application of the LBP-TOP operator on each frame-block were subsequently concatenated to form the feature vector of each region of interest.

The major drawback of the video feature extraction process was its reliance on the face detection at the frame level. The latter affected the accurate localization of the facial landmarks and consequently the robust computation of the regions of interest, needed for the extraction of the appearance-based features. Since the participants were allowed to move their head freely, detecting the facial region using just a frontal camera turned out to be a difficult task. As a result of this, the system was occasionally unable to accurately detect the facial region in every frame. Therefore, the frames without detected faces were replaced by duplicates of frames where the face was successfully detected. Both geometric-based and appearance-based features were extracted after the completion of the previous step.

 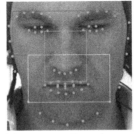 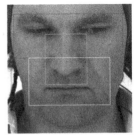

(a) Facial Landmarks and distance features (red)   (b) Regions Computation   (c) Computed Regions

Fig. 2. Regions Of Interest For Appearance-based Features

4 Continuous Prediction of Pain Intensity

The next step following the feature extraction process is the classification. The present work focuses on leave-one-participant-out experiments. The generalization capabilities of a trained model is therefore tested on unseen participants. This can be considered a highly relevant scenario in practice, as no additional learning phase adapts the system to the unknown person.

To combine predictions of different modalities, a number of fusion strategies [8, 12] are employed. Since the focus does not lie on fusion techniques, the setting is restricted to *early* and *late* fusion with the same classifier for each modality. Early fusion denotes the combination of features prior to learning while late or decision fusion denotes the combination of classification results (compare [8]). For the following experiments, a Random Forest (RF) has been chosen as classifier, because it gave the best results in initial experiments (in comparison to, for example, Support Vector Regression).

5 Experimental Validation

The goal of the first experiment is to assess the discriminative power of the computed feature sets, especially in comparison to earlier work on the same dataset ([8, 16, 17]).

A classification experiment similar to the one in [8] was carried out. For each modality, a Random Forest was trained using 500 trees [1].

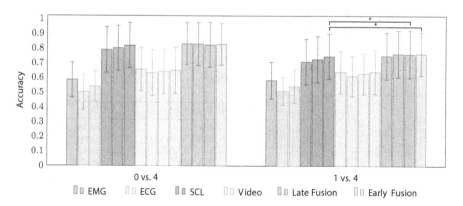

Fig. 3. Binary classification results. To make the results comparable, the two class pairings 0 vs 4 and 1 vs 4 are analysed. Significance (indicated with asterisk) is computed using a Wilcoxon signed rank test with a significance level of 5%. More detailed channel descriptions can be found in Figure 4 (the order is kept the same).

---

[1] Here, only part A of the dataset is used. See [8] for details.

Late fusion was conducted using fixed mappings (here: maximum and product) and a trainable mapping in the form of a pseudo-inverse. Early fusion, consequently, by concatenating the features before training. As the results in Figure 3 suggest, the recognition system is able to generalize well on unseen subjects. Both fusion variants improve the final results in comparison to the individual channels. Furthermore, the results show an improvement over the best results of competing approaches. In [17], the authors achieved 80.6% accuracy using a Random Forest with bio-physiological, facial expression and head movement features in the setting 0 vs. 4. Another study investigated fusion strategies in combination with Random Forests and obtained 80.2% on the same dataset [8].

Using late fusion, the accuracy reaches 83.1% with the computed features, which slightly outperforms early fusion (82.7%) and the SCL channel. The results indicate that SCL is the strongest modality with 81.9% accuracy. ECG seems to be the channel with the lowest performance. Both early and late fusion are able to improve the results for both experiments, for 1 vs 4 even significantly in relation to the best single modality. For details the reader is referred to Figure 3.

By switching from a classification to a regression problem, the binary class pairings are removed and the classifier is directly trained on the intensity levels. While the performance for the pairings of the highest pain level with a very low one (i.e. 0 or 1) was relatively high, now the regression has to deal with all the levels at once and distinct very similar ones like 1 and 2 or 2 and 3 from each other. Performance is now measured as root mean squared error (RMSE) averaged over the 86 participants. Again the results for single modalities are presented as well as early fusion. Since late fusion relies on class probabilities it is omitted in this experiment. The results show that the setting is more challenging with a mean RMSE of more than 1 for the individual channels and 0.98 for early fusion. Again SCL is the best performing modality followed by video. It seems to be evident, that the SCL channels carries more information about applied painful stimuli than ECG or EMG.

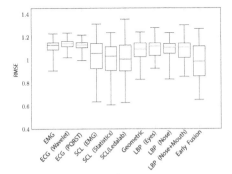

Feature	RMSE	MAE
EMG	1.11 (±0.06)	0.97 (±0.06)
ECG (Wavelet)	1.14 (±0.05)	1.00 (±0.04)
ECG (PQRST)	1.12 (±0.05)	0.98 (±0.05)
SCL (EMG)	1.04 (±0.13)	0.87 (±0.12)
SCL (Statistics)	1.01 (±0.14)	0.85 (±0.13)
SCL (Ledalab)	1.01 (±0.15)	0.85 (±0.13)
Geometric	1.08 (±0.09)	0.94 (±0.08)
LBP (Eyes)	1.09 (±0.07)	0.95 (±0.07)
LBP (Nose)	1.08 (±0.08)	0.94 (±0.08)
LBP (Nose+Mouth)	1.09 (±0.08)	0.94 (±0.08)
Early Fusion	**0.98** (±0.14)	**0.84** (±0.13)

Fig. 4. Regression results. The RMSE values of the different feature sets of the various modalities suggest that SCL is the most robust modality on its own, followed by video. Again, ECG has the worst performance. Early fusion is able to slightly outperform the unimodal results with an error of 0.98. The standard deviation is given in parentheses.

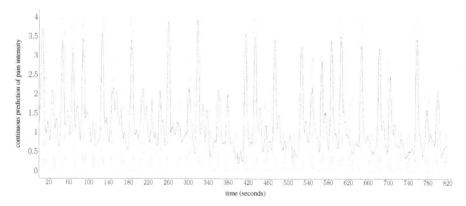

Fig. 5. Continuous estimation of pain intensity. The stimuli (especially the high ones) can be distinguished by the learned model. Smaller ones are more difficult to discern.

To visualize the possible quality of the estimation, the above mentioned features are extracted over the whole run of the experiment with a window length of 5.5 and an offset of 0.1 seconds. Training is done on the remaining 85 participants. Note, that the baseline is excluded from the training process. That means only levels 1-4 are used. The result of the continuous prediction can be seen in Figure 5. As can be seen, the high intensity pain is clearly discernible. It is more difficult to predict pain levels 1 and 2. It is interesting to see that despite excluding the baseline, the predictor learns a level that seems to be lower than the lowest level seen during training. The prediction has been slightly rescaled because the person-independent training led to a concentration of the predicted values in the middle of the spectrum at about 2.2. Note that the scaling factor can be determined using a hold-out set in case of person independence or additional snippets (if available) of the person to be tested for the best results (e.g. in a production setting). While person independent recognition seems to work well in this scenario, it will be interesting to see how the model performs in settings with a different (or omitted) manual calibration phase.

6   Conclusion

The presented findings suggest that person-independent estimation of pain intensity from physiological and video features is possible. Especially strong pain can robustly be recognized. Additionally, it was found that the SCL channel exhibits a very good performance with the computed features. As previously mentioned, the robustness of the classification process highly depends on reliable face and facial feature detection. Since other works showed that the video channel can be a highly effective modality, we plan to further investigate how to improve the prediction based on it. Furthermore, future work includes the investigation of the applicability of the presented approach for real-time purposes.

Acknowledgments. This paper is based on work done within the Transregional Collaborative Research Centre SFB/TRR 62 *Companion-Technology for Cognitive Technical Systems* funded by the German Research Foundation (DFG). Markus Kächele is supported by a scholarship of the Landesgraduiertenförderung Baden-Württemberg at Ulm University.

References

1. Artemiadis, P.K., Kyriakopoulos, K.J.: An EMG-based robot control scheme robust to time-varying EMG signal features. Trans. Info. Tech. Biomed. **14**(3), 582–588 (2010)
2. Ashraf, A.B., Lucey, S., Cohn, J.F., Chen, T., Ambadar, Z., Prkachin, K.M., Solomon, P.E.: The painful face - pain expression recognition using active appearance models. Image and Vision Computing **27**(12), 1788–1796 (2009)
3. Benedek, M., Kaernbach, C.: Decomposition of skin conductance data by means of nonnegative deconvolution. Psychophysiology **47**(4), 647–658 (2010)
4. Cao, C., Slobounov, S.: Application of a novel measure of EEG non-stationarity as 'Shannon- entropy of the peak frequency shifting' for detecting residual abnormalities in concussed individuals. Clinical neurophysiology : Official Journal of the International Federation of Clinical Neurophysiology **122**(7), 1314–1321 (2011)
5. Chen, W., Zhuang, J., Yu, W., Wang, Z.: Measuring complexity using FuzzyEn, ApEn, and SampEn. Medical Engineering & Physics **31**(1), 61–68 (2009)
6. Hammal, Z., Cohn, J.F.: Automatic detection of pain intensity. In: Proceedings of the International Conference on Multimodal Interaction, ICMI 2012, pp. 47–52. ACM (2012)
7. Kächele, M., Schels, M., Schwenker, F.: Inferring depression and affect from application dependent meta knowledge. In: Proceedings of the 4th International Workshop on Audio/Visual Emotion Challenge, AVEC 2014, pp. 41–48. ACM (2014)
8. Kächele, M., Werner, P., Al-Hamadi, A., Palm, G., Walter, S., Schwenker, F.: Biovisual fusion for person-independent recognition of pain intensity. In: Schwenker, F., Roli, F., Kittler, J. (eds.) MCS 2015. LNCS, vol. 9132, pp. 220–230. Springer, Heidelberg (2015)
9. Kaltwang, S., Rudovic, O., Pantic, M.: Continuous pain intensity estimation from facial expressions. In: Bebis, G., Boyle, R., Parvin, B., Koracin, D., Fowlkes, C., Wang, S., Choi, M.-H., Mantler, S., Schulze, J., Acevedo, D., Mueller, K., Papka, M. (eds.) ISVC 2012, Part II. LNCS, vol. 7432, pp. 368–377. Springer, Heidelberg (2012)
10. Lucey, P., Cohn, J.F., Prkachin, K.M., Solomon, P.E., Matthews, I.: Painful data: The UNBC-McMaster shoulder pain expression archive database. Image, Vision, and Computing Journal, 197–205 (2012)
11. Phinyomark, A., Limsakul, C., Phukpattaranont, P.: A novel feature extraction for robust EMG pattern recognition. Journal of Computing **1**(1), 71–80 (2009)
12. Schwenker, F., Dietrich, C.R., Thiel, C., Palm, G.: Learning of decision fusion mappings for pattern recognition. International Journal on Artificial Intelligence and Machine Learning (AIML) **6**, 17–21 (2006)
13. Tkach, D., Huang, H., Kuiken, T.A.: Research study of stability of time-domain features for electromyographic pattern recognition. J. Neuroeng Rehabil **7**, 21 (2010)
14. Treister, R., Kliger, M., Zuckerman, G., Aryeh, I.G., Eisenberg, E.: Differentiating between heat pain intensities: The combined effect of multiple autonomic parameters. Pain **153**(9), 1807–1814 (2012)

15. Walter, S., Gruss, S., Ehleiter, H., Tan, J., Traue, H., Werner, P., Al-Hamadi, A., Crawcour, S., Andrade, A., Moreira da Silva, G.: The BioVid heat pain database data for the advancement and systematic validation of an automated pain recognition system. In: 2013 IEEE International Conference on Cybernetics (CYBCONF), pp. 128–131, June 2013

16. Walter, S., Gruss, S., Limbrecht-Ecklundt, K., Traue, H.C., Werner, P., Al-Hamadi, A., Diniz, N., Silva, GMd, Andrade, A.O.: Automatic pain quantification using autonomic parameters. Psychology & Neuroscience **7**, 363–380 (2014)

17. Werner, P., Al-Hamadi, A., Niese, R., Walter, S., Gruss, S., Traue, H.C.: Automatic pain recognition from video and biomedical signals. In: International Conference on Pattern Recognition, pp. 4582–4587 (2014)

18. Zhao, G., Pietikainen, M.: Dynamic texture recognition using local binary patterns with an application to facial expressions. IEEE Transactions on Pattern Analysis and Machine Intelligence **29**(6), 915–928 (2007)

19. Zhao, Q., Zhang, L.: ECG feature extraction and classification using wavelet transform and support vector machines. In: International Conference on Neural Networks and Brain, ICNN B 2005, vol. 2, pp. 1089–1092, October 2005

Classification

Time Series Forecasting in Cyberbullying Data

Nektaria Potha and Manolis Maragoudakis[✉]

Department of Information and Communication Systems Engineering,
University of the Aegean, 83200 Karlovasi, Greece
{nekpotha,mmarag}@aegean.gr

Abstract. The present article deals with sexual cyberbullying, a serious subject that has gained significant attention throughout recent years of emerging social media platforms. The detection of sexual predation is one of the most important and challenging tasks in our days. Using real-world data, we follow a time series modeling approach, in which predator's posts (i.e. questions) are associated with numeric labels, according to the style of the attack (e.g. attempts for physical approach, grooming, retrieval of personal information, etc.). Upon modeling the domain as time series, in order to allow for forecasting the severity of a future question of a predator (i.e. the class label), a sliding window method was adopted. Two well-known methods that have been traditionally applied in time series problems, namely Support Vector Machines and Neural Networks, were utilized for forecasting. Simultaneously, since text processing is almost certain to derive a large number of input features, an additional method for reducing dimensionality of the original dataset was applied, implemented with Singular Value Decomposition. We demonstrate that the use of SVM classifier is more appropriate for our data and we show that it is able to provide accurate results that surpass current state-of-the-art outcomes, by using both the original feature set as well as the reduced SVD dimensions.

1   Introduction

After the Internet was invented, people have received a lot of benefits from it. On the other hand, the widespread use of the Internet and social media generally has created many problem and has affected us badly in many ways. Cyberbullying is the type of bullying that takes place using electronic means. It is defined as a person tormenting, threatening, harassing or embarrassing another person using the Internet or other technologies, like cell phones. In its most basic form, bullying involves two individuals, a bully or intimidator or predator and a victim. Sexual predation is a special case of cyberbullying and refers the sexual exploitation of children via the Internet. The Internet has become a prime place where sexual predators can lure children and teens into dangerous situations. Generally, a sexual predator is a person seen as obtaining or trying to obtain sexual contact with another person in a metaphorically "predatory" or abusive manner by using social media in order to lure children and adolescents from their families. Using chat rooms, email, and instant messaging can put a child at risk to encounter a sexual predator. Examples of sexual cyberbullying include mean text messages or

© Springer International Publishing Switzerland 2015
L. Iliadis and C. Jayne (Eds.): EANN 2015, CCIS 517, pp. 289–303, 2015.
DOI: 10.1007/978-3-319-23983-5_27

emails with sexual content, embarrassing pictures and videos posted on social network-ing sites. It is an undeniable fact that the use of Information and Communications Technology (ICT) and the Internet with specific strategy by sexual offenders, in order to exploit vulnerable children or adolescents for sexual or other abusive purposes tar-geting on victimization of minors, is a daily and very dangerous phenomenon which must be stopped. One main factor that have intensified this social menace is the ano-nymity of technology which allows cyber predators to constantly reach their targets (potential victims) to do so more intensely than regular bullying with a lessened sense of responsibility. In the online world, kids develop a sense of anonymity, which leads them to develop trust and confidence quickly with other online members. In this way, sexual predators prey upon these children by developing online relationships designed to lead to inappropriate discussions and meetings. This relatively new phenomenon of sexual cyberbullying has proven to be difficult for researchers to study due to its "in-tangible, no corporeal nature, much like other forms of cyber deviance". This intangi-ble, non-corporeal nature makes electronic mediums ideal for bullies because of the existence of the ease with which one may disguise themselves as another or remain entirely anonymous while online, with the ability to "hide" behind the computer screen. Temporary e-mail accounts and pseudonyms in chat rooms, instant messaging programs, and other Internet venues can make it very difficult for adolescents to de-termine the identity of aggressors (e.g. an unknown nickname is often used). Having the ability of hiding behind some form of anonymity by using their personal computer or mobile phone, an individual may bully another individual [1].

In the present paper, data mining is applied to the problem of sexual predation de-tection. The proposed method is utilizing a dataset of real dialogues (i.e. pairs of ques-tions and answers between cyber predator and the victim), in which each question of predator is manually annotated in terms of severity using a numeric label. In each online conversation, we only examine the set of questions of each predator, thus mod-eling this set as a multivariate time series. The feature space is represented as a bag of words, experimenting with the classic vector space model, known as term frequency-inverse document frequency representation (*tf-idf*), analytically is explained in section 3.6. In order to reduce dimensionality the Singular Value Decomposition (SVD) tech-nique was additionally incorporated. The goal of the present work is to capture the linguistic aspects of each predator's defensive or aggressive questions in order to predict the severity of the upcoming question. An interesting aspect of this dataset is the degree of severity, which is captured by signs of swearing or offending words. The prediction of the numeric label (category) according to the content of the pre-vious questions in each predator's question set is of great importance, thus not only gives the possibility of capturing the strategy followed by a predator but also provides the opportunity of generalizing the time series forecasting to unknown real-life conversation. In this way these predictions could be used as a guide or basis for deci-sion making. For instance, since the victims of sexual cyberbullying are mostly inno-cent or naïve children and adolescents, a system that could predict the severity of an upcoming question of the predator and thus, notify the victim would make the latter more careful and more self-conscious about the purposes of the other side of the di-alogue. Recall that in almost all cases, the final goal of the predator is to physically approach the victim. The main contribution of this work is the consideration of a varying size window of predators' previous posts in order to forecast the severity of their following post, using state-of-the-art time series analysis algorithms.

The rest of the paper is structured as follows: Section 2 briefly reviews previous work in cyberbullying detection while Section 3 describes the proposed forecasting approach. In Section 4, the experimental results are presented using the online conversations between predators and victims as corpora (as they are downloaded from Prevented Justice organization) and Section 5 includes the main conclusions drawn from this study and discusses future work directions.

2   Previous Work

Cyberbullying is a growing problem in the social web and is becoming a major threat to teenagers and adolescents. Cyberbullying is a well-studied problem from a social perspective while few studies have been dedicated to automatic cyberbullying detection, this is not an easy task. Textual contents in web environment is often unstructured, informal, and even misspelled, making undoubtedly difficult and even impossible in many times the identification of cyberbullying behavior. As a result, very few research teams are working on the detection of cyberbullying. At PAN-2012, a Sexual Predator Identification took place for first time [2]. Given a set of chat logs, the participants had to identify the predators among all users in the different conversations or the part (the lines) of the conversations which are the most distinctive of the predator behavior. In conclusion, it is impossible to identify predators using a unique method but it is necessary the use of different approaches. Moreover, the most effective method for identifying distinctive lines of the predator behavior in a chat log appeared to be those based on filtering on a dictionary or LM basis. In a recent study, Kontstathis et al., in order to trace instances of cyberbullying proposed a "bag-of-words" language model, which based on the text in online collection of posts [3]. Their corpus was taken by the website Formspring.me. Their goal was the identification of the most commonly used cyberbullying terms. Moreover, they exploited a supervised machine learning method called Essential Dimensions of LSI (EDLSI) in order to select additional terms of cyberbullying in Formspring.me data. The data was labeled using a web service, Amazon's *Mechanical Turk*. The *Mechanical Turk* (MTurk) is a crowdsourcing Internet marketplace that enables individuals or businesses (known as *Requesters*) to co-ordinate the use of human intelligence to perform tasks that computers are currently unable to do. It is one of Amazon's Wed Services. Yin et al., proposed a supervised learning approach for detecting harassment. Combining local features, sentiment features, and contextual features training a model for detecting harassing posts in chat rooms and discussion forums [4]. By employing a SVM classifier with the linear kernel and combining *tf-idf* measure as local features, sentiment features, and contextual features of documents proved that identification of online harassment provide significantly improved performance when *tf-idf* is supplemented with sentiment and contextual feature attributes. Furthermore, there are studies which have not only focused on the content of the conversations (dialogues between participants, usually the victim and the predator), but in the features and characteristics of those involved. In 2012, Dadvar et al., proposed cyberbullying detection based on gender information categorizing users into male and female groups. They hypothesized that incorporation of the users' information, such as age and gender, could improve the accuracy of cyberbullying detection [5].

3   Proposed Methodology

Time series forecasting is the process of training a model in order to provide predictions for future events based on known past events. Time series data has a natural temporal ordering. This differs from typical data mining applications, where each data point is an independent example of the concept to be learned, and the ordering of data points within a data set does not matter. The process examined in the current paper is depicted in Figure 1. As this figure illustrates, raw data taken from XML-format transcripts of 31 cases found in Perverted Justice [6] were converted to vectors using the *tf-idf* representation scheme. Upon creating input vectors, a multivariate time series is generated using the severity of the predator's question as a magnitude of the signal and the lexical terms as additional dimensions. Furthermore, we form our training set by applying a sliding window approach that sets the severity of the next question as the class of the problem. We experiment using either the multivariate time series from the original question set or also by transforming all terms into a reduced vector using the Singular Value Decomposition (SVD) method [7]. Upon that, we apply either Neural Nets or SVM in order to forecast the target variable and thus predict the label of the forthcoming predator's question [8].

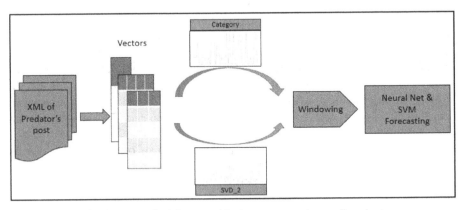

Fig. 1. The methodology used in the current study as a flowchart

By considering various values for the size of the window, we conduct cross-validation evaluation, in order to measure the performance of each classifier for each representation method, either the original *tf-idf* or the reduced SVD one. Finally, a closer inspection of the question set revealed that n-grams of term could also be exploited in order to gain in forecasting accuracy, therefore experiments also include variations of the size of the n-grams. The utilization of SVM as classifier could also be exploited by domain experts in order to have a better insight of which terms of phrases are closely associated with each severity label by studying the so-called "support vectors" which are characteristic instances that are close to the decision boundaries of each label. Figure 2 portrays our strategy for dealing with textual information and transform it to multivariate time series, using a sliding window validation. As mentioned before the collection of questions asked by predators is utilized as dataset in the present work. However, as the severity of each question is of course not available in a real life conversation, it is necessary to take the textual content of each predator's previous questions into consideration, in order to infer on the severity of the next question. Initially, the question set of each predator is represented as a vector based on the *tf-idf* representation. Each dimension corresponds to a separate term (single word) in case of a unigram approach or two or more neighboring terms in case of n-grams. By considering a window size of two (which was found to be the most accurate in the experimental evaluation), we calculate the *tf-idf* scores of all terms presented in the two foregoing questions (named as "current" and "previous" questions). For example, for the question belonging to category 200 (named as "next" in the blue box), we take into account the two previous questions marked as current and previous respectively. The input vector is created by calculating the *tf-idf* scores on the words "pictures", "naked" and "meet" (assuming that stopwords such as "I", "any", "have", etc. have been removed). Note that the category attributes (named as "label") in these two questions is not taken into account, since in a real-case scenario, this information is not available. In addition, for the question being part of category 900 (named as next in the brown box), we take into consideration the two previous questions marked as current (category of 200) and previous (category of 900) correspondingly. The input vector is constructed by estimating the *tf-idf* scores on the word "meet", the only word belonging to the previous questions (presuming that stopwords such as "ok" and "wanna" have been removed). The column, on the left of the figure 2, contains a time series representation of a real dialogue between a predator and a victim (only the questions posed by the predator). The vertical boxes at the data column depict the size of the windows were choosing. The transformation of the sliding window gives a generic data set (the horizontal data set) is used as the example set (or row data) for the predictive model where be applied, with the label variable shown in the green box in figure below. The green vertical box contains the label (target) variable that should be predicted by classifiers.

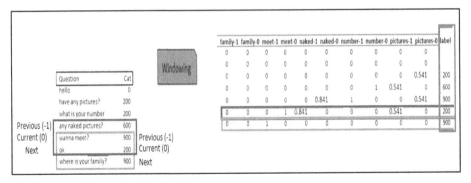

Fig. 2. The current time series data and its transformed structure according to sliding window method.

3.1 Time Series Forecasting

A time series is a sequence of values or readings ordered by a time parameter. In general consecutive samples of a time series are dependent on each other to an extent dictated by the process in question. Because of this dependency the prediction of the future of the series is possible. There are several algorithm that capable in predicting or forecasting time series data. In this work we apply the most widely used as time prediction classifiers, Support Vector Machine (SVM) and Multi-Layer Perceptron (MLP) neural net. In some past works, it was quite often that whenever feature selection was performed using linear SVM, the same algorithm also served as the classifier. Nevertheless, there are also previous researches that distinguish the two processes and use another algorithm for classification. In our setting, linear as well as non-linear SVM were measured and the performance was benchmarked against that of a Multi-Layer Perceptron (MLP) neural net. In the following section, analytic results tabulate this difference. For the forecasting task, a window size of K previous posts was chosen and the horizon was set to one (i.e. the following post).

3.2 Neural Networks

Neural Networks have been widely used as time series prediction. The MLP architecture was chosen to contain one hidden layer perceptron with the following data processing function:

$$Y_t^* = \Phi_{output}\left(\sum_{j=1}^{h} u_j \Phi_{hidden}\left(\sum_{l=1}^{K}\sum_{i=1}^{m} w_{m(l-1)+i}X_{it-l} + w_0\right) + u_0\right)$$

where Y_t^* is the prediction of the tth post, X_i is the data for current post i, K is the window size, m is the dimensionality of the input vector, w_i are the MLP hidden layer coefficients, v_j represent the output layer coefficients, n is the number of inputs, h is the number is neurons within the hidden layer and $\varphi(x)$ are the activation functions. As regards to the output layer, the function (1) is utilized and for hidden layer, the traditional sigmoid function (2) is used. In order to estimate the perceptron's weights, a genetic algorithm was applied, following the approach of Korning PG. study [9].

$$\Phi_{output}(x) = \left(\frac{1}{1 + e^{-x}} - 0.5\right) \quad (1) \qquad \phi_{hidden}(x) = \frac{1}{1 + e^{-x}} \quad (2)$$

3.3   Support Vector Machines (SVM)

A new, learning and powerful algorithm for any classification problem is SVM (Support Vector Machines). Support vector machines derived from statistical learning theory by Vapnik and Chervonenkis [10]. An SVM model is a representation of the examples as points in space, mapped so that the examples of the separate categories are divided by a clear gap that is as wide as possible. New examples are then mapped into that same space and predicted to belong to a category based on which side of the gap they fall on. A support vector machine constructs a hyperplane or set of hyperplanes in a high- or infinite-dimensional space, which can be used for classification, regression, or other tasks. Intuitively, a good separation is achieved by the hyperplane that has the largest distance to the nearest training data point of any class (so-called functional margin), since in general the larger the margin the lower the generalization error of the classifier. SVMs maximize the margin around the separating hyperplane (between the two different classes). As margin of the separator is called the width of separation between support vectors of classes. Support vectors are the examples closet to the hyperplane. In a linear hypothesis space all training data is at least distance 1 from the hyperplane [11].

$$x_i + b \geq 1 \ if \ y_i = 1$$
$$W^T x_i + b \leq -1 \ if \ y_i = -1$$

Linear hyperplane defined by support vectors. For support vectors, the distance from the hyperplane is:

$$r = \frac{W^T x + b}{\|W\|} = \frac{1}{\|W\|}$$

3.4   Singular Value Decomposition (SVD)

An indexing and retrieval method is Latent semantic indexing (LSI) that uses a mathematical technique called singular value decomposition (SVD) to identify patterns in the relationships between the terms and concepts contained in an unstructured collection of text. LSI is based on the principle that words that are used in the same contexts tend to have similar meanings. The goal of LSI is the extraction of the "meaning" of words by using their co-occurrences with other words that appear in the documents of a corpora. A key feature of LSI is its ability to extract the conceptual content of a body of text by establishing associations between those terms that occur in similar contexts. LSI begins by constructing a term-document matrix A to identify the occurrences of the m unique terms within a collection of n documents. In a term-document matrix, each term is represented by a row, and each document is represented by a column, with each matrix cell $a_{i,j}$. The singular value decomposition (SVD) is a factorization of a real or complex matrix, with many useful applications in signal processing and statistics. SVD is an alternative for feature selection approach in order to alleviate the issue of high-dimensionality of data. The main idea behind is that SVD can reveal "concepts" i.e. latent relationships between features in an ordered manner, so that one could choose the degree of dimensionality reduction. The first SVD dimension that results upon transforming the n*m matrix represents the level of interrelations between the rows (i.e. the questions of the predator) and the columns (i.e. the terms), as set by trivial metrics such as the count of terms and the length of each vector (i.e. how many terms were found in a single question). Therefore, this dimension is of no use for the task at hand. The second, the third and so forth dimensions capture latent semantic relations in a decreasing order of significance [7].

3.5   Data Description

In this study, the dataset consists of a set of chat transcripts in text files as collected form Prevented Justice. Prevented Justice is an American organization which investigates, identifies, and publicizes the content of online sexual conversations between adults and adults posing as minors. These files contain entire conversations between adults posing as young teens and convicted predators. In all questions asked by the predator, a numeric class label was assigned, manually annotated. The values of the class label are from the set {0,200,600,900}. Zero is assigned to posts where no cyberbullying activity or intention is identified. Questions which contain personal information are described as category 200. In this category, indicators such as personal information (age, hobbies). Category 600 is assigned as grooming and characterizes the posts which involve the use of sexual terminology (e.g., sex, orgasm, kiss, naked, etc.). Any attempt of the predator to physically approach the victim is labeled as category 900 and includes terms that are associated with such a behavior, such as approach verbs (e.g., come, meet), families nouns (e.g., mom, divorce), etc. In this category, predator intends to meet the victim, obviously try to isolate the victim from its support network of family, friends, etc., for example, the use of a noun such as divorce could represent a predators attempt to gain insight on the physical location of the victims family [12],[13]. According to the Perverted Justice organization the complete unedited chat logs are posted to the website only after the person's legal case has been resolved. The current follow-up process consists of notifying a community of the offender's status once a person has been arrested and convicted. In the present work, we examine the questions set of each predator. As previously mentioned, the goal of the current work is to capture the textual aspects of each predator's defensive or aggressive questions and create a predictor that will be based on the content of a number of previous questions (in our experiments we found that number to be equal to 2), in order to predict the severity (also mentioned as category in the article) of his next question to the potential victim. To the best of our knowledge, time series have never been used before in cyberbullying data. Therefore, we cannot directly compare our results with those of previous studies of cyberbullying. Despite the fact that we utilize exactly the same data set as in work of Kontostathis et al., [3], we present a completely different approach.

3.6   Preprocessing

Having all predator posts as a set of XML transcripts, we parsed and transformed all this collection into a vector set. Analysis of these transcripts revealed that there are certain acronyms and text shorthand used in web chat situations that need to be taken into consideration and would be neglected if one applied a standard preprocessing procedure. Some characteristic examples of such cases include the use of "121", which is a well-known shorthand for the phrase "one to one", "182", which is used to represent "I hate you", "ADIDAS" for "All Day I Dream About Sex" and many others. For a full list of acronyms, visit: *http://www.netlingo.com*. As a result, the only preprocessing steps that we applied was a tokenization based on the space character a stop word removal and a case of transformation, in contrast with the most text mining approaches that there are some standard preprocessing steps such as stop words removal, tokenization, stemming and Part-Of-Speech tagging. The feature representation of text documents plays a critical role in many applications of data-mining. Each document is represented by a high dimensional sparse vector, using the *tf-idf* weighting scheme. The *tf-idf* representation is a way to score and rank the importance of words (or "terms") in a document. The importance of a term depends of the frequency of its appearance in a document or across multiple documents. A word (term) receives a high score (weight) when appears frequently across multiple documents. This word is considered of great importance in its contextual document corpus. The size of the resulted matrix of each document equals $n \times (k*m+1)$, where n is the number of questions of each predator (i.e. the rows of the matrix) and m is the number of terms obtained upon the text-processing phase (i.e. the columns). Since we utilize a windowing approach, the total size of features equals to $k*m$, where k is the window size. The +1 is for the class label, which is the numeric score of the level of assault, set by the annotators according to the rules as explained previously at the data description section. Therefore, each set of questions asked by a predator was modeled as a time series ts_j, with j ranging from 1 to 33 (i.e. the number of transcribed logs from the aforementioned dataset) and each ts_j is represented as an $n \times (k*m+1)$ matrix. It is evident that using the specific data representation format, m should be quite large. For our case, k was set to various sizes from 2 to 5 and 2 was found to be the most accurate.

4   Experimental Study

4.1   Forecasting Performance

In order to validate the outcome of the proposed methodology, was measured the forecasting performance of the category label for each predator, utilizing two different prediction algorithms, SVM using a Gaussian Kernel and MLP Neural net. While experimenting with learning algorithms, it is important not to test the prediction of an estimator on the data used to fit the estimator as this would not be evaluating the performance of the estimator on new, unseen data. Therefore, we start by splitting the data into a train (60%) and test set (40%), then a common basis for evaluating the performance of a classifier is the confusion matrix. Specifically, in the current work, four labels were considered within the class namely {0,200,600,900}. The evaluation criteria used in order to demonstrate the robustness of our predictive models are described in detail bellow. The reference on the performance of the present predictive models indicates the model's ability to correctly predict the classes.

The confusion matrix shows how the predictions are made by the model for each question set per predator. It provides a summary of classification outcomes and permits the inspection of the number of correct and incorrect predictions in each class of each signal. Finally, we estimate the value of precision and recall measures respectively for each predator (signal) taking the average of all classes. Precision is a measure of the accuracy provided that a specific class has been predicted. It is defined by function (3), where TP and FP are the number of true positive and false positive predictions for the considered class. Recall is a measure of the ability of a prediction model to select instances of a certain class from a dataset. It is commonly also called sensitivity and corresponds to the true positive rate. It is defined by formula (4), where TP and FN are the numbers of true positive and false negative predictions for the considered class. Furthermore, one of the most common metric has been considered, namely the Root Mean Square Error. The Root Mean Square Forecasting Error (RMSFE) of an algorithm i is calculated by the equation (5), where $P_{(ij)}$ is the value that algorithm i forecasted for the sample j (from a set of examples) and T_j is the value of the "target" value for the jth example. For an ideal forecast, $P_{(ij)} = T_j$ and $RMSFE_i = 0$. So, the error indicator varies from 0 to infinity, with 0 to correspond to the ideal prediction [14].

$$ \text{Precision} = \frac{TP}{TP + FP} \quad (3) \quad \text{Recall} = \frac{TP}{TP + FN} \quad (4) \quad RMSFE_{i=} \sqrt{\frac{\sum_{j=1}^{n} |P_{(ij)} - T_j|^2}{n}} \quad (5) $$

4.2   Experimental Results

This section presents the experimental outcome of the time series forecasting process, using, the SVM and Neural Net classifiers as predictive models, for the original and the SVD reduced signals. The comparison of the results, obtained by the two prediction models, is shown on the following tables. Specifically, Table 1 decomposes the prediction performance for both original and SVD signals of each predator, using the SVM classifier. We observe that the prediction of SVM classifier is very high and more effective when implementing the SVD methodology than the original signals, with a varying difference from 0% to 3.7%. Note that the best scores are shown in boldface letters for each case studied. Note that the experimental results shown express a statistical significance of 95% ±1.7.

Table 2 tabulates the forecasting performance of both original and SVD signals for each predator using the Neural Net classifier. Neural Net classifier gives better classification performance in SVD than the original signals, however, this difference is now smaller than before. It is noticeable that SVD signals depict better performance when using either classifier compared with the original signals. This is somehow expected since high dimensionality data and scarcity pose significant challenges to both classifiers, because the performance of these algorithms may substantially deteriorate due to high dimensionality and existence of many noisy features in these data.

By comparing the results of the Tables 1 and 2 the best results are obtained utilizing the SVM classifier for SVD dimension signals. SVM outperforms Neural Net on performance prediction by a factor of 4 % and 6% for the implementation of the original and SVD signals respectively. Neural Nets are also showing satisfactory forecasting performance for both SVD and original signals. The effect of SVD is slightly worse in Neural Nets than in SVM, which could be attributed to the fact the former does not use kernel functions to cope with multi-dimensional data as the latter do.

Table 1. Analytical performance evaluation (RMSFE, PRECISION, and RECALL) per predator for both original and SVD dimension Signals, using: Window size=2 for SVM classifier.

Name	Original			SVD		
	precision	recall	RMSFE	precision	recall	RMSFE
ArmySgt196	1.000	1.000	0.000	1.000	1.000	0.000
arthinice	0.910	0.910	0.000	1.000	1.000	0.000
asian_kreationz	0.922	0.922	0.000	0.989	0.890	0.037
aticloose	0.899	0.899	0.032	0.999	0.910	0.030
corazon23456	0.878	0.888	0.039	0950	0.988	0.010
crazytrini85	0.897	0.857	0.031	0.930	0.973	0.021
flxnonya	0.970	0.870	0.029	0.992	0.902	0.011
fotophix	0.900	0.930	0.041	0.992	0.962	0.011
ghost27_73	1.000	1.000	0.000	1.000	1.000	0.000
hiexcitement	0.929	0.877	0.012	0.998	0.939	0.010
i_8u_raw	0923	0873	0.019	0.988	0.918	0.023
icepirate53	0900	0980	0.029	0.997	0.971	0.019
italianlover37	0.880	0.840	0.067	0.908	0.999	0.036
jleno9	0.970	0.870	0.030	0.876	0.976	0.031
jon_raven2000	0.940	0.890	0.050	0.867	0.907	0.039
lee_greer74	0.928	0.988	0.010	0.999	0.899	0.010
manofdarkneedsl951	0.883	0.943	0.020	0.988	0.928	0.020
marc_00_48089	0.992	0.912	0.020	0.897	0.917	0.027
needinit1983	0.992	0.892	0.011	0.960	0.990	0.024
sebastian_calif	0.930	0.921	0.011	1.000	1.000	0.000
sjklanke	0.901	0.890	0.015	1.000	1.000	0.000
sphinx_56_02	0.900	0.939	0.039	0.967	0.997	0.029
spongebob_giantdick	0.949	0.999	0.038	0.999	0.899	0.028
stylelisticgrooves	0.950	0.950	0.047	0.960	0890	0.038
sugardavis	0.930	0.990	0.038	0.970	0.890	0.018
sweet_jason002	0.919	0.999	0.028	0.990	0.899	0.028
texassailor04	0.910	0.880	0.079	0.950	0.990	0.039
the_third_storm	0.942	0.912	0.059	0.972	0.912	0.031
thedude420xxx	0.912	0.940	0.060	0.980	0.940	0.039
tunnels12000	0.984	0.983	0.029	1.000	1.000	0.000
user194547	0.970	0.971	0.025	1.000	1.000	0.000

Table 2. Analytical performance evaluation (RMSFE, PRECISION, and RECALL) per predator for both original and SVD dimensions signals, using: Window size=3 for Neural Net MLP classifier.

Name	Original			SVD		
	precision	recall	RMSFE	precision	recall	RMSFE
ArmySgt196	0.978	0.910	0.024	0.980	0.900	0.020
arthinice	0.931	0.980	0.039	0.929	0.977	0.023
asian_kreationz	0.850	0.899	0.037	0.852	0.889	0.010
aticloose	0.940	0.940	0.020	0.949	0.944	0.031
corazon23456	0.988	0.850	0.019	0.979	0.849	0.020
crazytrini85	0.933	0.901	0.021	0.943	0.905	0.021
flxnonya	0.892	0.992	0.031	0.902	0.992	0.011
fotophix	0.992	0.912	0.031	0.992	0.922	0.011
ghost27_73	0.890	0.930	0.029	0.888	0.933	0.018
hiexcitement	0.978	0.900	0.033	0.990	0.990	0.015
i_8u_raw	0.970	0.910	0.028	0.970	0.900	0.035
icepirate53	0.890	0.860	0.043	0.900	0.870	0.041
italianlover37	0.950	0.900	0.030	0.948	0.911	0.032
jleno9	0.990	0.980	0.011	**1.000**	**1.000**	**0.000**
jon_raven2000	0.859	0.899	0.050	0.844	0.880	0.043
lee_greer74	0.999	0.839	0.010	0.999	0.849	0.010
manofdarkneedsl951	0.928	0.948	0.020	0.930	0.950	0.020
marc_00_48089	0.940	0.800	0.023	0.930	0.800	0.026
needinit1983	0.960	0.910	0.027	0.967	0.920	0.029
sebastian_calif	0.900	0.920	0.010	0.900	0.930	0.015
sjklanke	**1.000**	**1.000**	**0.000**	**1.000**	**1.000**	**0.000**
sphinx_56_02	0.929	0.859	0.039	0.978	0.978	0.039
spongebob_giantdick	0.959	0.899	0.028	0.962	0.902	0.028
stylelisticgrooves	0.960	0.911	0.027	0960	0920	0.017
sugardavis	0.831	0.930	0.071	0.830	0.930	0.034
sweet_jason002	0.899	0.999	0.028	0.900	0.990	0.028
texassailor04	0.920	0.888	0.051	0.960	0.920	0.031
the_third_storm	0.942	0.822	0.054	0.942	0.830	0.039
thedude420xxx	0.940	0.910	0.050	0.930	0.819	0.042
tunnels12000	**1.000**	**1.000**	**0.000**	**1.000**	**1.000**	**0.000**
user194547	**1.000**	**1.000**	**0.000**	**1.000**	**1.000**	**0.000**

Upon examining the experimental outcomes, an interesting observation is that it is possible to model the set of questions posed by a predator as a time series and unlike the majority of approaches that follow a typical regression analysis, incorporate textual information in a multivariate time series and achieve better classification outcomes using popular classifiers such as Neural Net MLP and SVM. We have to mention that all the predators are convicted and the total set of these dialogs were published by Prevented Justice, so the perpetrators' pseudonyms are not considered personally identifiable information, as they have already used in other published works [13],[15]. Table 3 illustrates the output of the forecasting performance of the models applied in a more aggregated manner. Initially, we consider the original set of questions and experiment with various sizes of n for the term n-grams using either SVM or MLP as classifiers. Then, we also consider the transformation of the original dataset into the SVD representation, using the first 5 singular values. Again, the two classifiers mentioned above were taken into consideration. Each cell holds the average value of   precision and recall measures for all predators (signals) and the average of all class labels. An important observation is that the use of word n-grams representation provides very high classification accuracy, demonstrating that when the textual documents are cleaned as much as possible from the useless information that they contain, the forecasting is more effective.

Table 3. he performance evaluation (RMSFE, PRECISION, RECALL) for various sizes of the terms n-grams: Window size=3, Learners SVM &Neural Net.

n-grams	Original			SVD		
	precision	recall	RMSFE	precision	recall	RMSFE
SVM						
n=1	0.890	0.780	0.047	0.920	0.893	0.031
n=2	0.969	0.956	0.019	0.987	0.933	0.010
n=3	0.920	0.870	0.024	0.977	0.944	0.022
Neural Net MLP						
n=1	0.850	0.800	0.050	0.867	0.816	0.036
n=2	0.941	0.934	0.023	0.969	0.901	0.029
n=3	0.930	0.901	0.034	0.944	0.910	0.028

5   Concluding Remarks

The present article presents our study on time series forecasting in sexual cyberbullying data and evaluation of the strategies used by online sexual predators in their efforts to develop relationships with minors using the Internet. For the purposes of our research, 31 real world transcriptions have been used as source data, obtained from a well-known American organization (i.e. Perverted Justice), which inspects, identifies, and publicizes the conduct of adults who solicit online sexual conversations with adults posing as minors. Based on the questions asked by each predator and a manual assignment of a category label, denoting the type of attack, analysis was performed in order to model the predator's tactics. We described how we can transform each question set as a multivariate time series forecasting process, by utilizing a sliding widow of size equal to two and demonstrate that the prediction of the SVM classifier is more satisfactory than MLP. Furthermore, in order to reduce the dimensionality of our data, an alternative experimental phase was run, this time using the same classifiers but data have been significantly reduced using the well-known SVD algorithm. Experiments show that the implementation of the SVD representation on the original signals leads to a better classification accuracy notably improving the outcome of the two classifiers and reducing the error indicator between of the real and the forecasting value, both in SVM and Neural Net MLP implementation. By reducing the size of the attributes set, we reinforce the view that larger number of weak features gives inferior performance. Furthermore, the representation based on word n-grams performs high results on the prediction of the label of the next question in the aforementioned classifiers, showing up that the two word sequence of words is the best option and confirming that the sequence of the words is powerful and useful. It is well-known that at the majority of cases, an online conversation is informal, short ranging from a few words to a few lines and its content is not well structured. The inclusion of text-preprocessing such as removal of stopwords and *tf-idf* filtering also helped in reducing the bias posed by many irrelevant words such as idioms, slang, etc. Moreover, the present study showed that it is possible to accurately predict label of the following predator's post in an online chat transcript. The main strength of our approach remains the use of the complete dialogs courses and their modeling as time series. As time series are used in main real-life cases, we are of the belief that cyberbullying investigators can gain positive insights when dealing with the temporal dimension. However, there are also some limitations such as the need for a large number of dialogue interactions so that our model can work correctly and in some cases with numerous examples, the method can be computationally expensive. In the future, we plan to utilize an extension of our previous experimental setup in order to support multi-labels and a more fine-grained categorization.

The importance of the results drawn by the aforementioned approach is based on the argument that cyberbullying identification remains a critical problem. More specifically, a potential use of such a system could be related to minors, who may be potential victims of sexual cyberbullying attack during an online dialogue. By having a system that warns them in order to be more careful and pay attention to suspicious signs such as vulgar languages, sexual content insults, etc., these minors could be more aware of the actual intentions of the predator. Finally, the prediction of the severity of the upcoming post in an online conversation, helps cybercrime investigation experts to detect specific behavioral patterns.

References

1. Cyberbullying, The National Crime Prevention. http://www.ncpc.org/cyberbullying
2. Inches, G., Crestani, F.: Overview of the international sexual predator identification competition at pan-2012. In: Forner, P., Karlgren, J., Womser-Hacker, C. (eds.) CLEF 2012 Evaluation Labs and Workshop - Working Notes Papers, Rome, Italy (2012)
3. Kontostathis, A., Reynolds, K., Garron, A., Edwards, L.: Detecting cyberbullying: query terms and techniques. In: Proceedings of the 5th Annual ACM Web Science Conference, pp. 195–204 (2013)
4. Yin, Z.X., Hong, L., Davison, B.D., Kontostathis, A., Edwards, L.: Detection of harassment on web 2.0. In: CAW 2.0 2009: Proceedings of the 1st Content Analysis in Web 2.0 Workshop, Madrid, Spain (2009)

5. Dadvar, M., de Jong, F., Ordelman, R., Trieschnigg, D.: Improved cyberbullying detection using gender information. In: Proceedings of the 12th -Dutch-Belgian Information Retrieval Workshop (DIR 2012), Ghent, Belgium (2012)
6. http://www.Perverted-Justice.com (2008)
7. Kalman, D.: Asingularly valuable decomposition: the SVDof a matrix. College Math. J. **27**, 2–23 (1996)
8. http://www.simafore.com/blog/bid/109051/Time-Series-Forecasting-comparing-RapidMiner-and-R-for-analysis
9. Korning, P.G.: Training neura networks by means of genetic algorithms working on very long chromosomes. International Journal of Neural Systems **6**(3), 299–316 (1995)
10. Vapnik, V.: The Natural Of Statistical Learning Theory. Springer, New York (1995)
11. Chang, Y.-W., Lin, C.-J.: Feature ranking using linear SVM. J. Machine Learning Res. **3**, 53–64 (2008)
12. Kontostathis, A., West, W., Garron, A., Reynolds, K., Edwards, L.: Identify predators using chatcoder 2.0 - notebook for pan at clef 2012. In: Forner, et al. (eds.)
13. Kontostathis, A., Edwards, L., Leatherman, A.: ChatCoder: toward the tracking and categorization of internet predators. In: Proceedings of Text Mining Workshop 2009 Held in Conjunction with the Ninth SIAM International Conference on Data Mining (SDM 2009) (2009)
14. Davis, J., Goadrich, M.: The relationship between precision-recall and ROC curves. In: Proceedings of the International Conference on Machine Learning (ICML) (2006)
15. McGhee, I., Bayzick, J., Kontostathis, A., Edwards, L., McBride, A., Jakubowski, E.: Learning to Identify Internet Sexual Predation. International Journal of Electronic Commerce **15**(3), 103–122 (2011)

Classification of Binary Imbalanced Data Using A Bayesian Ensemble of Bayesian Neural Networks

Marcelino Lázaro[1]([✉]), Francisco Herrera[2], and Aníbal R. Figueiras-Vidal[1]

[1] Department of Signal Theory and Communications,
Carlos III University of Madrid, 28911 Leganés, Spain
`mlazaro@tsc.uc3m.es`
[2] Department of Computer Science and Artificial Intelligence,
University of Granada, 18071 Granada, Spain

Abstract. This paper presents a new method to deal with classification of imbalanced data. A Bayesian ensemble of neural network classifiers is proposed. Several individual neural classifiers are trained to minimize a Bayesian cost function with different decision costs, thus working at different points of the Receiver Operating Characteristic (ROC). Decisions of the set of individual neural classifiers are fused using a Bayesian rule that introduces a "balancing" parameter allowing to compensate the imbalance of available data.

Keywords: Imbalanced data · Classification · Neural networks · Bayes risk

1 Introduction

Pattern classification is the act of taking in raw data and making a decision to assign it to a category or class [1]. There are many real world classification tasks in industry, business and science. In many cases, classification task has to be learned from an available labeled data set, containing samples of the objects to be classified along with their corresponding class labels. Machine learning methods, and specifically neural networks, have been extensively used to solve this kind of classification problems [2,3].

In last years, the classification of imbalanced data has attracted a great attention. Imbalanced data means that the number of instances of each class are very different. Focusing in a binary classification problem, learning from imbalanced data has the difficulty of representing the minority class, that can be shrouded in the thicker cloud of samples of the majority class. This is a serious concern in applications where the importance of detecting the minority class is high, such as medical or fraud detection applications.

This work has been partially supported by Research Grant S2013/ICE-2845 (CASI-CAM-CM), DGUI - Comunidad de Madrid

© Springer International Publishing Switzerland 2015
L. Iliadis and C. Jayne (Eds.): EANN 2015, CCIS 517, pp. 304–314, 2015.
DOI: 10.1007/978-3-319-23983-5_28

Different techniques have been used to face the problem of classification of imbalanced data: standard algorithms modified to compensate imbalance; pre-processing techniques that modify the data set to balance it (removing samples of the majority class, or introducing synthetic samples of the minority class); or ensembles of classifiers to improve the accuracy of individual classifiers. A detailed review of several of these methods can be found in [4,5]. In particular, ensembles of classifiers have shown excellent results [5,6].

Bayesian theory [7] allows to specify different costs for errors classifying each class, and at the same time to take into account the prior probability of each class. This formulation can be usefull to deal with imbalanced data.

In this paper, we will introduce a new ensemble method to deal with imbalanced binary classification problems. A "balancing" parameter is introduced in the design of the individual classifiers and in the design of the ensemble. This "balancing" parameter, based on the Bayesian formulation, allows to specify the relative importance of errors in the decision for each class.

2   Bayes Risk and Training of Neural Networks

A classification problem consists in assigning a D-dimensional pattern $\mathbf{x}$ (instance or sample) to one out of a known set $\mathcal{H}$ of possible classes or hypotheses. In a binary classification problem only two classes are possible, namely $\mathcal{H} = \{0, 1\}$. Bayesian formulation considers the a priori class probabilities along with the different costs of each possible decision for samples of every class. The goal of a Bayesian classifier is to minimize the Bayesian risk function, which includes the statistical average of these costs [7,8]

$$\mathcal{R} = \sum_{t \in \mathcal{H}} \sum_{d \in \mathcal{H}} \pi_t \, c_{d,t} \, p_{\hat{\mathcal{H}}|\mathcal{H}}(d|t) \tag{1}$$

where π_t denotes the probability of hypothesis t, $c_{d,t}$ is the cost of deciding hypothesis d when the true hypothesis is t, and $p_{\hat{\mathcal{H}}|\mathcal{H}}(d|t)$ denotes the conditional probability of this decision

$$p_{\hat{\mathcal{H}}|\mathcal{H}}(d|t) \equiv P(\text{Decide } d \mid t \text{ is true}) \tag{2}$$

The classifier minimizing the Bayesian risk is defined by the likelihoods (conditional distributions of input pattern under both hypothesis, $f_{\mathbf{x}|\mathcal{H}}(\mathbf{x}|t)$, for $t \in \{0, 1\}$), as well as the a priori class probabilities and decision costs. The optimal decision rule minimizing Bayesian risk is

$$\Lambda(\mathbf{x}) = \frac{f_{\mathbf{x}|\mathcal{H}}(\mathbf{x}|1)}{f_{\mathbf{x}|\mathcal{H}}(\mathbf{x}|0)} \underset{\hat{\mathcal{H}}=0}{\overset{\hat{\mathcal{H}}=1}{\gtrless}} \frac{(c_{1,0} - c_{0,0})}{(c_{0,1} - c_{1,1})} \frac{\pi_0}{\pi_1} = \gamma \, (> 0) \tag{3}$$

i.e., a test comparing the likelihood ratio (LR) with a threshold γ given by costs $c_{d,t}$ and prior probabilities π_t [7,8].

A binary classifier is usually characterized by the false alarm and the miss probabilities (probabilities of erroneous decisions under both hypothesis)

$$p_{FA} = p_{\hat{\mathcal{H}}|\mathcal{H}}(1|0) \equiv P(\text{Decide } 1 \mid 0 \text{ is true}) \qquad (4)$$

$$p_M = p_{\hat{\mathcal{H}}|\mathcal{H}}(0|1) \equiv P(\text{Decide } 0 \mid 1 \text{ is true}) \qquad (5)$$

Obviously, different values of γ originate different pairs of values for p_{FA} and p_M, which show a compromise: reducing one of them means increasing the other. The Receiver Operating Characteristic (ROC) is a curve that represents $p_D = 1 - p_M$ vs. p_{FA} for $0 \leq \gamma < \infty$, i.e., draws the working points (p_{FA}, p_D) from $(0,0)$ to $(1,1)$ [9]. The Area Under this Curve (AUC) is a parameter that serves to compare different classifier designs when $c_{d,t}$ or π_t are not given [10,11].

In binary classification, Bayesian formulation can be simplified assuming $c_{1,1} = c_{0,0} = 0$, and parameterizing the other two costs by means of a single parameter α as follows

$$c_{1,0} = \frac{\alpha}{\pi_0}, \ c_{0,1} = \frac{1-\alpha}{\pi_1} \qquad (6)$$

Using this parameterization, the Bayesian risk becomes

$$\mathcal{R}(\alpha) = \alpha \, p_{FA} + (1-\alpha) \, p_M \qquad (7)$$

and decision threshold is now given by $\gamma = \alpha/(1-\alpha)$. Parameter α establishes the relative importance given to errors for samples of both classes.

In practice, in many real problems distributions involved in (3) are unknown, and classifiers have to be designed from a set of labeled observations, $\{\mathbf{x}_k, y_k\}$, $k \in \{1, 2, \cdots, N\}$, with binary class labels y_k. In this case, a machine architecture with a number of trainable parameters, $\mathbf{w}$, can be used to obtain a classification rule. Machine learning methods, and in particular neural networks, can be used in several different ways to solve binary classification problems [3]. Here, architecture to implement the classifier is constrained to neural networks with a single output and a threshold based decision.

For these networks, typically labels $y_k = -1$ and $y_k = +1$ are associated to samples belonging to hyphotesis $\mathcal{H} = 0$ and $\mathcal{H} = 1$, respectively. The soft output of the network is

$$z_k = g(\mathbf{x}_k, \mathbf{w}) \qquad (8)$$

where $g(\mathbf{x}, \mathbf{w})$ is the parameter-dependent non-linear transfer function of the neural network, which depends on the specific network architecture. Finally, the decision rule of the neural classifier, in terms of the given binary labels, is

$$\hat{y}_k = \text{sgn}(z_k) \qquad (9)$$

where $\text{sgn}(\cdot)$ is the well known sign function.

The design of the neural classifier consists in obtaining the values of neural network parameters, $\mathbf{w}$. Frequently, $\mathbf{w}$ is obtained by imposing the minimization of some appropriate cost function $J(\mathbf{w})$ that measures the difference between z_k and y_k. If cost is conveniently selected, network output z_k can provide an estimate of the posterior probabilities for each class and there is a relationship with Bayesian formulation (see [3,12] for more details). This is the case of the Mean Squared Error (MSE) cost function

$$J^{MSE}(\mathbf{w}) = \frac{1}{N} \sum_{k=1}^{N} (y_k - z_k)^2 \tag{10}$$

probably the most used cost function because it can be used to minimize the probability of error (Bayesian risk for $c_{0,1} = c_{1,0}$, or equivalently $\alpha = \pi_0$). This cost is useful for balanced problems, but for imbalanced problems, with $\pi_0 >> \pi_1$, reducing p_{FA} has a bigger impact in cost than reducing p_M, thus resulting in a poor performance for the minority class.

Typically, the cost function is iteratively minimized by using a gradient descent method to adapt parameters from iteration $(i - 1)$ to iteration i

$$\mathbf{w}^{(i)} = \mathbf{w}^{(i-1)} - \mu \frac{\partial J(\mathbf{w})}{\partial \mathbf{w}} \tag{11}$$

It is interesting to remark that gradient expression in updating equation (11) can be written as

$$\frac{\partial J(\mathbf{w})}{\partial \mathbf{w}} = \sum_{k=1}^{N} \frac{\partial J(\mathbf{w})}{\partial z_k} \frac{\partial z_k}{\partial \mathbf{w}} \tag{12}$$

Term $\frac{\partial z_k}{\partial \mathbf{w}}$ depends on network architecture, in particular on the dependence of neural network output on network parameters. Term $\frac{\partial J(\mathbf{w})}{\partial z_k}$ is the one including the dependence on the cost function. For MSE cost function

$$\frac{\partial J^{MSE}(\mathbf{w})}{\partial z_k} = -\frac{2}{N} (y_k - z_k) \tag{13}$$

3   Proposed Methods

In this communication we propose a classifier based on the ensemble of several binary neural network classifiers. The neural classifiers as well as the ensemble classifier will be based on the Bayes risk. In this section, first the individual classifiers will be presented, and then the ensemble classifier will be formulated.

3.1   Individual Bayesian Neural Network Classifiers

In the design of the individual classifiers that will compose the ensemble, the aim is to find the network parameters minimizing an estimate of Bayes risk (7) for a given value of parameter α when decision rule is subject to be provided by a neural network like the one presented in Section 2, i.e., given by (8) and (9).

Considering decision rule (9), false alarm (false positive) and miss (false negative) probabilities defining risk (7) for the neural classifier are

$$p_{FA} = \int_{0}^{\infty} f_{Z|\mathcal{H}}(z|0) \, dz, \quad p_M = \int_{-\infty}^{0} f_{Z|\mathcal{H}}(z|1) \, dz \tag{14}$$

In general, conditional distributions on $\mathcal{Z}$, which models network output (8), are unknown. The proposed training method will estimate these distributions from available data using Parzen window estimator [13]. If sets $\mathcal{S}_0$ and $\mathcal{S}_1$ contain indexes for data corresponding to hypothesis $\mathcal{H} = 0$ and $\mathcal{H} = 1$, respectively

$$\mathcal{S}_0 = \{k : y_k = -1\} \text{ and } \mathcal{S}_1 = \{k : y_k = +1\} \tag{15}$$

and N_0 and N_1 denote the number of samples in each set, Parzen window estimate for conditional distribution on $\mathcal{Z}$ given $\mathcal{H} = t$ is

$$\hat{f}_{\mathcal{Z}|\mathcal{H}}(z|t) = \frac{1}{N_t} \sum_{k \in \mathcal{S}_t} K_\sigma(z - z_k), \ t \in \{0, 1\} \tag{16}$$

The window or kernel $K_\sigma(z)$ is any valid probability density function (PDF) with parameter σ controling its width. Typically, symmetric zero mean distributions, such as Gaussians, are used for estimation. Replacing $f_{\mathcal{Z}|\mathcal{H}}(z|t)$ in (14) by its Parzen estimation (16) to estimate p_{FA} and p_M

$$\hat{p}_{FA} = \int_0^\infty \hat{f}_{\mathcal{Z}|\mathcal{H}}(z|0) \, dz, \ \hat{p}_M = \int_{-\infty}^0 \hat{f}_{\mathcal{Z}|\mathcal{H}}(z|1) \, dz \tag{17}$$

and inserting now these estimates in (7), the proposed cost function is

$$J^{Bayes}(\mathbf{w}) = \alpha \, \hat{p}_{FA} + (1 - \alpha) \, \hat{p}_M \tag{18}$$

By defining function $L_\sigma(x)$ from integration of Parzen kernel function,

$$L_\sigma(x) = \int_x^{+\infty} K_\sigma(z) \, dz \tag{19}$$

and taking into account that for symmetric zero mean PDFs $K_\sigma(x)$

$$\int_0^{+\infty} K_\sigma(x - z) \, dx = L_\sigma(-z) \text{ and } \int_{-\infty}^0 K_\sigma(x - z) \, dx = L_\sigma(+z) \tag{20}$$

the proposed cost function can be written as

$$J^{Bayes}(\mathbf{w}) = \frac{\alpha}{N_0} \sum_{k \in \mathcal{S}_0} L_\sigma(-z_k) + \frac{1 - \alpha}{N_1} \sum_{k \in \mathcal{S}_1} L_\sigma(+z_k) \tag{21}$$

It is straightforward to obtain the derivative of this cost function with respect to network output

$$\frac{\partial J^{Bayes}(\mathbf{w})}{\partial z_k} = a_k \, K_\sigma(z_k), \text{ with } \begin{cases} \alpha/N_0, & \text{if } y_k = -1 \\ (\alpha - 1)/N_1, & \text{if } y_k = +1 \end{cases} \tag{22}$$

Finally, gradient with respect to network parameters is given by

$$\frac{\partial J^{Bayes}(\mathbf{w})}{\partial \mathbf{w}} = \sum_{k=1}^N \frac{\partial J^{Bayes}(\mathbf{w})}{\partial z_k} \frac{\partial z_k}{\partial \mathbf{w}} \tag{23}$$

Again, note that $\frac{\partial z_k}{\partial \mathbf{w}}$ depends on the neural network specific architecture. From an implementation viewpoint, it is important to remark that with respect to the training of a neural network using MSE cost function, the only difference in the proposed training algorithm is that (13) has to be replaced by (22) in the computation of gradient for each iteration.

3.2 Ensemble Bayesian Classifier

The proposed classifier is a Bayesian ensemble of N_c individual neural network classifiers. Each individual classifier will be trained with the procedure presented in Section 3.1, for a different value for parameter α. The values for α used to train each individual classifier will be denoted as $\alpha^{(j)}$, with $j \in \{1, 2, \cdots, N_c\}$ This means that each individual classifier will try to work at a different point of the ROC of the classification problem, which is given by a different pair of probabilities of false alarm and detection, thus having N_c different points

$$\{(p_{FA}^{(j)}, p_D^{(j)})\} \text{ for } j \in \{1, 2, \cdots, N_c\} \tag{24}$$

This provides diversity in the decisions of individual classifiers. After training each individual neural classifier, these probabilities can be estimated from the training set, or from a validation set if one is available.

The proposed classifier will fuse the decisions of the previous N_c individual classifiers using a Bayesian rule. Therefore, to classify a given pattern $\mathbf{x}_k$, the input for this ensemble classifier is the vector containing the N_c decisions provided by the individual neural network classifiers for this pattern. If $\hat{y}_k^{(j)}$ denotes the binary decision of the j-th individual classifier for input pattern $\mathbf{x}_k$, the input of the ensemble for pattern $\mathbf{x}_k$ is

$$\mathbf{x}_k^E \equiv \mathbf{x}_k^E(\mathbf{x}_k) = \left[\hat{y}_k^{(1)}, \hat{y}_k^{(2)}, \cdots, \hat{y}_k^{(N_c)}\right] \tag{25}$$

A likelihood ratio can be defined for this input of the ensemble. Conditional distributions of the decisions of the individual classifiers are given by

$$p_{\hat{y}|\mathcal{H}}^{(j)}(\hat{y}_k^{(j)}|0) = \begin{cases} p_{FA}^{(j)}, & \text{if } \hat{y}_k^{(j)} = +1 \\ 1 - p_{FA}^{(j)}, & \text{if } \hat{y}_k^{(j)} = -1 \end{cases} \tag{26}$$

and

$$p_{\hat{y}|\mathcal{H}}^{(j)}(\hat{y}_k^{(j)}|1) = \begin{cases} p_D^{(j)}, & \text{if } \hat{y}_k^{(j)} = +1 \\ 1 - p_D^{(j)}, & \text{if } \hat{y}_k^{(j)} = -1 \end{cases} \tag{27}$$

Using these conditional distributions, and assuming conditional independence between the output of the N_c classifiers, the likelihood ratio for $\mathbf{x}_k^{(E)}$ is given by

$$\Lambda\left(\mathbf{x}_k^E\right) = \prod_{j=1}^{N_c} \frac{p_{\hat{y}|\mathcal{H}}^{(j)}(\hat{y}_k^{(j)}|1)}{p_{\hat{y}|\mathcal{H}}^{(j)}(\hat{y}_k^{(j)}|0)} = \prod_{j=1}^{N_c} \frac{p_D^{(j)} \, \delta[\hat{y}_k^{(j)} - 1] + \left(1 - p_D^{(j)}\right) \, \delta[\hat{y}_k^{(j)} + 1]}{p_{FA}^{(j)} \, \delta[\hat{y}_k^{(j)} - 1] + \left(1 - p_{FA}^{(j)}\right) \, \delta[\hat{y}_k^{(j)} + 1]} \tag{28}$$

where the discrete-time delta function $\delta[n]$ is used to provide a compact expression for (26) and (27). The final decision of the ensemble for input $\mathbf{x}^E$ will be obtained comparing the likelihood ratio with the threshold associated with the value of parameter α defined for the Bayesian risk of the ensemble, α^E

$$\Lambda(\mathbf{x}^E) \underset{\hat{\mathcal{H}}=0}{\overset{\hat{\mathcal{H}}=1}{\gtrless}} \gamma^E, \text{ with } \gamma^E = \frac{\alpha^E}{1-\alpha^E} \tag{29}$$

4   Experiments

This section presents the results obtained with the proposed method in the classification of several imbalanced databases.

4.1   Databases

We have tested the proposed method with several imbalanced real-world databases obtained from KEEL data-set repository [14]. Data and information about these data sets can be found in the Website http://www.keel.es/dataset.php. Data sets in [14] are organized in different k-fold partitions for training and test data. Here, we have worked with the 5-fold partition provided in KEEL-dataset repository, thus making easier to compare results. Results obtained for these 5 folds will be averaged. Table 1 shows the main characteristics of the tested databases.

4.2   Implementation Details

The goal of this paper is to demonstrate the intrinsic potential of the proposed method to work with different data. Therefore, instead of looking for the best configuration for each database, a generic setup has been used for all databases.

The only pre-processing for input data is normalization, forcing zero mean and unit variance for each dimension. Multilayer perceptrons (MLP) with a single hidden layer of N_n neurons, a single neuron in the output layer, and hyperbolic

Table 1. Description of the databases used for experiments.

Database	Number of patterns	Dimensionality	Imbalance Ratio
Yeast05679vs4	528	8	9.35
Ecoli067vs5	220	6	10.00
Glass2	214	9	10.39
Led7digit	443	7	10.97
Cleveland0vs4	177	13	12.62
Yeast4	1484	8	28.41
Yeast5	1484	8	32.78
Yeast6	1484	8	39.15

tangent activation functions, are used for the individual classifiers of the ensemble. Transfer function (8) and gradient expressions $\frac{\partial z_k}{\partial \mathbf{w}}$ in (23) are well known for this architecture [15,16], and have not been included due to lack of space. An adaptive step-size μ has been used in gradient updating equation (11). After each epoch, cost $J^{Bayes}(\mathbf{w}^{(i)})$ is evaluated and compared with $J^{Bayes}(\mathbf{w}^{(i-1)})$

- If $J^{Bayes}(\mathbf{w}^{(i)}) < J^{Bayes}(\mathbf{w}^{(i-1)})$: step size is increased, $\mu = c_I \, \mu$
- If $J^{Bayes}(\mathbf{w}^{(i)}) \geq J^{Bayes}(\mathbf{w}^{(i-1)})$: step size is decreased, $\mu = \mu/c_D$, and weights $\mathbf{w}^{(i)}$ are re-computed with the new step size.

$c_I = 1.05$ and $c_D = 2$ have been used in all experiments, with initial value $\mu = 1$. Relatively small networks have been used: MLP networks with $N_n = 4$ neurons in the hidden unit. Parzen window used in these experiments (to define cost and to compute (22)) has been, with $\sigma = 1$ in all cases

$$K_\sigma(z) = \begin{cases} \frac{1}{2\sigma}\left(1 + \cos\left(\frac{\pi}{\sigma}|z|\right)\right), & |z| \leq \sigma \\ 0, & |z| > \sigma \end{cases} \tag{30}$$

An ensemble of 9 individual neural network classifiers has been implemented, with the goal of showing advantage against ensembles with a higher number of elements. Parameters defining the Bayes risk objective for each one of the individual neural networks are $\alpha^{(j)} = 0.1 \times j$, for $j \in \{1, 2, \cdots, 9\}$. Network parameters $\mathbf{w}$ are randomly initialized, with values drawn from independent Gaussian distributions, with zero mean and variance 0.1.Not a single validation mechanism has been considered for training of individual neural classifiers, whereas 2000 epochs have been used for training every individual network. This generic setup will allow to assess the capability of the proposed method to deal with problems with a reduced number of samples, where a validation set can not be available.

4.3   Experimental Results

100 independent Monte Carlo simulations, starting with different initial parameters for individual neural classifiers, have been performed for each one of the 5 folds of every database. Average results obtained in the 5 folds, along with standard deviations, will be presented.

Table 2 compares the AUC of the ensemble with the AUC given by individual classifiers. AUC for ensemble has been obtained varying parameter γ^E from 0 to ∞ (in practice to maximum value for $\Lambda(\mathbf{x}_k^E)$ in samples of the test sets). AUC for individual classifiers has been computed by means of the trapezoidal integral of the 9 points (24) along with trivial ROC points $(0,0)$ and $(1,1)$. AUC for ensemble is higher than AUC obtained with individual classifiers in all databases, which shows the advantage of combining classifiers.

AUC gives an idea of the whole capability of the classifier to work at different tradeoffs for p_{FA} and p_M. Therefore, it is an appropriate figure of merit for a method designed to be able to work at any specified tradeoff. To evaluate the ability of the method to work at a specified tradeoff for errors of both classes,

Table 2. AUC for the ensemble and for the 9 working points (24) of the individual neural classifiers, along with trivial $(0,0)$ and $(1,1)$ ROC points.

Database	Individual classifiers	Ensemble
Yeast05679vs4	0.8286 ($\pm$ 0.0609)	0.8618 ($\pm$ 0.0546)
Ecoli067vs5	0.8935 ($\pm$ 0.0602)	0.9226 ($\pm$ 0.0712)
Glass2	0.7934 ($\pm$ 0.1542)	0.9096 ($\pm$ 0.0723)
Led7digit	0.9312 ($\pm$ 0.0441)	0.9474 ($\pm$ 0.0450)
Cleveland0vs4	0.8803 ($\pm$ 0.1201)	0.9531 ($\pm$ 0.0558)
Yeast4	0.8573 ($\pm$ 0.0234)	0.8968 ($\pm$ 0.0179)
Yeast5	0.9439 ($\pm$ 0.0503)	0.9734 ($\pm$ 0.0309)
Yeast6	0.8903 ($\pm$ 0.0730)	0.9125 ($\pm$ 0.0699)

Table 3. Average success probability $p_S = 1 - (p_{FA} + p_M)/2$ for best method in [5], individual neural classifier with $\alpha^{(j)} = 1/2$, and ensemble of 9 classifiers with $\alpha^E = 1/2$.

Database	Best in [5]	Individual	Ensemble
Yeast05679vs4	**0.8144**	0.7721 ($\pm$ 0.0527)	0.7940 ($\pm$ 0.0532)
Ecoli067vs5	0.8900	0.8736 ($\pm$ 0.0539)	**0.8963** ($\pm$ 0.0656)
Glass2	0.8045	0.7787 ($\pm$ 0.1775)	**0.8675** ($\pm$ 0.0917)
Led7digit	0.8880	0.8581 ($\pm$ 0.0491)	**0.8911** ($\pm$ 0.0838)
Cleveland0vs4	0.8280	0.8014 ($\pm$ 0.1598)	**0.9164** ($\pm$ 0.0712)
Yeast4	**0.8489**	0.7856 ($\pm$ 0.0532)	0.8219 ($\pm$ 0.0186)
Yeast5	**0.9661**	0.9343 ($\pm$ 0.0548)	0.9506 ($\pm$ 0.0481)
Yeast6	0.8678	0.8517 ($\pm$ 0.0819)	**0.8771** ($\pm$ 0.0712)

Table 3 compares the average probability of a successful classification of samples of both classes, measured as

$$p_S = 1 - \frac{p_{FA} + p_M}{2} \tag{31}$$

This means to assign the same importance to errors in both classes, which in the Bayesian risk corresponds to $\alpha = 1/2$. Therefore, $\alpha^E = 1/2$ will be used in the ensemble. Table also includes results obtained using a single individual neural classifier using $\alpha^{(j)} = 1/2$, and the best result provided by all the methods compared in [5]. This work compares performance obtained using ensembles of classifiers (10 to 40 classifiers) along with several preprocessing techniques used to balance data sets before constructing the ensemble. The proposed ensemble method provides better results than the individual neural classifier for all databases. Compared with the best method in [5], it provides better results in 5 of the 8 databases. We want to remark that the proposed method does not modify the data set before constructing the ensemble, to balance the number of samples of each class, as some methods in [5] do. The classifier is constructed using the original data.

5   Discussion

A new method, designed to classify imbalanced data, has been presented. The method combines several classifiers, each one designed to give a different relative importance to errors in majority and minority classes (p_{FA} and p_M) through parameters $\alpha^{(j)}$. This provides the ensemble with decisions for different working points in the ROC curve. Using these diverse decisions, the ensemble has the capability of working efficiently at different points of the ROC, allowing to select different compromise for both classes by varying parameter γ^E (or α^E).

The proposed method has shown competitive results in several imbalanced databases using a generic and simple setup, without validation nor an specific tuning for each database, and without data pre-processing to balance data sets. Lack of balance is handled with a proper choice for parameter α^E. This demonstrates its intrinsic potential to work with imbalanced data sets.

References

1. Duda, R.O., Hart, P.E., Stork, D.G.: Pattern classification, 2nd edn. John Wiley & Sons (2001)
2. Widrow, B., Rumelhard, D.E., Lehr, M.A.: Neural networks: Applications in industry, business and science. Communications of the ACM **37**(3), 93–105 (1994)
3. Zhang, G.P.: Neural networks for classification: A survey. IEEE Transactions on Systems, Man, and Cybernetics **30**(4), 451–462 (2000)
4. He, H., Garcia, E.A.: Learning from imbalanced data. IEEE Transactions on Knowledge and Data Engineering **21**(9), 1263–1284 (2009)
5. Galar, M., Fernández, A., Barrenechea, E., Herrera, F.: EUSBoost: Enhancing ensembles for highly imbalanced data-sets by evolutionary undersampling. Pattern Recognition **46**, 3460–3471 (2013)
6. Galar, M., Fernández, A., Barrenechea, E., Bustince, H., Herrera, F.: A review on ensembles for the class imbalance problem: bagging-, boosting-, and hybrid-based approaches. IEEE Transactions on Systems, Man, and Cybernetics - Part C: Applications and Reviews **42**(4), 463–484 (2012)
7. Kay, S.M.: Fundamentals of statistical signal processing: detection theory. Prentice-Hall Inc., Upper Saddle River (1998)
8. Scharf, L.S.: Statistical signal processing: detection, estimation, and time series analysis. Addison-Wesley (1991)
9. Van Trees, H.L.: Detection, Estimation, and Modulation Theory: Part I. John Wiley and Sons, New York (1968)
10. Fawcett, T.: An introduction to ROC analysis. Pattern Recognition Letters **27**(8), 861–874 (2006)
11. Bradley, A.P.: The use of the area under the ROC curve in the evaluation of machine learning algorithms. Pattern Recognition **30**(7), 1145–1159 (1997)
12. Cid-Sueiro, J., Arribas, J.I., Urbán-Muñoz, S., Figueiras-Vidal, A.R.: Cost functions to estimate a posteriori probabilities in multiclass problems. IEEE Transactions on Neural Networks **10**(3), 645–656 (1999)
13. Parzen, E.: On the estimation of a probability density function and the mode. Annals of Mathematical Statistics **33**, 1065–76 (1962)

14. Alcalá-Fdez, A., Fernández, A., Luengo, J., Derrac, J., García, S., Sánchez, L., Herrera, F.: KEEL data-mining software tool: data set repository, integration of algorithms and experimental analysis framework. Journal of Multiple-Valued Logic and Soft Computing **17**(2–3), 255–287 (2011)
15. Rumelhart, D.E., Hinton, G.E., Willians, R.J.: Learning representations by back-propagating errors. Nature (London) **323**, 533–536 (1986)
16. Widrow, B., Lehr, M.A.: 30 years of adaptive neural networks: perceptron, madaline and backpropagation. Proceedings of the IEEE **78**(9), 1415–1441 (1990)

Inferring Users' Interest on Web Documents Through Their Implicit Behaviour

Stephen Akuma[✉], Chrisina Jayne, Rahat Iqbal, and Faiyaz Doctor

Department of Computing and the Digital Environment, Coventry University, Coventry, UK
akumas@uni.coventry.ac.uk, {ab1527,aa0535,aa9536}@coventry.ac.uk

Abstract. This paper examines the correlation of implicit and explicit user behaviour indicators in a task specific domain. An experiment was conducted and data was collected from 77 undergraduate students of Computer science. Users' implicit features and explicit ratings of document relevance were captured and logged through a plugin in Firefox browser. A number of implicit indicators were correlated with user explicit ratings and a predictive function model was derived. Classification algorithms were also used to classify documents according to how relevant they are to the current task. It was found that implicit indicators could be used successfully to predict the user rating. These findings can be utilised in building individual and group profile for users of a context-based recommender system.

Keywords: Implicit indicators · Explicit rating · Context based · Recommender system

1 Introduction

An average user searching on the Internet is normally faced with the problem of information overload due to the volumes of information updated on a daily basis. The present challenge in the field of Information Retrieval (IR) is to help users find relevant documents for their current needs. This challenge is due to the fact that large amount of data are updated every day on the web and only a single document in the large collection might be relevant to a user [3]. A document is said to be of interest to a user if it is relevant to what the user is searching for. Query inputs are used for representing users' interest in an IR system but they do not adequately represent users' interest. Advances have been made to supplement user queries with additional sources of information obtained from user interaction with the system [14]. Implicit features can be obtained from mouse activity, key activity, dwell time, eye-tracking and psychological measures. To effectively improve the retrieval system through recommendation of relevant web resources to users, the interest of the users must be captured implicitly or explicitly. The explicit approach is the most consistent but it is expensive and it alters user-browsing patterns [6]. Implicit approaches unobtrusively capture users browsing behaviour through some special indicators called implicit indicators. It is cost effective and large amount of data can be captured anytime.

© Springer International Publishing Switzerland 2015
L. Iliadis and C. Jayne (Eds.): EANN 2015, CCIS 517, pp. 315–324, 2015.
DOI: 10.1007/978-3-319-23983-5_29

Previous studies [1], [6], [11], [13] on information retrieval have examined correlation between implicit indicators and users' explicit relevant judgment and they inferred that user implicit indicators could be used to predict explicit ratings.

This study focuses on studying user behaviour in a task specific context and investigating the correlation between implicit indicators and explicit ratings. It provides a novel task specific approach for deriving a predictive model that combines the implicit indicators. In a preliminary study, we found dwell time to correlate positively with user rating [2]. The preliminary study was specific to reading and not searching. In this study, users were allowed to freely surf the Internet and rate documents according to their interest. We performed two types of analysis. First we investigate the user search behaviour by correlating their implicit generated indicators with their explicit ratings. Secondly, we derive a predictive model and classify the implicit indicators based on user ratings for document relevance. This study differs from previous research in implicit indicators [6], [9], [15] in two ways. First it is task specific and it involves more than one type of tasks. Secondly, large numbers of participants are involved in the experiments.

The paper is organised as follows: Section 2 covers related work, Section 3 presents the user study, Section 4 shows the results, Section 5 gives the discussion of the results and Section 6 describes the conclusion.

2   Related Work

Previous research on implicit feedback tries to infer user's interest from their browsing behaviour. Morita and Shinoda [9] introduced the use of reading time as a behavioural characteristic for creating user profile and data filtration. They found that users spend longer time on articles they find interesting and the length of the article does not have any significant effect on the reading time. They did not also see any correlation between the readability of an article and reading time. Morita & Shinoda's research was based on a single implicit indicator (reading time) and users were restricted to certain news group thereby limiting their true web experience.

Nichols [10] explored and examined other implicit indicators that can be used as a source of implicit feedback. He introduced a list of additional observable behaviour that can be a source for implicit feedback. Oard and Kim [12] extended Nichols work by focusing on three observable behaviours: examination, retention and reference. The Curious Browser is a web browser developed by Claypool et al.[6] to study predictive strength of some given indicators. They captured implicit data as user's browse while the explicit data was collected through a five point rating scale. Their result show that the time spent on a document, the amount of scroll and the combination of time spent and amount of scroll are good predictive indicators, which have a stronger correlation with user explicit ratings. Claypool et al. research did not use information seeking task to engage participants, which may create uncertainty in applying the findings.

Most studies of implicit indicators in the context of information retrieval have been based on Search Engine Result Page (SERP) and very few studies [4], [7] have focus on users post click behaviour. This work uses task-based approach with focus on the actual documents visited by users. It revisits some of the implicit indicators used in the studies above but instead of the multiple domain approach of data collection, information tasks were used to limit user's goals in a particular domain. The following implicit indicators are evaluated in this work: Users Dwell time, Mouse Movement, Mouse Distance, Mouse Velocity, Mouse Clicks, Amount of Scroll, Keystroke and Amount of Copy.

3   User Study

Unlike existing studies, which are either controlled or naturalistic, both methods were employed for the study to make up for the disadvantages that exist in using one of the methods. The study was for 45minutes for every group and the participants were given an option to either perform the experiment in a controlled environment or to perform it at home in a natural setting. Those that participated in the controlled study performed it in selected computer laboratories of the University. A consent form was given to them to complete and sign; this was followed by a short tutorial about the task. A zipped Mozilla Firefox Portable with an embedded JavaScript plugin was uploaded to Google drive for participants to download and extract to their computers. After the extraction, they open the Firefox browser in the folder and begin the search for answers to the given task. They were asked to enter their own query in a search engine to find documents or enter the URL address of web documents directly. They were asked to read the documents for information relating to their task then close the current tab (web document). On closing the tab, they were prompted to state whether they were familiar with the web document and whether it was difficult to understand. They were also asked to explicitly rate the webpage according to how they consider it to be relevant to their task. The rating was on a 6-point scale. The participants were asked to visit and read a minimum of 7 web documents and to prepare a 200 words report of the solution to the given task. The participants recruited for the study were 77 undergraduate students from Computing department, Coventry University. Only students above 18 years were allowed to participate.

3.1   Experimental System

The system used for this research was an instrumented web browser that stores data in a central server. A JavaScript plugin developed to automatically capture user web activities was injected in a Mozilla Firefox Portable for easy and remote use. The injected plugin captured users' behavioural features, explicit ratings for document familiarity, document difficulty and relevance for each document visited. URLs from Google result page, Facebook and Yahoo were prevented from being logged to ensure that only documents related to the task at hand were recorded. The data collected were then sent to central server, and then recorded to MySQL database for storage. Figure 1 depicts how data was automatically captured and stored.

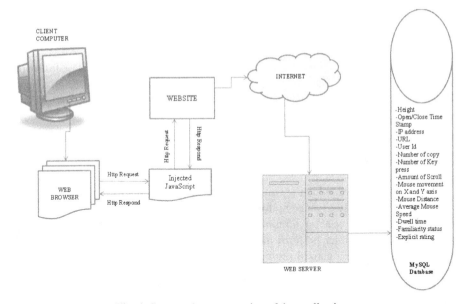

Fig. 1. Structural representation of data collection

3.2   Behavioural Features Captured

The following behavioural measures were captured as users visited web documents. The events captured were grouped into Interactive and Site structural data.

Interactive Events Captured

1. *Active Time Spent on the Document (TS):* The Active Time Spent (TS) is the accumulated time in seconds spent by a user on an active page during browsing. It is also called dwell time or reading time.
2. *Distance of Mouse Movement (DMM):* The Euclidean distance of mouse movement is calculated by its x and y coordinates on the monitor in every 100ms. The formula is given in equation (1):

$$MD = \sum_{i=1}^{n} \sqrt{(x - x_i)^2 + (y - y_i)^2} \tag{1}$$

Where x and y are the mouse location along the x and y coordinates.

3. *Total Mouse Movement (TMM):* This is the total mouse movement calculated by its x and y coordinates on the monitor. The count is incremented by one for x and y as the mouse hovers.
4. *Mean Mouse Velocity (MMV):* This is the mean speed covered by the mouse on the monitor as shown in equation (2):

$$\sum_{i=1}^{n} \sqrt{(x - x_i)^2 + (y - y_i)^2} / (t) \tag{2}$$

Where x and y are the mouse locations along the x and y coordinates. And t is the number of mouse distance interval recorded in every 100ms.

5. *Number of Mouse Clicks (NMC):* This is the total amount of mouse clicks on a page. The number of mouse click is incremented every time the mouse is clicked by a user.

6. *Number of Scrollbar Clicks (NSC):* Most web pages are longer in length than the monitor height. When readers are interested in a page, they scroll the page. The scrolling is normally done by either clicking or dragging the scroll bar. Any time a user clicks the scrollbar up or down, the count is incremented.

7. *Number of Key UP and Down (NK):* This is the total number of keystroke on a document. The "Page Up", "Page down", "up", "left", "right" and "down" keys or any other key can be used in place of the mouse in moving the document. This is incremented when the user strikes a key.

8. *Amount of Copy (AC):* This is the number of times texts are copied to the clipboard on a document. It is incremented by one any time a text from a particular document is copied.

9. *Mouse Duration Count (DC):* This is the total number of 100ms interval that occurred while the mouse moved on the screen.

10. *Explicit Ratings (ER)* This is the actual rating of the web document by the user. The Firefox plugin attaches a six scale-rating button on each of the webpages. After reading a webpage, the user rates it by clicking on any of the six scale buttons where 5 – means very relevant, 4 – means more relevant, 3 – means moderate relevant, 2 – means slightly relevant, 1– means very low relevance, 0 – means not relevant.

11. *Document Familiarity:* This captures user's familiarity with the current web document on a Boolean scale.

12. *Document Difficulty:* This captures on a Boolean scale of how difficult the current web document is to the user.

Site Structural Data

1. *Page Height:* This is the document height. It is the length of the entire document.
2. *Time Stamp:* This is the time and date in GMT when a document is loaded and when a document is closed.
3. *IP Address (IP):* This is the Internet protocol address of a user. It represents the User's location.
4. *URL:* This is the http address of any web document visited by a user.

3.3   Tasks

The tasks used for this research were simulated work situations, which follows [5] recommendation. The tasks were designed in consideration of the subjects' background to encourage them to search the web as they naturally do. A classification scheme [8] was used to group the tasks components into a single scheme. The tasks

320 S. Akuma et al.

domain for the study was in the area of Computer Science. Two simulated search tasks were composed of two parts – the simulated work task situation and indicative request.

Simulated Work Task Situation 1 (Mixed Task)
GIG Software Development Company employed you as a consultant to provide a solution to the Company's pressing problem of developing customized software within a minimal time frame. Some professional software developers achieved this by using the Rational Unified Process while others used the waterfall model.

Indicative Request 1
Which of the approach would you consider for a small project of few lines of code (LOC) and what stages of the software life cycle do you consider to be the most important? State the reason for your answer in your report.

Simulated Work Task Situation 2 (Factual Task)
Google is looking for young and ambitious students of Computer science for internship to work under the Company's Service Management Department. Consider that you are shortlisted for an interview among 2000 applicants and you are asked to search the Internet and find answers to questions related to Information Technology Infrastructure Library (ITIL):

Indicative Request 2
a) What are the five stages of the ITIL lifecycle?
b) What are the differences between ITIL v1, v2 and v3 (2007)?
c) What are ITIL processes?
d) What are ITIL functions?
e) Who should use ITIL?
f) When should ITIL be used?
g) What are the differences between ITIL and ISO/IEC?

4   Results

Web documents used during participants training and short tutorial were manually filtered out. 343 web documents were extracted for analysis. Every participant visited at least one web document while performing the assigned task. The maximum number of website visited by a participant in single task was 11 and the minimum was one. Pearson correlation was used in finding correlation between the user explicit ratings and the implicit indicators. Independent T-Test was used to determine statistical significance of the mean differences of the variables based on user ratings for relevance. A confidence interval of 95% and a statistical significant coefficient, $p < 0.05$ was used for analysing the dataset in terms of correlation and the independent T-test.

4.1   Correlation of Implicit and Explicit Relevance Ratings

A Pearson correlation was run to access the relationship between the implicit indicators and explicit relevance rating. Initial analysis for linearity shows that the implicit variables were linearly related with the explicit relevance ratings. There was positive correlation between the explicit relevance rating and the mouse clicks ($r = 0.211$), amount of copy ($r = 0.28$), the amount of scroll ($r = 0.123$), the mouse movement along X ($r = 0.225$) and Y axis ($r = 0.261$), the dwell time ($r = 0.285$), the mouse distance ($r = 0.254$) and the mouse duration count ($r = 0.238$) with the p-values as 0.000, 0.000, 0.013, 0.000, 0.000, 0.000, 0.000 and 0.000 respectively as shown in Table 1. In order to find whether there is a statistical difference between documents rated as relevant from documents rated as non-relevant, the 6 point scale explicit ratings were merged into two: relevant and non-relevant. Rating 0, 1 and 2 were merged as non-relevant while rating 3, 4 and 5 were merged as relevant. Independent T-Test was used to compare the mean of the two groups of relevancy based on the implicit parameters. There was a significance difference ($p < 0.05$) in the means for the following variables: mouse clicks ($p=0.000$), number of copy ($p=0.013$), the amount of scroll ($p=0.000$), the mouse movement along X ($p=0.000$) and Y ($p=0.000$) axis, the dwell time ($p=0.000$), the mouse distance ($p=0.000$) and the mouse duration count ($p=0.000$). The mean of the indicators for the relevant group was higher than the non-relevant group as can be seen in Table 1. This signifies that the users focused on documents perceived to be relevant than those perceived to be non-relevant. This supports the assertion that the more relevant the documents, the more the user generated implicit indicators.

Table 1. Pearson correlation between the implicit indicators and explicit relevant ratings and T-Test comparison of implicit indicators based on relevancy groups

Implicit Indicators	Pearson Correlation (r) with User Explicit Rating	Significant coefficient level (p)	Mean Not Relevant(N =79)	Mean Relevant (N=264)	T-Test(p)
Clicks	*0.211*	*0.000*	**0.59**	**2.0**	0.000
Height	0.012	0.819	4494.96	4370.76	0.665
Copy	*0.28*	*0.000*	**0.05**	**0.66**	0.000
Scroll	*0.123*	*0.023*	*140.12*	*204.27*	0.013
Mouse movement X	*0.225*	*0.000*	1827.35	4632.80	0.000
Mouse movement Y	*0.261*	*0.000*	2785.24	5540.16	0.000
Dwell time	*0.285*	*0.000*	46.96	164.24	0.000
Mouse distance	*0.254*	*0.000*	3980.39	8684.45	0.000
Mouse duration count	*0.238*	*0.000*	53.19	144.82	0.000
Mean mouse speed	-0.73	0.180	927.82	853.74	0.201
Keystroke	-0.18	0.742	1.38	2.29	0.473

4.2   Consistency in Explicit Relevance Ratings

Common documents visited by the users were extracted from the pool of documents visited. 21 users viewed the most common document visited and two users viewed the least. An investigation was carried out to see if documents commonly visited are relevant. The mean of the explicit relevance ratings of the common documents visited was calculated and a mean of 3.205 was obtained. This suggests that common documents visited by users are relevant and common consistent implicit indicators can be used to infer relevance among a community of users.

4.3   Multiple Linear Regression Analysis

To obtain the most predictive indicators of interest, multiple linear regression analysis was used to estimate the user behavioural features that best represent the user explicit interest. Stepwise method was used to select indicators with the most predictive power. The stepwise regression included only dwell time (total active time) and copy (amount of copy) parameters. The rest of the parameters were excluded and the function below was obtained:

Explicit Relevance ratings = 2.978 + 0.281(Copy) + 0.002(Dwell Time)
Correlation coefficient   0.364

4.4   Classification Analysis

The ratings for relevance were merged into relevant and non-relevant. The rating for 0, 1 and 2 were merged as non-relevant and rating 3, 4 and 5 were merged as relevant. WEKA library was used for the classifiers with the following indicators as the input: Dwell time, Mouse Movement, Mouse Distance, Mouse Clicks, Amount of Scroll, Mean Mouse Speed, Keystroke and Amount of Copy. The classifiers were run on the dataset to classify the data according to the two groups of relevancy. 10-folds cross-validation was carried out and 70% of the data (N=240) was used for the training, while 30% (103) of the data was used for the testing. The classifiers results are shown in table 2:

Table 2. The predictive power of selected classifiers

S/N	Classifier	Accuracy
1	Logistic Regression	68.2%
2	K-Nearest Neighbours	70.85%
3	Naïve Bayes	73.12%
4	Multilayer Perceptron	73.8%
5	J48 Decision	76.97%

Since the entropy for the J48 decision tree is lower and it gives the best accuracy, it may be used in practice to classify documents based on relevancy.

5   Discussion

Apart from the keystroke and mean mouse speed, all other implicit indicators measured correlate significantly with the explicit relevance ratings. The correlation between implicit indicators and explicit relevance ratings obtained in this work is higher than those obtained in previous research. For instance, [7] estimated document relevance from dwell time, cursor movements and other post-click behaviour and the authors obtained a correlation of 0.167 for dwell time, 0.101 for mouse movement along X axis, 0.172 for mouse movement along Y axis, -0.143 for mouse speed along the X axis and -0.124 for mouse speed along the Y axis and -0.008 for amount of vertical scroll. They described the correlation for dwell time as moderate. The T-Test in our experiment show that documents that were rated as relevant have higher Dwell time, Mouse Movement, Mouse Distance, Mouse Clicks, Amount of Scroll, and Amount of Copy than documents that were rated not-relevant. This indicates that the higher the presence of this user generated indicators on a document, the more relevant is the document.

Our results also showed that documents that are commonly viewed by users are relevant and can be recommended to users of a particular domain through a collaborative feedback system. Reasonable accuracy of prediction was obtained when the User generated features (Dwell time, Mouse activity and Key stroke activity) were classified based on relevant and non-relevant explicit ratings. Since all the classifiers used produced accuracy above 65%, it indicates that a good fit model could be obtained which validates our hypothesis that user generated implicit indicators can be used to predict documents that are perceived to be relevant from those that are not. These findings can be applied in the area of recommending relevant documents to a community of users based on their browsing behaviour.

6   Conclusion

The study explored the correlation of implicit and explicit user behaviour indicators in a task specific context. Logistic Regression, Decision Tree, Naïve Bayes, K-Nearest Neighbours and Multilayer Perceptron were used to model the relationship between the implicit and explicit indicators. The experimental results for these models show that with relatively high accuracy, document relevance can be predicted based on user implicit behaviour. These findings contribute to the understanding of user behaviour in a task specific context and deducing the best implicit indicators that can be used in a recommender system. Future work will include validating the interest indicators obtained in this study with gaze-generated features. A framework for context-based recommender system will be developed.

References

1. Akuma, S.: Investigating the Effect of Implicit Browsing Behaviour on Students' Performance in a Task Specific Context. International Journal of Information Technology and Computer Science (IJITCS) **6**(5), 11–17 (2014)
2. Akuma, S., Jayne, C., Iqbal, R., Doctor, F.: Implicit predictive indicators: mouse activity and dwell time. In: Iliadis, L. (ed.) AIAI 2014. IFIP AICT, vol. 436, pp. 162–171. Springer, Heidelberg (2014)
3. Alhindi, A., Kruschwitz, U., Fox, C., Albakour, M.: Profile-Based Summarisation for Web Site Navigation. ACM Transactions on Information Systems **33**(1), 1–40 (2015)
4. Balakrishnan, V., Zhang, X.: Implicit user behaviours to improve post-retrieval document relevancy. Computers in Human Behavior **33**, 104–112 (2014)
5. Borlund, P.: The IIR evaluation model: A framework for evaluation of interactive information retrieval systems. Information Research **8**(3) (2003)
6. Claypool, M., Le, P., Wased, M., Brown, D.: Implicit interest indicators. In: Proceedings of the International Conference on Intelligent User Interfaces, IUI, pp. 33–40 (2001)
7. Guo, Q., Agichtein, E.: Beyond dwell time: estimating document relevance from cursor movements and other post-click searcher behavior. In: WWW 2012 - Proceedings of the 21st Annual Conference on World Wide Web, pp. 569–578 (2012)
8. Liu, J., Liu, C., Belkin, N.: Examining the effects of task topic familiarity on searchers' behaviors in different task types. In: Proceedings of the ASIST Annual Meeting, vol. 50(1) (2013)
9. Morita, M., Shinoda, Y.: Information filtering based on user behaviour analysis and best matchtext retrieval. In: Proceedings of SIGIR Conference on Research and Development, pp. 272–281 (1994)
10. Nichols, D.M.: Implicit ratings and riltering. In: Proceedings of the 5th DELOS Workshop on Filtering and Collaborative Filtering, Budapaest, Hungary, pp. 10–12 (1997)
11. Núñez-Valdéz, E.R., Cueva Lovelle, J.M., Sanjuán, O., García-Díaz, V., De Pablos, P.O., Montenegro Marín, C.E.: Implicit feedback techniques on recommender systems applied to electronic books. Computers in Human Behavior **28**(4), 1186–1193 (2012)
12. Oard, D., Kim, J.: Implicit feedback for recommendation systems. In: Proceedings of the AAAI Workshop on Recommender Systems, pp. 81–83 (1998)
13. Teevan, J., Dumais, S.T., Horvitz, E.: Potential for personalization. ACM Transactions on Computer-Human Interaction **17**(1) (2010)
14. White, R.W., Kelly, D.: A study on the effects of personalization and task information on implicit feedback performance. In: Proceedings of the International Conference on Information and Knowledge Management, pp. 297–306 (2006)
15. Zemirli, N.: WebCap: inferring the user's interests based on a real-time implicit feedback. In: 7th International Conference on Digital Information Management, ICDIM 2012, pp. 62–67 (2012)

One-Class Classification for Microarray Datasets with Feature Selection

Beatriz Pérez-Sánchez$^{(\boxtimes)}$, Oscar Fontenla-Romero,
and Noelia Sánchez-Maroño

Laboratory for Research and Development in Artificial Intelligence (LIDIA),
Department of Computer Science, Faculty of Informatics,
University of A Coruña, Campus de Elviña, S/n, 15071 A Coruña, Spain
bperezs@udc.es

Abstract. Microarray data classification is a critical challenge for computational techniques due to its inherent characteristics, mainly small sample size and high dimension of the input space. For this type of data two-class classification techniques have been widely applied while one-class learning is considered as a promising approach. In this paper, we study the suitability of employing the one-class classification for microarray datasets while the role played by feature selection is analyzed. The superiority of this approach is demonstrated by comparison with the classical approach, with two classes, on different benchmark data sets.

1 Introduction

In the last years there has been a boom in the acquisition of biomedical data. The application of innovative technologies in the field of molecular genetics - such as deoxyribonucleic acid (DNA) microarrays - provides a global view of the cell enabling the measurement of simultaneous expression of tens of thousands of genes. Thus, microarrays allow for creating data sets that represent a system which may be of interest from a biological or clinical point of view. Moreover, due to the intrinsic characteristics of these data, in recent years they also have become a challenge for the scientific community in terms of machine learning and bioinformatics. Microarray data present a large dimensionality of the feature space (often reaching several thousands of genes) compared to the small number of samples available (usually less than a hundred). The high dimensionality of the feature space degrades the performance of the classifier and increases its computational complexity. As a consequence, the application of conventional statistical and machine learning techniques for classification purposes is heavily limited.

Several studies have shown that most genes measured in a DNA microarray experiment are not relevant in the accurate classification of different classes of the problem [1]. To overcome the problem known as *curse of dimensionality* or Hughes effect [2] that appears when with a fixed number of training samples, the predictive power of the learner reduces as the feature dimensionality increases, dimensionality reduction techniques plays a crucial role. Among dimensionality

© Springer International Publishing Switzerland 2015
M. Ali et al. (Eds.): IEA/AIE 2015, LNAI 9101, pp. 325–334, 2015.
DOI: 10.1007/978-3-319-19066-2_1

reduction methods, feature selection identifies and discards irrelevant features from the training data as a previous step of the classification stage. Thus, the learning algorithm can focus on those useful aspects of data improving its performance. There are usually three varieties of feature selection methods: wrappers, filters and embedded methods. Wrapper models involve optimizing a predictor as part of the selection process. Filter methods rely on the general characteristics of the training data to select feature independent of any predictor. And finally, embedded techniques generally use machine learning models for classification and then an optimal subset of features is built by the classifier algorithm. Both wrapper and embedded methods have an important computational cost due to the interaction with the classifier. In the case of microarray data classification the most employed methods fall into the filter category [3]. In [4] the authors review the most up-to-date feature selection methods developed in this field (filters and one embedded method) and a comparative among them is introduced.

Machine learning has been widely applied for handling microarray data sets being supervised machine learning a promising approach. Several surveys about machine learning in microarray were published over time [5–7]. To date, two-class classification techniques have mainly been applied and Support Vector Machines are among the most popular classifiers used for this task. In fact, there is a clear tendency in the literature to use SVM versus other classifiers. Many microarray data are used for cancer diagnosis. In this context, some datasets present unbalanced classes as healthy patients normally predominate. For that reason, the one-class classification (OCC) paradigm also has been employed to treat this kind of data but its use is not so extended as the two-class perspective. In [8] the author proposes to use a one-class approach to classify microarrays because of these models rely only on objects coming from single class distribution. In this case two approaches can be considered since training of the classifier can be focused on the majority class or on the minority one. Despite of having less information to distinguish between classes, one-class models can easily learn the specific properties of a given data set and are robust to intrinsic difficulties of the data.

Thus, the aim of this paper is to compare the behavior of the two-class classification versus the one-class paradigm when dealing with microarray data sets and analyze the role played by feature selection.

2   One-class Classifiers

In a multi-class classification problem, data from several categories are available and the decision boundary is chosen thanks to objects from each class. The main goal is to categorize an unknown object as belongs to one specific class from the more broad set of classes, two in the simplest case of binary classification. Nevertheless, occasionally the classification task does not consist just in categorizing an object into one specific class but deciding if this object fits to a particular class or not. OCC paradigm shows a favorable perspective to solve these kinds of problems since in OCC only instances from one of the classes are available or considered. The objects from this category will be called the *target* objects while all other are the *outlier* ones.

OCC problems are usual in the real world, there are different situations where normal examples are widely accessible but outliers are expensive or even unfeasible to collect. For example, for identifying failures in a machine, data about the regular working of the machine can be easily obtained. However, obtaining examples about failures is expensive and sometimes impossible since some of them would not have taken place. Another scenario refers to the diagnosis of a disease, as in the previously commented case of cancer. Although several methods to solve the one-class classification problem have been proposed we have selected the Support Vector Data Description [9]. This algorithm is one of the most up-to-date one-class classifiers applied to handle the type of data of interest. This method is briefly introduced as follows.

2.1 Support Vector Data Description.

A one-class classification technique based on Support Vector Machines (SVMs) [10] was proposed by Tax [9] and it is known as Support Vector Data Description (SVDD). This approach establishes a closed boundary around the target data by means of a small hypersphere which encloses all target data. As SVDD is based on SVMs, its decision boundary is described by a certain target data called support vectors. The hypersphere is characterized by a center $\mathbf{c}$ and a radius R ($R > 0$) around the dataset which has minimal volume [11] as it can be seen in Fig. 1. The error function to minimize is given below,

$$\min_{\mathbf{c},R} R^2$$

where

$$\|\mathbf{x}_i - \mathbf{c}\|^2 \leq R^2, i = 1, 2, \ldots, n.$$

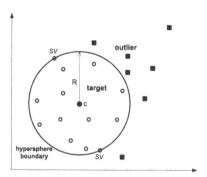

Fig. 1. The hypersphere boundary includes all target data. The objects which are on the boundary are known as the support vectors (SV).

As SVDD is a variant of the SVM, also it has to solve a quadratic optimization problem during its training. Therefore a dual formulation, in terms of inner products, can be derived. Replacing the normal inner products $(\mathbf{x}_i \cdot \mathbf{x}_j)$ by a kernel function $K(\mathbf{x}_i, \mathbf{x}_j)$ the flexibility of the model is increased. Data are mapped to a high dimensional feature space without much extra computational cost. The most commonly used kernel function is Gaussian kernel. The distance from the sphere center to a test object $\mathbf{x}$ is calculated by means of the following equation

$$d_{SVDD}(\mathbf{x}, X_t) = K(\mathbf{x}, \mathbf{x}) - 2 \sum_i \alpha_i K(\mathbf{x}, \mathbf{x}_i) + \sum_{i,j} \alpha_i \alpha_j K(\mathbf{x}_i, \mathbf{x}_j)$$

where the parameters α_i are obtained by quadratic optimization. To obtain a more efficient description there is the possibility of including outliers examples provided when these are available into the training procedure. In this case, the distance from $\mathbf{x}_i$ to the center $\mathbf{c}$ should not be strictly smaller than R^2, but larger distances should be penalized. Therefore, slack variables ξ_i are introduced and the new minimization problem is,

$$\min_{\mathbf{c}, R} R^2 + C_i \sum_i \xi_i$$

where

$$\|\mathbf{x}_i - \mathbf{c}\|^2 \leq R^2 + \xi_i, \ \xi_i \geq 0, \ \forall i$$

taking into account that the parameter C controls the trade-off between the volume and the errors. In this way a more robust classifier can be obtained.

3   Experimental Setup

In this section, we test the suitability of the one-class learning approach for microarray datasets and compare the results with those obtained by the two-class approach, specifically the SVM algorithm. It is worth mentioning that the OCC is addressed by using both minority and majority class as target concept. Furthermore, we study the role played by feature selection as a previous step to the classification stage. Next, we establish certain considerations which have been taken into account in the experimental study.

– The input dataset was previously normalized to have zero mean and a standard deviation of 1.
– In order to obtain statistically significant results, 30 simulations were run with the cross-validation technique to tune the parameters of each method.
– For the implementation of classifiers two different toolboxes for Matlab was used. The data description toolbox, DDtools library[12], for SVDD and the Statistics and Machine Learning toolbox for SVM.

- Regarding the parameters associated with the SVDD classifier, the fraction of the target set which will be rejected was established to $0,01$, and the width parameter in the radial basis function kernel was selected by means of cross-validation technique. In case of SVM, the kernel function was getting by cross-validation and the percentage of the target object that can be considered as outliers was established as 0.
- With reference to the feature selection methods, all techniques are available in the well-known Weka tool [13], except for mRMR filter, whose implementation is available for *Matlab*.
- To evaluate the goodness of the selected set of genes in terms of accuracy of the classifier it is necessary to have an independent test set with data which have not been seen by neither the feature selection method nor the classifier. The selected data sets come originally distributed into training and test sets, so the training set was employed to perform the feature selection process and posterior classification while the test set was used to evaluate the appropriateness of the selection and the posterior classification.
- Finally, a statistical study was conducted to determine whether the results are statistically different. First at all, the normality conditions of each distribution are checked by means of Kolmogorow Smirnov test. As in any case, normal conditions are verified then the non parametric Kruskal-Wallis test was applied.

The remainder of this section is devoted to the analysis of the characteristics of the selected datasets and also to introduce the feature selection methods and the evaluation measures.

3.1   Datasets Characteristics

For this experimental study, we have considered two widely used binary microarray dataset which are available for download at [14,15]. Both data sets come originally separated in training and test thus, Table 1 summarizes the main characteristics of both partitions. For each set, we introduce its number of examples (# Ex.), attributes (# Atts.) and some information for majority and minority classes (number of examples/percentage of examples). Moreover, we provide the imbalance ratio (IR) defined as, the number of outlier examples that are divided by the number of normal examples. A value of 1 indicates balance whereas a large value denotes a high imbalance. As it can be seen in Table 1 both datasets present more imbalance in the test set especially in prostate data set. This may be caused by the problem known as *dataset shift* [16]. It occurs when testing (unseen) data experience a phenomenon that leads to a change in the distribution of a single feature, a combination of features, or the class boundaries. As a result, the common assumption that the training and testing data follow the same distributions is often violated in real world application and scenarios. In this regard, prostate dataset poses a big challenge for machine learning methods since the test dataset was extracted from a different experiment. It is possible that some classifiers, whose features are selected according to the training set, assign all samples to the majority class.

Table 1. Description of the train and test binary datasets used in the experimental study.

Dataset	# Atts.	Train				Test			
		# Ex.	Min. class	Maj. class	IR	# Ex.	Min. class	Maj. class	IR
Breast	24.481	78	34/43,59%	44/56,41%	1,29	19	7/36,84%	12/63,16%	1,71
Prostate	12.600	102	50/49,02%	52/50,98%	1,04	34	9/26,47%	25/73,53%	2,78

3.2 Feature Selection Methods

In this study we choose seven classical feature selection methods widely used by the researchers in this field. Moreover, for the sake of comparison, these methods are those used in [4], where a thorough study can be found. Such methods are: Correlation-based Feature Selection (CFS) [17], Fast Correlation-Based Filter (FCBF) [18], INTERACT algorithm [19], Information Gain (IG) [20], ReliefF [21], minimum Redundancy Maximum Relevance (mRMR) [22] and Suppor Vector Machine based on Recursive Feature Elimination (SVM-RFE) [23]. All these methods are in the filter category except SVM-REF, the most famous embedded method to specifically deal with gene selection for cancer classification. The three first feature selection methods (CFS, FCBF and INTERACT) return a subset of features, the number of selected ones for each dataset is shown in Table 2. Remaining techniques (IG, ReliefF, mRMR and SVM-REF) provide an ordered ranking of the features, for these one we show the performance when the top 10 and top 50 features are retained.

Table 2. Total number of features and selected number by subset methods.

	Features			
	Total	CFS	FCBF	INT
Breast	24.481	130	99	102
Prostate	12.600	89	77	73

3.3 Evaluation Measures

Most of performance measures for a binary class problem are built over the classical confusion matrix from which four measures can be directly obtained. TP and TN denote the number of positive and negative cases correctly classified, while FP and FN refer to the number of misclassified positive and negative examples, respectively. *Accuracy*, defined as $Acc = (TP + TN) / (TP + FN + TN + FP)$, is the most common metric for assessing the performance of learning systems. Moreover, the *true negative rate* or *specificity* $Sp = TN/ (TN + FP)$, is the percentage of correctly classified negative examples (e.g. the rate of healthy patients who are correctly classify as not having cancer). Analogously, the *true positive rate*, also called *recall* or *sensitivity*, $Se = TP/ (TP + FN)$ is the percentage of correctly classified positive instances (e.g. the rate of cancer patients who are correctly identified as having cancer). A perfect predictor would be described as 100% sensitive and 100% specific. Regardless of the class (majority or minority) used as target class in the OCC approach, it should be mentioned that sensitivity and specificity measures are always calculated on the same criteria. We consider as positive the cancer samples and as negative the healthy ones.

4   Experimental Results

In this section we present the experimental results achieved on the Breast and Prostate datasets previously introduced. Tables 3 and 4 show the performance obtained by SVM and SVDD classifiers over both test datasets. In case of SVDD classifier we introduce the results reached by using both classes (minority and majority) as the target concept in training process. Each column represents one of the three performance measures while rows indicate the feature selection methods. Note that, for the sake of comparison, last row shows the results achieved using the whole set of features, i.e., no feature selection is applied. To facilitate the analysis of the results, in both tables the results corresponding to the best values (statistically speaking) of each performance measures for each dataset are marked in boldface type.

Firstly we focus on Breast dataset whose results are shown in Table 3. At first glance it seems that for all cases the results obtained by SVDD classifier class are better than those achieved by the SVM and statistical tests confirm this assumption. Only for FCBF and INT filters the SVM obtain a higher value in the specificity measure, however in both cases SVDD achieves the best value of accuracy and specificity and also balanced values for sensitivity and specificity. Furthermore, it is worth mentioning the importance of using feature selection methods (see last row in Table 3) because they help prevent overfitting.

Consider now the Prostate dataset whose results are shown in Table 4. As it can be observed, the results obtained by SVM and SVDD on this dataset follows along the same line as the previous one. The one-class approach overcomes the results obtained by SVM, showing important differences in all cases. As it was stated earlier, Prostate dataset suffers the dataset shift problem since the test distribution differs significantly from the train distribution. In this situation it is possible that classifiers assign the vast majority of samples to one of the classes such it shows for SVM. However, SVDD seems not to suffer this problem and very good results are reached. Although the results obtained when no feature selection is applied are good, it should take into account the important needs both computational and time to manage the original datasets.

Finally, another point to be borne in mind is that SVDD presents an important advantage respect to SVM. Although in provided tables the results are not statistical different, SVDD allows us to use minority or majority class as the target class in the training process and remain the best results depending on the specific application. The ideal situation would be obtain a classifier 100% sensitive and 100% specific but this fact is not easy. Therefore, a trade-off becomes

Table 3. Results for **SVM** and **SVDD** classifiers on Breast dataset.

	Acc			Se			Sp		
	SVM	OCCmin	OCCmaj	SVM	OCCmin	OCCmaj	SVM	OCCmin	OCCmaj
CFS	0,5295	**0,6116**	**0,6167**	0,3352	**0,4756**	**0,4772**	0,6428	**0,8448**	**0,8476**
FCBF	0,6109	**0,6937**	**0,6940**	0,0714	**0,7400**	**0,7067**	**0,9256**	0,6667	0,6724
INT	0,5846	**0,7091**	**0,7116**	0,1400	**0,7294**	**0,7400**	**0,8439**	0,6743	0,6629
IG-10	0,5284	**0,6733**	**0,6726**	0,4886	**0,6411**	**0,6450**	0,5517	**0,7286**	**0,7200**
IG-50	0,5435	**0,7329**	**0,7442**	0,3762	**0,7956**	0.8128	0,6411	0,6257	0,6267
RelieF-10	0,4958	**0,7905**	**0,7898**	0,5533	**0,7500**	**0,7500**	0,4622	**0,8600**	**0,8581**
RelieF-50	0,4705	**0,7351**	**0,7337**	0,5267	**0,6811**	**0,6739**	0,4378	**0,8276**	**0,8362**
SVM-REF-10	0,4944	**0,8849**	**0,8881**	0,5790	**0,8944**	**0,9006**	0,4450	**0,8686**	**0,8667**
SVM-REF-50	0,4821	**0,8351**	**0,8403**	0,6638	**0,8333**	**0,8450**	0,3751	**0,8381**	**0,8324**
mRMR-10	0,4944	**0,7614**	**0,7467**	0,4981	**0,7711**	**0,7556**	0,4922	**0,7448**	**0,7314**
mRMR-50	0,5151	**0,7586**	**0,7617**	0,4105	**0,8011**	**0,8028**	0,5761	**0,6857**	**0,6914**
no FS	0,4979	**0,6432**	**0,6358**	**0,5505**	0,4806	0,4705	0,4672	**0,9219**	**0,9190**

OCCmin corresponds to the test results obtained by training with minority class
OCCmaj corresponds to the test results obtained by training with majority class

Table 4. Results for **SVM** and **SVDD** classifiers on Prostate dataset.

	Acc			Se			Sp		
	SVM	OCCmin	OCCmaj	SVM	OCCmin	OCCmaj	SVM	OCCmin	OCCmaj
CFS	0,5909	**0,9747**	**0,9745**	0,3029	**0,9656**	**0,9653**	0,6947	**1,0000**	**1,0000**
FCBF	0,6216	**0,9245**	**0,9225**	0,1681	**0,9011**	**0,8989**	0,7848	**0,9896**	**0,9881**
INT	0,6549	**0,9590**	**0,9571**	0,1259	**0,9443**	**0,9416**	0,8453	**1,0000**	**1,0000**
IG-10	0,5976	**0,9508**	**0,93925**	0,2844	**0,9416**	**0,9269**	0,7104	**0,9763**	**0,9733**
IG-50	0,6470	**0,9933**	**0,9904**	0,2185	**0,9909**	**0,9869**	0,8013	**1,0000**	**1,0000**
RelieF-10	0,6141	**0,9309**	**0,9284**	0,2563	**0,9061**	**0,9067**	0,7429	**1,0000**	**1,0000**
RelieF-50	0,6937	**0,9571**	**0,9584**	0,1437	**0,9416**	**0,9437**	0,8917	**1,0000**	**0,9992**
SVM-REF-10	0,6369	**0,8696**	**0,8718**	0,2733	**0,8589**	**0,8592**	0,7677	**0,8993**	**0,9067**
SVM-REF-50	0,6278	**0,9598**	**0,9602**	0,2229	**0,9453**	**0,9459**	0,7736	**1,0000**	**1,0000**
mRMR-10	0,6384	**0,9486**	**0,9594**	0,2193	**0,9389**	**0,9496**	0,7893	**0,9756**	**0,9867**
mRMR-50	0,6153	**0,9363**	**0,9369**	0,2178	**0,9133**	**0,9141**	0,7584	**1,0000**	**1,0000**
no FS	0,5400	**0,9012**	**0,8976**	0,3348	**0,8656**	**0,8608**	0,6139	**1,0000**	**1,0000**

OCCmin corresponds to the test results obtained by training with minority class
OCCmaj corresponds to the test results obtained by training with majority class

a good option. For example, in case of cancer diagnosis purpose should take into account that a low value of Sensitivity (cancer patients who are correctly identified as suffering the disease) is more critic than a low value of specificity (healthy patients who are correctly classify as not suffering the disease).

5 Conclusions

Microarray data classification is a difficult challenge for learning systems due to its intrinsic characteristics. Machine learning has predominantly been employed for this kind of data and two-class classification techniques have been widely applied. Recent research indicates that the one-class approach is suitable to handle this kind of data because of it relies only on object coming from single class distribution. Despite of having less information to distinguish between classes, one-class models can easily learn the specific properties of a given dataset and are more robust to intrinsic difficulties of the data. In this paper, we demonstrate the suitability of applying one-class learning to handle microarray datasets. We made an experimental study to analyze and compare the behavior of one and two class classifiers, SVDD and SVM respectively, on two microarray datasets. At the same time the effect of applying feature selection techniques is considered, denoting its importance to reduce overfitting. The experimental results allow us to prove the superiority of the one-class classification. Therefore, we can confirm that one-class approach is a good technique to handle this kind of data offering a fine global performance and a good trade-off between sensitivity and specificity measures. Moreover, it offers the possibility of selecting one of the two available class as target concept in the learning process and remain the best results depending on the specific application. As lines of future research, we will conduct a study that includes other feature selection methods, since the tendency is toward focusing on new combinations (such as hybrid or ensemble methods). Moreover, we will incorporate both new microarray datasets as other one-class classification methods.

References

1. Golub, T.R., Slonim, D.K., Tamayo, P., Huard, C., Gaasenbeek, M., Mesirov, J.P., Coller, H., Loh, M.L., Downing, J.R., Caligiuri, M.A., Bloomfield, C.D.: Molecular classification of cancer: class discovery and class prediction by gene expression monitoring. Science **286**(5439), 531–537 (1999)
2. Hughes, G.: On the mean accuracy of statistical pattern recognizers. IEEE Transactions on Information Theory **14**(1), 55–63 (1968)
3. Guyon, I., Gunn, S., Nikravesh, M., Zadeh, L.A.: Feature Extraction: Foundations and Applications. Studies in Fuzziness and Soft Computing. Springer-Verlag New York, Inc. (2006)
4. Bolón-Canedo, V., Sánchez-Maroño, N., Alonso-Betanzos, A., Benítez, J.M., Herrera, F.: A review of microarray datasets and applied feature selection methods. Information Sciences **282**, 111–135 (2014)

5. Valafar, F.: Pattern recognition techniques in microarray data analysis: a survey. Annals of the NewYork Academy of Sciences **980**, 41–64 (2002)
6. Larrañaga, P., Calvo, B., Santana, R., Bielza, C., Galdiano, J., Inza, I., Lozano, J.A., Armañanzas, R., Santafé, G., Pérez, A., Robles, V.: Machine learning in bioinformatics. Briefings in Bioinformatics **7**(1), 86–112 (2006)
7. Yip, W.K., Amin, S.B., Li, C.: A survey of classification techniques for microarray data analysis. In: Handbook of Statistical Bioinformatics. Springer Handbooks of Computational Statistics, pp. 193–223 (2011)
8. Krawczyk, B.: Combining one-class support vector machines for microarray classification. In: Proc. Federated Conference on Computer Science and Information Systems (FedCSIS 2013), pp. 83–89 (2013)
9. Tax, D.M.J., Duin, R.P.W.: Support vector domain description. Pattern Recognition Letters **20**(11), 1191–1199 (1999)
10. Vapnik, V.: Statistical Learning Theory. Wiley (1998)
11. Tax, D.M.J., Duin, R.P.W.: Support vector data description. Machine Learning **54**, 45–66 (2004)
12. Tax, D.M.J.: DDtools, the data description toolbox for matlab. Delft University of Technology (2005)
13. Hall, M., Frank, E., Holmes, G., Pfahringer, B., Reutemann, P., Witten, I.H.: The weka data mining software: an update. ACM SIGKDD Explorations Newsletter **11**(1), 10–18 (2009)
14. Kent Ridge Bio-Medical Dataset. http://datam.i2r.a-star.edu.sg/datasets/krbd (online; accessed January 2015)
15. Microarray Cancers, Plymouth University. http://www.tech.plym.ac.uk/spmc/links/bioinformatics/microarray/microarray_cancers.html (online; accessed January 2015)
16. Moreno-Torres, J.G., Raeder, T., Alaiz-RodríGuez, R., Chawla, N.V., Herrera, F.: A Unifying View on Dataset Shift in Classification. Pattern Recognition **45**(1), 521–530 (2012)
17. Hall, M.: Correlation-Based Feature Selection for Machine Learning, PhD. Thesis (1999)
18. Yu, L., Liu, H.: Feature selection for high-dimensional data: A fast correlation-based filter solution, pp. 856–863 (2003)
19. Zhao, Z., Liu, H.: Searching for interacting features. In: Proceedings of the International Joint Conference on Artifical Intelligence, pp. 1156–1161 (2007)
20. Hall, M., Smith, L.: Practical feature subset selection for machine learning. Computer Science **98**, 181–191 (1998)
21. Kononenko, I.: Estimating attributes: analysis and extensions of RELIEF. In: Bergadano, Francesco, De Raedt, Luc (eds.) ECML 1994. LNCS, vol. 784, pp. 171–182. Springer, Heidelberg (1994)
22. Peng, H., Long, F., Ding, C.: Feature selection based on mutual information: criteria of max-dependency, max-relevance, and min-redundancy. IEEE Transactions on Pattern Analysis and Machine Intelligence **27**, 1226–1238 (2005)
23. Guyon, I., Weston, J., Barnhill, S., Vapnik, V., Cristianini, N.: Gene selection for cancer classification using support vector machines. Machine Learning **46**(1–3), 389–422 (2002)

Financial Intelligent Modeling

Intuitionistic Fuzzy Neural Network: The Case of Credit Scoring Using Text Information

Petr Hájek[✉] and Vladimír Olej

Institute of System Engineering and Informatics, Faculty of Economics and Administration,
University of Pardubice, Pardubice, Czech Republic
{Petr.Hajek,Vladimir.Olej}@upce.cz

Abstract. Intuitionistic fuzzy inference systems (IFISs) incorporate imprecision in the construction of membership functions present in fuzzy inference systems. In this paper we design intuitionistic fuzzy neural networks to adapt the antecedent and consequent parameters of IFISs. We also propose a mean of maximum defuzzification method for a class of Takagi-Sugeno IFISs and this method is compared with the center of area and basic defuzzification distribution operator. On credit scoring data, we show that the intuitionistic fuzzy neural network trained with gradient descent and Kalman filter algorithms outperforms the traditional ANFIS method.

Keywords: ANFIS · Intuitionistic fuzzy sets · Intuitionistic fuzzy inference systems of Takagi-Sugeno type · Intuitionistic fuzzy neural networks · Defuzzification methods

1   Introduction

Fuzzy inference systems (FISs) are one of the most widely used tools to model highly nonlinear systems with uncertainty. By incorporating imprecision into the models, FISs have provided good generalization performance in various applications. In addition, using linguistic terms makes the models descriptive and interpretable in their respective domains. However, determining the precise membership functions (both in antecedents and consequents of if-then rules) is problematic due to uncertainties associated with linguistic uncertainties, disagreement among experts, or noise in the data [1,2]. Therefore, researchers and practitioners call for additional freedom in the design of membership functions, making it possible to minimize the effects of the mentioned uncertainties [3]. As a result, two main categories of FIS generalizations have been proposed, interval-valued FISs [4] and intuitionistic FISs [5-8]. Currently, algorithms are being developed to optimize additional uncertainty in FISs' parameters.

Fuzzy neural networks have shown the ability to model non-linear systems at high accuracy, representing an implementation of FIS to adaptive networks for developing fuzzy if-then rules with suitable membership functions. For example, ANFIS [9,10] identifies a set of parameters through several learning rules: gradient descent algorithm, hybrid algorithm (combining back-propagation gradient descent and least squares method), and Kalman filter.

© Springer International Publishing Switzerland 2015
L. Iliadis and C. Jayne (Eds.): EANN 2015, CCIS 517, pp. 337–346, 2015.
DOI: 10.1007/978-3-319-23983-5_31

The concept of intuitionistic fuzzy sets (IF-sets) [11,12] was developed as an alternative approach to define a fuzzy set in cases where available information is not sufficient for the definition of an imprecise concept by means of a conventional fuzzy set, but the concept can be more naturally approached by separately envisaging positive and negative instances [13]. As a result, loosely related membership and non-membership functions can be defined in intuitionistic FISs (IFISs). This approach has been preferred in various application domains such as control [14], time series analysis [15], air pollution prediction [5], or bankruptcy forecasting [8]. However, these applications raised several questions regarding the adaptation of IFISs, most noticeably (1) which algorithms are effective for setting and fine-tuning parameters of IFISs and (2) how to defuzzify the IF-sets. This paper seeks to address the questions in the following way: (1) an intuitionistic fuzzy neural network (IFNN) is developed to adapt the antecedent and consequent parameters of IFIS, and (2) the mean of maximum (MOM) defuzzification method is proposed for a class of Takagi-Sugeno IFIS.

As an application, we investigate whether IFNNs can be effectively employed in credit scoring. Credit scoring has become an important application domain of soft computing [16] because it is based on a group of expert decision makers and, moreover, the determinants of credit scoring and their weight are associated with a high degree of uncertainty. Recently, it has been demonstrated that credit scoring prediction can be performed based on textual analysis of documents related to assessed companies [17]. Therefore, we first preprocessed textual data to obtain inputs to IFNN and then we trained IFNN to accurately predict the corresponding credit score.

The remainder of this paper has been organized in the following way. Section 2 presents the design of IFNN which is based on the IFIS of a first-order Takagi-Sugeno type. Section 3 describes the credit scoring data set and the preprocessing of input textual data and section 4 presents the results of credit scoring prediction using IFNN with various learning algorithms and defuzzification methods. Section 5 concludes the paper and discusses possible future research directions.

2   Intuitionistic Fuzzy Neural Network

Let a set X be a non-empty fixed set. An IF-set A in X is an object having the form $A=\{\langle x,\mu_A(x),v_A(x)\rangle|x\in X\}$, where the function $\mu_A:X\to[0,1]$ defines the degree of membership function $\mu_A(x)$ and the function $v_A:X\to[0,1]$ defines the degree of non-membership function $v_A(x)$, respectively, of the element $x\in X$ to the set A, $A\subset X$. For every $x\in X$, $0\leq\mu_A(x)+v_A(x)\leq1$, $\forall x\in X$ must hold [11,12,18,19]. The amount $\pi_A(x)=1-(\mu_A(x)+v_A(x))$ is called the hesitation part (intuitionistic index, IF-index) and represents a measure of non-determinacy.

In the IFIS of a first-order Takagi-Sugeno type [6], the k-th if-then rule R_k, $k=1,2,$ ... ,N, is defined as follows

R_k: **if** x_1 is $A_{1,k}$ AND x_2 is $A_{2,k}$ AND ... AND x_i is $A_{i,k}$ AND ... AND x_n is $A_{n,k}$

then $y_k=f(x_1,x_2, ... ,x_n)=a_{1,k}x_1+a_{2,k}x_2+ ... +a_{i,k}x_i+ ... +a_{n,k}x_n+b_k,$ (1)

where $A_{1,k}, A_{2,k}, \dots, A_{i,k}, \dots, A_{n,k}$ represent IF-sets, y_k is the output of the k-th rule represented by a linear function f of inputs $x_1, x_2, \dots, x_n$, and $a_{1,k}, a_{2,k}, \dots, a_{n,k}, b_k$ are the coefficients. The firing weight of each if-then rule is obtained using the application of t-norm operators, see [7] for an overview. The final defuzzified output is usually calculated as the weighted average of each if-then rule's output. The design of IFNN is based on the ANFIS model. Typically, there exist six layers in this model (Fig. 1).

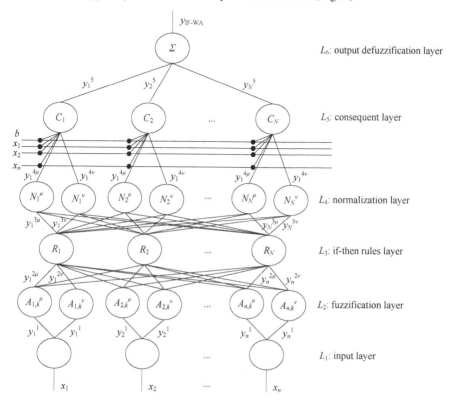

Fig. 1. IFNN model

Layer 1: This first layer is the input layer of the IFNN. Neurons in this layer transmit the external crisp input to the next layer. Namely, $x_1, x_2, \dots, x_n$ are the inputs and $y'_1, y'_2, \dots, y'_n$ are the outputs of the neurons in the first layer.

Layer 2: Neurons in the second layer represent antecedent IF-sets of if-then rules. Here, a fuzzification neuron receives an input and determines the degree to which this signal belongs to the neuron's IF-set. In this case, linguistic variables $A^\mu_{1,k}, A^\mu_{2,k}, \dots, A^\mu_{i,k}, \dots, A^\mu_{n,k}$ and $A^v_{1,k}, A^v_{2,k}, \dots, A^v_{i,k}, \dots, A^v_{n,k}$ determine membership functions $\mu(x_1), \mu(x_2), \dots, \mu(x_n)$ and non-membership functions $v(x_1), v(x_2), \dots, v(x_n))$, respectively, where for example $\mu(x_1) = 1 - v(x_1) - \pi(x_1)$, and $v(x_1) = 1 - \mu(x_1) - \pi(x_1)$.

If we have x_i^2 be the input and y_i^2 be the output signal of neuron i in the second layer, then we have $y_i^{2\mu} = f_2(x_i^{2\mu})$, $y_i^{2\nu} = f_2(x_i^{2\nu})$, where f_2 represents the activation function of neuron i, and is set to certain membership function $\mu(x)$ and non-membership function $\nu(x)$, usually Gaussian, bell, triangular, or trapezoidal. For example, Gaussian membership and non-membership functions can be defined as follows

$$\mu(x)=(1-\pi_a(x))e^{-\alpha}, \ \nu(x)=(1-\pi_b(x))-\mu(x)=1-\pi_b(x)-(1-\pi_a(x))e^{-\alpha}, \quad (2)$$

where $\alpha = (x-b)^2/2\sigma^2$ and $\pi_a(x)$ is the IF-index of center, and $\pi_b(x)$ is the IF-index of variance. Parameters b, σ, $\pi_a(x)$, and $\pi_b(x)$ are premise parameters that define the degrees of membership and non-membership, respectively.

Layer 3: Each neuron in the third layer corresponds to an if-then rule of the first-order Takagi-Sugeno type IFIS. An if-then rule neuron receives signals from the fuzzification neurons (involved in the antecedents of the if-then rule) and computes the firing weight of the if-then rule in the following way. Let w_k^μ and w_k^ν be firing weights which are computed using Gödel t-norm [20-22] as follows

$$w_k^\mu = \underset{k=1,2,\dots,N}{\text{MIN}} \ (\mu(x_1),\mu(x_2), \dots ,\mu(x_n)), \ w_k^\nu = \underset{k=1,2,\dots,N}{\text{MAX}} \ (\nu(x_1),\nu(x_2), \dots ,\nu(x_n)). \quad (3)$$

Then, if we have x_i^3 be the input and y_i^3 be the output signal of neuron i in the third layer, then we have $y_i^{3\mu} = f_3(x_i^{3\mu})$, $y_i^{3\nu} = f_3(x_i^{3\nu})$, where f_3 represents the MAX (MIN) operators of neuron i, and we obtain $y_i^{3\mu} = (w_1^\mu, w_2^\mu, \dots, w_N^\mu)$ and $y_i^{3\nu} = (w_1^\nu, w_2^\nu, \dots, w_N^\nu)$.

Layer 4: Neuron i in the fourth layer calculates the ratio of the i-th firing weight to the sum of all rules' firing weights. For convenience, the outputs of this layer are so-called normalized firing weights of the corresponding if-then rule.

If we have x_i^4 be the input and y_i^4 be the output signal of neuron i in the fourth layer, then we have $y_i^{4\mu} = f_4(x_i^{4\mu})$ and $y_i^{4\nu} = f_4(x_i^{4\nu})$, where f_4 represents the normalized function of neuron i. Then, $y_i^{4\mu} = (\overline{w}_1^\mu, \overline{w}_2^\mu, \dots, \overline{w}_N^\mu)$ and $y_i^{4\nu} = (\overline{w}_1^\nu, \overline{w}_2^\nu, \dots, \overline{w}_N^\nu)$ are normalized values.

Layer 5: The fifth layer, the fourth hidden layer, represents consequent parameters. Each neuron in this layer is connected to the if-then rule neurons in the fourth layer. A neuron in the fifth layer computes the weighted consequent value of a given rule in the following way

$$y_i^5 = x_i^5(x_1,x_2, \dots ,x_n) = x_i^5(a_{1,k}x_1 + a_{2,k}x_2 + \dots + a_{i,k}x_i + \dots + a_{n,k}x_n + b_k), \quad (4)$$

where x_i^5 is the input and y_i^5 is the output signal of neuron i in the fourth layer and $a_{1,k},a_{2,k}, \dots ,a_{n,k},b_k$ is a set of consequent parameters.

Layer 6: The sixth layer is both the output layer and the defuzzification layer. There is only one neuron in the layer, which calculates the weighted average of outputs of all neurons in the fifth layer and consequently produces the defuzzified output y_{IF_WA}. In related literature, defuzzification methods have been proposed for Mamdani type IFIS [8,23,24] such as center of area (IF_COA), basic defuzzification distribution operator

(IF_BADD), and mean of maximum (IF_MOM). IF_COA and IF_BADD have been adjusted to Takagi-Sugeno type IFISs recently as follows [8]

$$y_{\text{IF\_COA}} = \frac{\sum_{k=1}^{N} y_k (w_k^\mu - w_k^\nu)}{\sum_{k=1}^{N} (w_k^\mu - w_k^\nu)}, \quad y_{\text{IF\_BADD}} = \frac{\sum_{k=1}^{N} y_k (w_k^\mu - w_k^\nu)^\alpha}{\sum_{k=1}^{N} (w_k^\mu - w_k^\nu)^\alpha}, \text{ for } w_k^\mu - w_k^\nu > 0. \quad (5)$$

Here, we propose the IF_MOM method based on the weighted average defined by [23]. The defuzzified outputs can be defined as the weighted averages of all N if-then rules R_k. The weighted average IF_MOM can be defined as

$$y_{\text{IFWA3}} = \frac{\sum_{k=1}^{L} y_k^l}{2L}, \quad \text{for } y_k^l = \{ y \mid w_l^\mu - w_l^\nu = \max_k (w_k^\mu - w_k^\nu) \}. \quad (6)$$

In accordance with [9], the adaptation of IFNN (premise and consequent parameters) can be carried out, for example, using the two following algorithms: (1) gradient descent with one pass of LSE, where LSE is applied to obtain the initial values of the consequent parameters, and (2) sequential LSE, where IFNN is linearized w.r.t. the premise parameters and extended Kalman filter is used to adapt all parameters. Cluster centers b can be determined using a subtractive clustering algorithm [25]. In that case, the number of clusters is equal to the number of membership (non-membership) functions and, at the same time, to the number N of if-then rules.

3   Data Preprocessing

Credit scoring reduces information asymmetry between borrowers (issuers) and lenders (investors) by providing assessment of the credit risk of the borrowers. As output credit scores, we used the ratings provided by Standard and Poor's rating agency in 2011. We collected the ratings for 613 U.S. companies. The ratings AAA, AA+, AA, … ,C were transformed to credit scores 1,2,3, … ,21 (see Fig. 2 for their frequencies). As a source of input data, we used the annual reports of the companies, which are freely available on the U.S. Securities and Exchange Commission EDGAR System (www.sec.gov/edgar.shtml). Specifically, we used only the Management Discussion and Analysis section of the annual reports, where management discusses past and present corporate performance. The textual part of the annual reports is considered as an important indicator of future financial performance [17].

First, we preprocessed the documents using tokenization and lemmatization. Thus, potential term candidates were obtained. Second, the term candidates were compared with the positive (354 terms) and negative terms (2349 terms) developed specifically for financial domain [26]. Third, the *tf.idf* (term frequency–inverse document frequency) term weighting scheme was applied to calculate the importance of the terms obtained in the previous step. Fourth, the data were randomly divided into a training and testing set (4:1). This process was repeated five times. Fifth, a feature selection was performed using a correlation-based filter [27]. Note that feature selection was conducted after data division in order to prevent feature subset selection bias. On average, 29 terms were selected in training data sets. Table 1 shows all terms that were selected at least once.

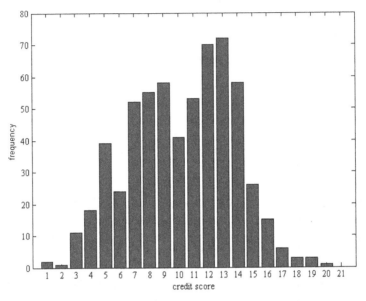

Fig. 2. Frequencies of corporate credit scores in the data set

Table 1. A set of input attributes

Category	Terms
Positive	abl, adequ, advantag, benefici, boom, compliment, construct, favor, good, charit, ideal, inventor, leadership, loyal, plenti, prestigi, progress, prosper, rebound, satisfi, solv, stabl, strength, success, win
Negative	abolish, acquiesc, alien, anticompetit, bribe, catastroph, confront, deeper, deleteri, demis, deter, disagr, disclos, dishonest, disincent, divert, drastic, embargo, erron, exculp, exoner, expos, feloni, grossli, harsh, impair, imped, impound, inabl, incapacit, incompet, incomplet, inconclus, indict, insurrect, know, limit, malfunct, miscalcul, misdemeanor, nonpay, nonrenew, object, obstacl, overdu, overestim, press, prevent, question, reckless, refin, riskier, stagnant, stoppag, uncorrect, unfeas, unlicens, unreli, unremedi, unresolv, unsold, unwant, vulner, weak, worst, wrongdo

4   Experimental Results

As mentioned above, the identification of IFNNs was carried out in two steps. In the first step, cluster centers were found using the subtractive clustering algorithm. Note that by using this algorithm, the number of cluster centers c is equal to the number of if-then rules N, $c=N$. Therefore, the radius of influence of a cluster r_a largely determines both the number of membership (non-membership) functions and the number of if-then rules. To control the complexity of the IFNN (and the potential over-fitting risk), we examined the radiuses varying from the set $r_a=\{0.50, 0.55, \ldots, 0.95\}$. As a result, we obtained on average $c=N=5.00\pm0.63$ if-then rules (and membership and non-membership functions) for the five training sets. In the second step, the premise and consequent parameters of the IFNN were optimized using (1) gradient descent with one pass of LSE, and (2) extended Kalman filter. The learning parameters of these algorithms were set as follows. The gradient algorithm was trained with the maximum number of epochs equal to 500, step size was set to 0.01, step increasing rate to 1.1, and step decreasing rate to 0.9. The parameters of the Kalman filter were data forgetting factor (set to 0.99) and increasing factor of data forgetting factor (set to 0.99). The quality of prediction was measured by RMSE on testing data owing to the ordinal character of the credit scores.

In the first set of experiments, we used the COA defuzzification method and compared the results with traditional ANFIS [9]. For the ANFIS, the initial setting of FIS was also determined using the subtractive clustering algorithm with the same procedure as for the IFNN. Hybrid and backpropagation algorithm are two commonly used algorithms to train ANFIS. The number of epochs was set to 10 for the hybrid algorithm and to 500 for the backpropagation algorithm. Fig. 3 shows that the IFNN trained with a gradient algorithm performed significantly best. To compare the algorithms, we performed a Friedman test at $p=0.05$.

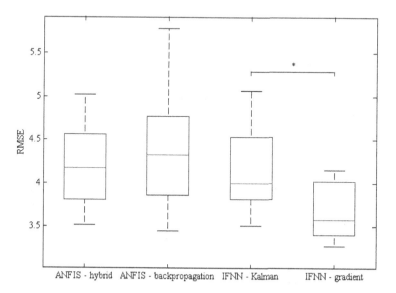

Fig. 3. RMSE on testing data – comparison between ANFIS and IFNN. COA defuzzification method was used for both ANFIS and IFNN. Median, lower, and upper quartile, and minimum and maximum RMSE are depicted in boxplots. IFNN trained with gradient performed significantly best (Friedman test at $p=0.05$).

In the second set of experiments, we compared the defuzzification methods used in the IFNNs, COA, BADD ($\alpha=2$), and MOM. Fig. 4 shows that the MOM performed significantly better than both BADD and COA in case of the Kalman learning algorithm. In addition, MOM and BADD performed significantly better than COA for the IFNN trained with gradient algorithm. Overall, the use of the MOM defuzzification method resulted in the decrease in RMSE, particularly for IFNN trained with Kalman algorithm. In contrast, COA was not an effective defuzzification method, irrespective of the IFNN learning algorithm. These findings suggest that the most accurate weighted consequent values were calculated using the functions with the maximum difference between w^μ_k and w^ν_k firing weights, in particular.

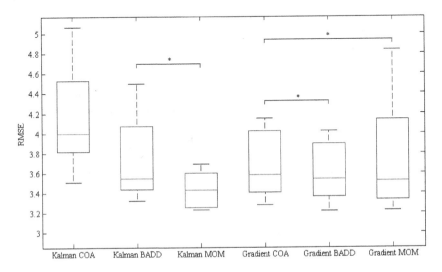

Fig. 4. RMSE on testing data – comparison between IFNN defuzzification methods. Median, lower, and upper quartile, and minimum and maximum RMSE are depicted in boxplots. Prediction performance was statistically compared using Friedman test at $p=0.05$.

5   Conclusion

Consistent with previous studies [3,28,29], we show that the introduction of non-membership functions may improve the prediction performance of FISs. However, substantially more experiments should be conducted to generalize our findings.

Taken together, we have demonstrated that IFNN trained by Kalman filter or gradient descent algorithms may significantly outperform ANFIS. In the case study of credit scoring using text information, hybrid and backpropagation algorithms were employed as two common algorithms to train ANFIS. The MOM defuzzification method, proposed in this study, has also shown promising results for use in the output layer of the IFNN. This class of IFNNs, see [30] for a generalized model, appears to provide better performance especially in the cases where high uncertainty and imprecision have to be captured in membership functions' design. Financial prediction seems to be a suitable application domain. Previous evidence with type-2 FISs supports this finding [3,28,29].

Future research should concentrate on the optimization of both the remaining parameters of the membership (and non-membership) functions and the base of if-then rules. Evolutionary algorithms have shown promising results for other generalizations of FISs [28]. Finally, we recommend further extensions of IFNNs corresponding to recent development in IF-sets, for instance temporal IF-sets [31].

The experiments in this study were carried out in Fuzzy Logic Toolbox, Matlab 2010b using the MS Windows 7 operation system.

Acknowledgments. This work was supported by the scientific research project of the Czech Sciences Foundation Grant No: 13-10331S.

References

1. Liang, Q., Mendel, J.M.: Interval Type-2 Fuzzy Logic Systems: Theory and Design. IEEE Transactions on Fuzzy Systems **8**(5), 535–550 (2000)
2. Hagras, H., Wagner, C.: Towards the Widespread Use of Type-2 Fuzzy Logic Systems in Real World Applications. IEEE Computational Intelligence Magazine **7**(3), 4–24 (2012)
3. Zarandi, F., et al.: A Type-2 Fuzzy Rules-based Expert System Model for Stock Price Analysis. Expert Systems with Applications **36**, 139–154 (2009)
4. Mendel, J.M.: Interval Type-2 Fuzzy Logic Systems Made Simple. IEEE Transactions on Fuzzy Systems **14**(6), 808–821 (2006)
5. Olej, V., Hájek, P.: IF-Inference systems design for prediction of ozone time series: the case of pardubice micro-region. In: Diamantaras, K., Duch, W., Iliadis, L.S. (eds.) ICANN 2010, Part I. LNCS, vol. 6352, pp. 1–11. Springer, Heidelberg (2010)
6. Olej, V., Hájek, P.: Comparison of fuzzy operators for if-inference systems of takagi-sugeno type in ozone prediction. In: Iliadis, L., Maglogiannis, I., Papadopoulos, H. (eds.) EANN/AIAI 2011, Part II. IFIP AICT, vol. 364, pp. 92–97. Springer, Heidelberg (2011)
7. Hájek, P., Olej, V.: Adaptive intuitionistic fuzzy inference systems of takagi-sugeno type for regression problems. In: Iliadis, L., Maglogiannis, I., Papadopoulos, H. (eds.) Artificial Intelligence Applications and Innovations. IFIP AICT, vol. 381, pp. 206–216. Springer, Heidelberg (2012)
8. Hájek, P., Olej, V.: Defuzzification methods in intuitionistic fuzzy inference systems of takagi-sugeno type. The case of corporate bankruptcy prediction. In: The 11th Int. Conf. on on Fuzzy Systems and Knowledge Discovery (FSKD 2014), Xiamen, China, pp. 240–244 (2014)
9. Shing, J., Jang, R.: ANFIS: Adaptive Network Based Fuzzy Inference System. IEEE Transactions on Systems, Man, and Cybernetics **23**(3), 665–685 (1993)
10. Loganathan, C., Girija, K.V.: Hybrid Learning for Adaptive Neuro Fuzzy Inference System. International Journal of Engineering and Science **2**(11), 6–13 (2013)
11. Atanassov, K.T.: Intuitionistic Fuzzy Sets. Fuzzy Sets and Systems **20**, 87–96 (1986)
12. Atanassov, K.T.: Intuitionistic Fuzzy Sets. Physica-Verlag, Heidelberg (1999)
13. Dubois, D., Prade, H.: Interval-valued fuzzy set, possibility theory and imprecise probability. In: Proc. of the Joint 4th Conf. of the European Society for Fuzzy Logic and Technology, EUSFLAT/ LFA, Barcelona, Spain, pp. 314–319 (2005)
14. Akram, M.S., et al.: Intuitionistic Fuzzy Logic Control for Washing Machines. Indian Journal of Science and Technology **7**(5), 654–661 (2014)

15. Castillo, O., et al.: An Intuitionistic Fuzzy System for Time Series Analysis in Plant Monitoring and Diagnosis. Applied Soft Computing **7**(4), 1227–1233 (2007)
16. Hájek, P.: Credit Rating Analysis using Adaptive Fuzzy Rule-Based Systems: An Industry Specific Approach. Central European Journal of Operations Research **20**(3), 421–434 (2012)
17. Hájek, P., Olej, V.: Evaluating sentiment in annual reports for financial distress prediction using neural networks and support vector machines. In: Iliadis, L., Papadopoulos, H., Jayne, C. (eds.) EANN 2013, Part II. CCIS, vol. 384, pp. 1–10. Springer, Heidelberg (2013)
18. Atanassov, K.T.: New operations defined over the intuitionistic fuzzy sets. Fuzzy Sets and Systems **61**(2), 137–142 (1994)
19. Szmidt, E., Kacprzyk, J.: Distances between intuitionistic fuzzy sets. Fuzzy Sets and Systems **114**(3), 505–518 (2000)
20. Klement, E.P., Mesiar, R., Pap, E.: Triangular Norms. Position Paper I: Basic Analytical and Algebraic Properties. Fuzzy Sets and Systems **143**, 5–26 (2004)
21. Barrenechea, E.: Generalized atanassov's intuitionistic fuzzy index. Construction Method. In: IFSA-EUSFLAT, Lisbon, pp. 478–482 (2009)
22. Deschrijver, G., Cornelis, C., Kerre, E.: On the Representation of Intuitionistic Fuzzy t-norm and t-conorm. IEEE Transactions on Fuzzy Systems **12**, 45–61 (2004)
23. Angelov, P.: Crispification: Defuzzification over Intuitionistic Fuzzy Sets. BUSEFAL **64**, 51–55 (1995)
24. Angelov, P.: Multi-Objective Optimisation in Air-Conditioning Systems: Comfort/Discomfort Definition by IF Sets. Notes on Intuitionistic Fuzzy Sets **7**(1), 10–23 (2001)
25. Chiu, S.: Fuzzy Model Identification based on Cluster Estimation. Journal of Intelligence and Fuzzy Systems **2**, 267–278 (1994)
26. Loughran, T., McDonald, B.: When is a Liability not a Liability? Textual Analysis, Dictionaries, and 10-Ks. The Journal of Finance **66**(1), 35–65 (2011)
27. Hall, M.A.: Correlation-based Feature Selection for Machine Learning. Doctoral dissertation, The University of Waikato (1999)
28. Bernardo, D., Hagras, H., Tsang, E.: A Genetic Type-2 Fuzzy Logic based System for the Generation of Summarised Linguistic Predictive Models for Financial Applications. Soft Computing **17**(12), 2185–2201 (2013)
29. Huarng, K., Yu, H.K.: A Type-2 Fuzzy Time Series Model for Stock Index Forecasting. Statistical Mechanics and its Applications **353**, 445–462 (2005)
30. Sotirov, S., Atanassov, K.: Intuitionistic fuzzy feed forward neural network. Cybernetics and Information Technologies **9**(2), 71–76 (2009)
31. Chen, L.H., Tu, C.C.: Time-validating-based Atanassov's Intuitionistic Fuzzy Decision-making. IEEE Transactions on Fuzzy Systems. doi:10.1109/TFUZZ.2014.2327989

Decision Making on Container Based Logistics Using Fuzzy Cognitive Maps

Athanasios Tsadiras[✉] and George Zitopoulos

Department of Economics, Aristotle University of Thessaloniki, Thessaloniki, Greece
{tsadiras,gzitopou}@econ.auth.gr

Abstract. Fuzzy Cognitive Maps is a well established decision making technique that combines Artificial Neural Networks and Fuzzy Logic. In this paper we present Fuzzy Cognitive Maps for making decisions regarding Container based Logistics, based on the knowledge extracted by a domain expert. Based on this knowledge, a model of the interactions and causal relations among various key Logistics factors is created. Having the FCM created, it is examined both statically and dynamically. A number of scenarios are introduced and the decision making capabilities of the technique are presented by simulating these scenarios and finding the predicted outcomes according to the model and expert's knowledge. FCM's predicted consequences of specific decisions can be valuable to Decision Makers since they can test their decisions and proceed with them only if the results are desirable.

Keywords: Decision making · Predictions · Fuzzy cognitive maps · Logistics

1 Introduction

Based on Axelord's work on Cognitive Maps [1], Kosko introduced in 1986, Fuzzy Cognitive Maps (FCMs) [2]. FCMs are considered a combination of Artificial Neural Networks and Fuzzy Logic and many researchers have made extensive studies on their capabilities (see for example [3-7]). FCMs are created as collections of concepts and the various causal relationships that exist between these concepts. The concepts are represented by neurons and the causal relationships by directed arcs/arrows between the neurons. Each arc is accompanied by a weight that defines the degree of the causal relation between the two concepts/neurons. The sign of the weight determines the positive or negative causal relation between the two concepts/neurons. An example of FCM concerning performance of RFID-enabled Reverse Logistics is given in Fig. 1, [8-9]. FCMs were used by various researchers for various Supply Chain Management and Logistics topics such as for forward–backward analysis of RFID-enabled supply chain [10] or for the development of a Material Management System [11].

In FCMs, although the degree of the causal relationships could be represented by a number in the interval [-1, 1], each concept, in a binary manner, could be either activated or not activated. Certainty Neuron Fuzzy Cognitive Maps (CNFCMs) are introduced [7], to provide additional representing capabilities to FCMs, by allowing each

© Springer International Publishing Switzerland 2015
L. Iliadis and C. Jayne (Eds.): EANN 2015, CCIS 517, pp. 347–357, 2015.
DOI: 10.1007/978-3-319-23983-5_32

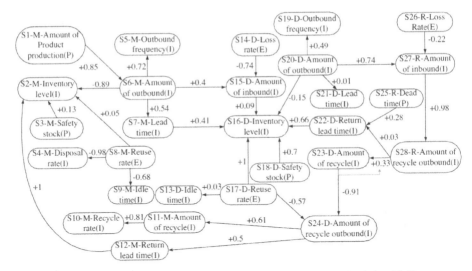

Fig. 1. FCM regarding performance of RFID-enabled reverse logistics, [8-9]

concept's activation to be activated just to a degree. To do that, a new updating function was proposed for the aggregation of the influences that each concept/neuron receives from other concepts/neurons. This was function f_M () that was initially used in MYCIN Expert System [12] for the handling of certainty factors. Certainty Neurons are defined as artificial neurons that use this function as their threshold function [13]. Using such neurons, the updating function of CNFCMs is the following:

$$A_i^{t+1} = f_M \left(A_i^t, S_i^t \right) - d_i A_i^t$$

where, A_i^{t+1} is the activation level of concept C_i at time step t+1,

$$S_i^t = \sum_j w_{ji} A_j^t$$ is the sum of the weight influences that concept C_i receives at

time step t from all other concepts,

d_i is a decay factor and

$$f_M(A_i^t, S_i^t) = \begin{cases} A_i^t + S_i^t(1 - A_i^t) = A_i^t + S_i^t - S_i^t A_i^t & if A_i^t \geq 0, \ S_i^t \geq 0 \\ A_i^t + S_i^t(1 + A_i^t) = A_i^t + S_i^t + S_i^t A_i^t & if A_i^t < 0, \ S_i^t < 0 \quad |A_i^t|, |S_i^t| \leq 1 \\ (A_i^t + S_i^t) / (1 - \min(|A_i^t|, |S_i^t|)) & otherwise \end{cases}$$

is the function that was used for the aggregation of certainty factors to the MYCIN expert system.

2   Development of a Fuzzy Cognitive Map for Containers based Logistics

To construct an FCM, specific rules should be followed to ensure its reliability. Because the FCM is created by the personal opinions and points of view of the expert(s) on the specific topic, the reliability of the model is heavily depended on the level of expertise of the domain expert(s). In our case, we used the Questionnaire method [14], which involves interviews and filling in of questionnaires by domain experts. The domain expert that we used is President of a Greek Logistics company, President of the Local Transport Business Association and First Vice President of the Local Chamber of Commerce. He is an expert on Containers based Logistics and provided us with the actors and factors of the FCM, as well as the possible alternative scenarios to be imposed to the developed system. The interested reader can find more on Containers based Logistics in [15].

After extended interviews and long discussions with the expert, which were time consuming due to the heavy schedule of the expert, the list of the concepts that are playing important role in Containers based Logistics and they should appear in the FCM, were identified. These concepts are the following:

Concept 1.   Logistics Services Prices
Concept 2.   Customer Demand for Logistics Services
Concept 3.   Revenue - Profits of Logistics Company
Concept 4.   Goods Insurance Costs
Concept 5.   Trust - Customer Satisfaction
Concept 6.   Empty Containers Safety Stock
Concept 7.   Containers Storage Size
Concept 8.   Security during Transportation
Concept 9.   Logistics Operating Costs
Concept 10. Delivery Times
Concept 11. Fleet Size
Concept 12. Road Traffic – Bad Weather
Concept 13. Fuel Prices
Concept 14. IT Systems - Technology
Concept 15. Strikes - Delays at Ports

After a structured procedure of filling in specific questionnaires by our domain expert, the causal relationships that exist between the concepts above were identified and the model presented in Fig. 2 was developed. This FCM can be studied both statically and dynamically.

3   Static Analysis of FCM

The static analysis of an FCM is based on studying the characteristics of the weighted directed graph that represent it, using graph theory techniques. The most important feature that should be studied is that of the feedback cycles that exist in the graph. Each cycle is accompanied by a sign, identified by multiplying the signs of the arcs that participate in the cycle. Positive cycle behaviour is that of amplifying any initial change, leading to a constant increase, in case an increase is introduced to the system. An example of a positive cycle is that of C2: Customer Demand for Logistics Services⇨ C11: Fleet Size ⇨ C5: Trust - Customer Satisfaction ⇨ C2: Customer Demand for Logistics Services. Through this cycle, an initial increase to "Customer's Demand for Logistics Services" will cause an additional increase to it.

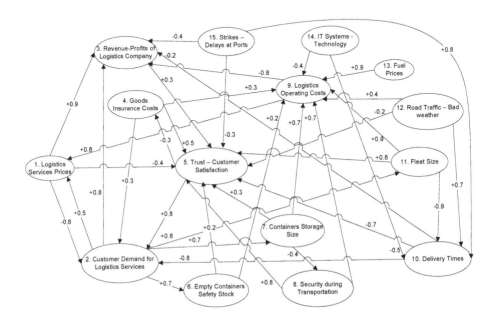

Fig. 2. FCM for Containers Based Logistics

Negative cycles on the other hand, counteract any initial change. This means that in the case where an increase is introduced in these cycles, they lead to decrease. An example of a negative cycle is that of   C1: Logistics Services Prices ⇨ C2: Customer Demand for Logistics Services (-) ⇨ C6: Empty Containers Safety Stock ⇨ C9: Logistics Operating Costs ⇨ C1: Logistics Services Prices. Through this cycle, an initial increase to the Logistics Services Prices will return to a decrease to them.

Examining the static characteristics of the FCM, it was found that the weighted directed graph of Fig. 2, contains a total of 23 cycles from which 12 are positive cycles and 11 are negative cycles. The existence of such a number of cycles implies strong and long interactions between the concepts of the FCM.

Another way to examine statically the FCM's graph is by calculating its density [16]. The density d is defined as

$$d = \frac{m}{n(n-1)}$$

where m is the number of arcs in the model and n is the number of concepts of the model. Density gives an indication of the complexity of the model. Typical values of density are in the interval [0.05, 0.3]. The density of the graph in Figure 2 is $38/(10*9) = 0.1809$ which is high and gives an indication of the complexity of the problem that it represents.

Graph Theory provides also the notion of node's importance [1] that assists the static analysis of FCMs. Node's importance gives an indication of the importance that the node/concept has for the model, by measuring the degree to which the node is central to the graph. The importance of a node i is evaluated as

$$imp(i) = in(i) + out(i)$$

where in(i) is the number of incoming arcs of node i and out(i) is the number of out-coming arcs of node i. According to this definition, the importance of the nodes of the FCM of Fig. 2 is given in Table 1. It is found that the most central/important concept is C5: Trust - Customer Satisfaction, followed by C9: Logistics Operating Costs.

Table 1. Importance of nodes

	C1	C2	C3	C4	C5	C6	C7	C8
In	2	4	5	1	10	1	1	1
Out	3	5	1	3	2	2	3	2
Imp	5	9	6	4	12	3	4	3

	C9	C10	C11	C12	C13	C14	C15
In	8	4	1	0	0	0	0
Out	2	3	3	3	1	2	3
Imp	10	7	4	3	1	2	3

4   Dynamic Study of FCM

The FCM model concerning Containers based Logistics was also simulated using the CNFCM technique that was mentioned in Section 1. The 15 concepts of the model and the 38 causal relationships that exist among these concepts were inserted to the CNFCM simulation program we developed in Excel using VBA programming. After that, various scenarios could be imposed to the system. The "what-if" scenarios that are presented in this paper were chosen in order to exhibit the decision making capabilities of the proposed method and of course much more can be imposed. The examined scenarios are shown in Table 2.

Table 2. Scenarios imposed to FCM concerning Containers based Logistics

Scenario	Description
#1	What-if there is a "small" increase to "Security during Transportation"? Activation of C8: "Security during Transportation" is set & kept to 0.25 (small increase).
#2	What-if there is a "large" increase to "Security during Transportation"? Activation of C8: "Security during Transportation" is set & kept to 0.75 (large increase).
#3	What-if there is a "moderate" increase to "Containers Storage Size"? Activation of C7: "Containers Storage Size" is set & kept to 0.5 (moderate increase).
#4	What-if there is a "moderate" increase to "Containers Storage Size"? Activation of C7: "Containers Storage Size" is set & kept to 0.5 (moderate increase). Additionally, a new causal relationship, weight $W_{7,14}=0.4$ is introduced.

In the first scenario, the decision maker is examining the impact of a small increase (0.25) of concept C8: "Security during Transportation" to the other concepts of the FCM model. The dynamical behavior of the system, as simulated by the CNFCM technique for this case, is shown in Fig. 3. From that figure, we can see that after an initial transition phase of 100 time steps with a lot of interactions, the concepts reach equilibrium at values that are presented in Table 3.

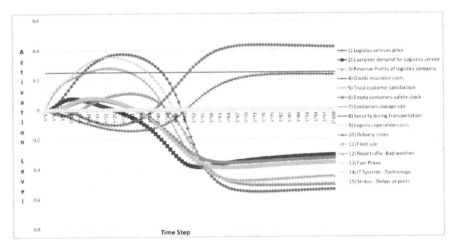

Fig. 3. Transition phase to equilibrium for Scenario #1 (C8: "Security during Transportation" is set to 0.25).

Table 3. Equilibrium Points of Scenarios #1 and #2

Activation level at equilibrium	1) Logistics Services Prices	2) Customer Demand for Logistics Services	3) Revenue-Profits of Logistics Company	4) Goods Insurance Costs	5) Trust-Customer Satisfaction	6) Empty containers Safety Stock	7) Containers Storage Size	8) Security during Transportation
#1-C8=0.25	-0.538	-0.307	-0.501	0.233	-0.449	-0.327	-0.327	0.250
#2-C8=0.75	0.639	0.433	0.491	-0.326	0.728	0.402	0.402	0.750
Activation level at equilibrium	9) Logistics operating costs	10) Delivery Times	11) Fleet Size	12) Road Traffic - Bad weather	13) Fuel Prices	14) IT Systems - Technology	15) Strikes - Delays at Ports	
#1-C8=0.25	-0.456	0.419	-0.357	0.000	0.000	0.000	0.000	
#2-C8=0.75	0.724	-0.465	0.435	0.000	0.000	0.000	0.000	

It can be noticed that scenario #1 is not good for the Logistics company because it predicts that the small increase of C8: "Security during Transportation" will lead to a moderate decrease of concepts C3: "Revenue-Profits of Logistics Company" (-0.501), C1 "Logistics Services Prices" (-0.538) and a small decrease to concept C2: "Customer Demand for Logistics Services (-0.307).

The Decision Maker, trying to improve the situation, can examine a different scenario, e.g. scenario #2. According to scenario #2, the activation of concept C8: "Security during Transportation" is changed from 0.25 to 0.75 that implies a "large" increase instead of a "small" increase. The transition of the system to the new equilibrium is shown in figure 4.

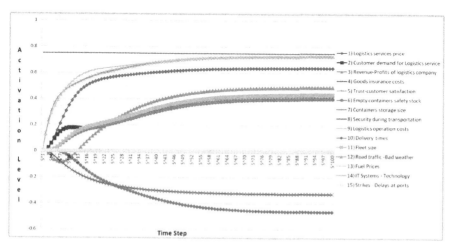

Fig. 4. Transition phase to equilibrium for scenario #2 (C8: "Security during Transportation" is set to 0.75).

The activation of each concept at the equilibrium state of scenario #2 is given Table 3. It can be noticed that for the Logistics company, scenario #2 is much better than that of scenario #1 because it predicts that the large increase of C8: "Security during Transportation" will lead to a moderate increase of concepts C3: "Revenue-Profits of Logistics Company" (0.491), C1: "Logistics Services Prices" (0.639) and C2: "Customer Demand for Logistics Services (0.433).

The decision maker that imposes the scenarios can draw the conclusion that scenario #2 should be preferred. He/she can interpret the results of scenario #2 the following way: The larger improvement of "Security during Transportation" that scenario #2 imposes compared to that of scenario #1, leads to a large increase in the "Trust-Customer Satisfaction". Subsequently, this causes the increase of "Customers Demand for Logistics Service". Due to the significant raise in "Customers Demand for Logistics Services", the "Empty Containers Safety Stock", the "Containers Storage size" and the "Fleet Size" are increased respectively. Furthermore, the "Logistics Services Prices" increases substantially due to the corresponding increase of "Customers Demand for Logistics Services" and "Logistics Operating Costs". The "Revenue - Profits of Logistics Company" increases despite the negative impact they receive from the "Logistics Operating Costs". This happens because the "Revenue - Profits of Logistics Company" receive powerful positive impacts from "Logistics Services Prices" and "Customers Demand for Logistics Services" that neutralize the negative impact. Finally, the "Goods Insurance Costs" and "Delivery Times" are reduced due to the increase of "Trust - Customer Satisfaction" and the "Fleet Size" respectively.

In the third scenario that was introduced, the effect of a moderate increase (0.50) of concept C7: "Containers Storage Size" is examined. Imposing this scenario to FCM, the system reaches a state of equilibrium with the transition of the system to equilibrium to be shown in Fig. 5. The concepts reach equilibrium at values that are presented in Table 4. From the values that appears in Table 4, it can be concluded that scenario #3 is not good for the Logistics company because it predicts that the moderate increase of C7: "Containers Storage Size" will lead to a moderate decrease of concepts C3: "Revenue-Profits of Logistics Company" (-0.582), C1: "Logistics Services Prices" (-0.528) and C2: "Customer Demand for Logistics Services (-0.432).

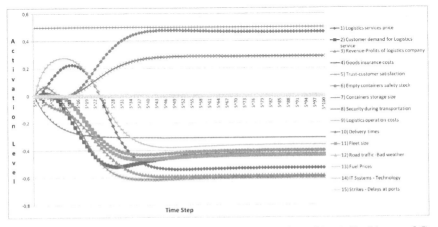

Fig. 5. Transition phase to equilibrium for scenario #3 (C7: "Containers Storage Size" is set to 0.5)

Table 4. Equilibrium Points of Scenarios #3 and #4

Activation level at equilibrium	1) Logistics Services Prices	2) Customer Demand for Logistics Services	3) Revenue-Profits of Logistics Company	4) Goods Insurance Costs	5) Trust-Customer Satisfaction	6) Empty containers Safety Stock	7) Containers Storage Size	8) Security during Transportation
#3-C7=0.5	-0.528	-0.432	-0.582	0.285	-0.598	-0.402	0.500	-0.308
#4 - C7=0.5 & W7.14=0.4	0.568	0.427	0.565	-0.264	0.538	0.400	0.500	-0.308

Activation level at equilibrium	9) Logistics operating costs	10) Delivery Times	11) Fleet Size	12) Road Traffic - Bad weather	13) Fuel Prices	14) IT Systems - Technology	15) Strikes - Delays at Ports	
#3-C7=0.5	-0.359	0.465	-0.435	0.000	0.000	0.000	0.000	
#4 - C7=0.5 & W7.14=0.4	0.472	-0.547	0.432	0.000	0.000	0.308	0.000	

The decision maker can make considerations about how the FCM should change, e.g. by the introduction of a new causal relationship (a new weight on the map), to avoid the negative results of scenario #3. To examine such a case, a new moderate positive causal relation between C7: "Containers Storage Size" and C14: "IT Systems-Technology" was added to the FCM model (scenario #4). In order to do that $W_{7,14}=0.4$ was added to the weight matrix of the FCM. By adding this new weight, the decision maker assumes that an increase to "Containers Storage Size" is causing a direct moderate increase to "IT Systems-Technology" (e.g. due to the additional need for computerization & automation). Simulating this new FCM for the same case with that of scenario #3 where C7:"Containers Storage Size" was moderately increased (C7=0.5), a new equilibrium is found for the system that appear at Table 4. The transition of the system to the new equilibrium is shown in Fig. 6.

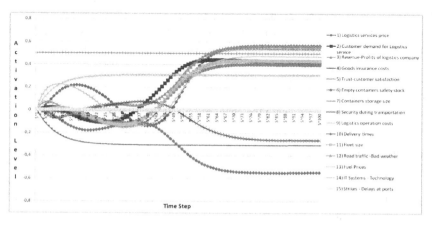

Fig. 6. Transition phase to equilibrium for scenario #4 (C7: "Containers Storage Size" is set to 0.5 and $W_{7,14} = 0.4$)

The introduction of the new weight $W_{7,14}$ led the system to a completely different equilibrium state from that of scenario #3, that is much better in regards of "Revenue - Profits of Logistics Company" (from moderate decrease -0.582, changes to moderate increase 0.565). The change of the equilibrium point can be interpreted in the following way: The system foresees that the moderate increase in the "Containers Storage Size" will result in a small decrease of "Security during Transportation" and in a relatively moderate improvement in the "IT systems-Technology". The improvement of "IT systems-Technology" leads to a reduction in the "Delivery times". Subsequently, this enhancement significantly increases the "Customer Demand for Logistics Services" and the "Trust - Customer Satisfaction". Moreover, the increase of "Customer Demand for Logistics Services" leads to a further increase in the "Empty Containers Safety Stock" and in the "Fleet Size". "Logistics Operating Costs" are moderately increased, which leads to a raise in the "Logistics Services Prices". Finally, the "Revenue - Profits of Logistics Company" will increase despite the negative effect they receive from the "Logistics Operating Costs". This happens because there are some powerful positive impacts from "Logistics Services Prices" and "Customer Demand for Logistics Services" that neutralize the negative impact.

5   Summary – Conclusions

After long discussions and extended interviews with a Logistic expert, a Fuzzy Cognitive Map is created that models the key factors of Container based Logistics and the various causal relationships that exist among them. The model was first examined statically. The density of model's graph was calculated and found particularly high, indicating the complexity of the case. The conceptual centralities of the concepts that exist in the model were also calculated and the most central, and consequently the most important concepts of the model were found.

After that, dynamic studies of the FCM model are made and the FCM technique is identified as an important and useful tool that can assist Logistics Decision Makers. This is because, it is capable of providing support by making predictions on various scenarios that are imposed to the Container based Logistics model that FCM creates. The novelty of the technique is apparent since it can also be used for studying structural changes to Logistics (new weights to the FCMs). This can be done by initially examining what such changes would cause to the various concepts of the corresponding FCM model and proceed with the changes only if the effects are the desirable.

Acknowledgments. The research of Athanasios Tsadiras has been co-financed by the European Union (European Social Fund – ESF) and Greek national funds through the Operational Program "Education and Lifelong Learning" of the National Strategic Reference Framework (NSRF) - Research Funding Program: Thales. Investing in knowledge society through the European Social Fund.

References

1. Axelrod, R.: Structure of Decision. The Cognitive Maps of Political Elites. Princeton University Press, Princeton (1976)
2. Kosko, B.: Fuzzy Cognitive Maps. Int. Journal of Man-Machine Studies **24**, 65–75 (1986)
3. Khan, M.S., Chong, A., Gedeon, T.: A Methodology for Developing Adaptive Fuzzy Cognitive Maps for Decision Support. Journal of Advanced Computational Intelligence **4**(6), 403–407 (2000)
4. Khan, M.S., Quaddus, M.: Group Decision Support Using Fuzzy Cognitive Maps for Causal Reasoning. Group Decision and Negotiation **13**, 463–480 (2004)
5. Stach, W., Kurgan, L., Pedrycz, W., Reformat, M.: Genetic learning of fuzzy cognitive maps. Fuzzy Sets and Systems **153**(3), 371–401 (2005)
6. Taber, R., Yager, R.R., Helgason, C.M.: Quantization effects on the equilibrium behavior of combined fuzzy cognitive maps. Int. Journal of Intelligent Systems **22**(2), 181–202 (2006)
7. Tsadiras, A.K., Margaritis, K.G.: Cognitive Mapping and Certainty Neuron Fuzzy Cognitive Maps. Information Sciences **101**, 109–130 (1997)
8. Trappey, A.J.C., Trappey, C.V., Wu, C.: Genetic algorithm dynamic performance evaluation for RFID reverse logistic management. Expert Systems with Applications **37**(11), 7329–7335 (2010)
9. Trappey, A.J.C., Trappey, C.V., Wu, C.-R., Hsu, F.-C.: Using fuzzy cognitive map for evaluation of RFID-based reverse logistics services. In: Proceedings of the 2009 IEEE International Conference on Systems, Man, and Cybernetics, pp. 1510–1515 (2009)
10. Kim, M.C., Kim, C.O., Hong, S.R., Kwon, I.H.: Forward–backward analysis of RFID-enabled supply chain using fuzzy cognitive map and genetic algorithm. Expert Systems with Applications **35**(3), 1166–1176 (2008)
11. Li, G., Peng, Q.: The development of the material management system based on ontology and fuzzy cognitive map. In: Jin, D., Lin, S. (eds.) CSISE 2011. AISC, vol. 106, pp. 431–436. Springer, Heidelberg (2011)
12. Buchanan, B.G., Shortliffe, E.H.: Rule-Based Expert Systems. The MYCIN Experiments of the Stanford Heuristic Programming Project. Addison-Wesley, Reading (1984)
13. Tsadiras, A.K., Margaritis, K.G.: Using Certainty Neurons in Fuzzy Cognitive Maps. Neural Network World **6**, 719–728 (1996)
14. Roberts, F.R.: Strategy for the energy crisis: the case of commuter transportation policy. In: Axelrod, R. (ed.) Structure of Decision, Princeton, pp. 142–179 (1976)
15. Hsu, W.-K.K.: Improving the service operations of container terminals. The International Journal of Logistics Management **24**(1), 101–116 (2013)
16. Hart, J.A.: Cognitive Maps of Three Latin American Policy Makers. World Politics **30**, 115–140 (1977)

Credit Prediction Using Transfer of Learning via Self-Organizing Maps to Neural Networks

Ali AghaeiRad[✉] and Bernardete Ribeiro

CISUC, Department of Informatics Engineering, University of Coimbra,
Polo II, Rua Silvio Lima, 3030-290 Coimbra, Portugal
{ali,bribeiro}@dei.uc.pt

Abstract. For financial institutions, the ability to predict or forecast business failures is crucial, as incorrect decisions can have direct financial consequences. Credit prediction and credit scoring are the two major research problems in the accounting and finance domain. A variety of pattern recognition techniques including neural networks, decision trees and support vector machines have been applied to predict whether borrowers are in danger of bankruptcy and whether they should be considered a good or bad credit risk.

In this paper a clustering and unsupervised method named Self Organizing Map (SOM) is used. We propose to label each cluster with voted method and improve labeling process by training a feedforward Neural Network (NN). The approach uses transfer of learning via SOM to the NN, is tested on the Australian Credit Approval financial data set. We compare both approaches and we will discuss which one is the best prediction model for financial data.

Keywords: Credit scoring · Self-organizing map · Neural networks · Financial risk analysis · Transfer learning

1 Introduction

Credit scoring has been introduced in the 1940s for the first time and over the decades had grown up significantly. In the 1960s, with the creation of credit cards, credit scoring in the credit granting process became more important for banks and other credit card issuers. In the 1980s, credit scoring was used for other purposes [6] such as an aid decision tool for approving personal loan applications.

Credit prediction models have evolved over time based on consumer behavior. Nowadays mortgage crisis and the economic downturn make credit prediction more important. Financial industry focuses on developing effective systems that can evaluate and manage credit. There are various methods of credit prediction regarding how exactly credit predicted to individuals for various purposes.

Before artificial intelligence has been used in credit prediction traditional statistical classification methods had been used for tackling this problem. In [2] a statistical model for discrete choice for consumer loan default and credit card

© Springer International Publishing Switzerland 2015
L. Iliadis and C. Jayne (Eds.): EANN 2015, CCIS 517, pp. 358–365, 2015.
DOI: 10.1007/978-3-319-23983-5_33

expenditure was derived. Later on [5] had explored the performance of credit scoring using two commonly discussed data mining techniques, classification and regression tree (CART) and multivariate adaptive regression splines (MARS). However, with the massive increase of applicants, which led to the overload of big data, it is often impossible to build such hand-craft models. Thus, with the advance in information and computer technology there is an ever-increasing need for Artificial Intelligence that is becoming a very popular alternative in credit scoring tasks. K-nearest neighbor method was proposed in [9] with an adjust version of Euclidean distance metric for assessing consumer credit risk. More recently, [10] presented an investigation of the use of supervised neural network models for credit risk evaluation under different learning schemes.

Companies are seeking credit prediction techniques with better performance so in [11] a thorough comparison the classification performance of credit score-card model, logistic regression model and decision tree model was performed. The authors made credit reports of individuals for financial companies to decide whenever customers have credit or not.

Self-organizing maps (SOMs) have been recognized as a powerful tool in credit prediction. Its capability and many supervised variants have been demonstrated in comparison with statistical and other intelligent methods. For example, in [8] SOM was used to determine the credit class through a visual exploration, while in [15] SOM is used for forecasting of credit classes. Despite the fact that SOM has been considerably used for credit prediction, in [12] & [14] SOM is used on bankruptcy trajectory analysis in finance field.

A recent approach in the area of Machine Learning is the combination of supervised and unsupervised methods to improve classification accuracy, although it has been seldom used in financial risk area. For example in [4] [7] [13] a fusion between supervised classification and clustering was used to improve the classification accuracy.

As SOM is a clustering technique, a common way for labeling the clusters of SOM is called voted method. Voted method labels each cluster based on the majority class in it. The purpose of this paper is to compare the capabilities of SOM that is labeled by voted method and SOM that is labeled by a feedforward Neural Network in forecasting of credit classes. The comparison has shown to perform well in a commonly used benchmark, the Australian dataset.

2   Background

2.1   Self-Organizing Maps

A Self-Organizing Map (SOM) is a type of artificial neural network (ANN) where neurons are set along an n-dimensional grid. Self-Organizing Maps are different from other ANN in the sense that they use a neighborhood function to preserve the topological properties of the input space. This data visualization technique invented by Teuvo Kohonen [1]. It has been created to help us understand high dimensional data because humans simply cannot visualize high dimensional data.

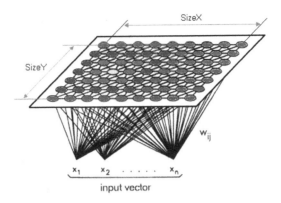

Fig. 1. Self-Organizing Map (SOM) [16]

SOM is able to reduce the amount of data and simultaneously projects data nonlinearly onto a lower dimensional array (see Figure 1). The neurons are distributed on a regular grid of usually two dimensions. Each neuron is related with a reference vector, reflecting the association strength for input vectors. A neighborhood kernel function is usually used to describe the topological relation of neurons. In each iteration of the training process, the reference vectors initialized at the beginning are updated. As a result, the neurons are topologically ordered on the grid, where neurons that have similar features in the input space will be located close to each other in the grid space.

2.2 Artificial Neural Network (ANN)

Information is processed by an ANN in a similar way to the human brain. There has been recently a renaissance of neural networks due their remarkable capabilities in performing tasks related to perception with many layers of processing information. Although deep neural networks are extremely good for computer vision and speech recognition models, shallow networks, have reduced resources needs, which convey for the problem handled in this work. In fact, a shallow ANN reduces memory requirements by storing a simple transfer function representing output values for multiple nodes. It typically consists of input nodes, output nodes, and hidden nodes.

To put it simply, a node can be referred as a data processing device that receives multiple inputs and generates a single output based on those inputs. During this process, input nodes receive one input, while both hidden nodes and output nodes receive several inputs. They are called hidden nodes because they do not receive any input data from sources outside the ANN and nor do they send output data to any devices outside the ANN. Because of this they are hidden from the existing universe outside the ANN.

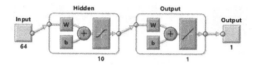

Fig. 2. Diagram corresponding to NN part of the proposed algorithm

A feedforward Neural Network is an artificial neural network where connections between the units do not form a directed cycle and are unilateral. It means that in this network information always moves one direction from the input nodes, through the hidden nodes and to the output node and never goes backwards [3].

3 Methodology

The financial dataset used in this paper does not need pre-processing steps for data preparation other than normalization. For this purpose we normalized the data set to zero mean and unit variance. The Australian dataset has 14 inputs and 2 outputs. It is fed to SOM as input data. SOM processed the input data with these parameters:

- Dimensions = 8×8 clusters, overall 64 clusters.
- Cover Steps = 100.
- Initial neighborhood = 5.
- Distance Functions= Manhattan distance.

The output of SOM is a map data of 8 by 8 clusters overall 64 clusters. Each input vector is mapped to a vector with 64 elements indicate clusters membership of input. The input layer of NN was connected to this output vector such that NN does not have any knowledge of 14 inputs of Australian dataset. Each subsequence layer has a connection from the previous layer. The row vector of hidden layer size is 10. The output layer produces weight of each input between 0 and 1. With threshold 0.5 final layer labels each input (see Figure 2) such that:

- Inputs with weight ≤ 0.5 is labeled as Bad Credit.
- Inputs with weight > 0.5 is labeled as Good Credit.

For learning in feed-forward NN each pair of input-output values are fed into the network for a number of epochs until the network 'learns' the relationship between the input and output. Levenberg-Marquardt backpropagation numerical optimization algorithm is used to optimize the performance function.
A description of the process can be gleaned from the layout in Figure 3 where the steps referred above are identified.

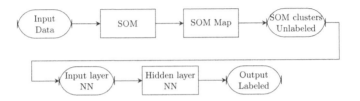

Fig. 3. Transfer learning labeling process

4 Results and Discussion

4.1 Dataset Description

Australian Credit Approval has used as financial datasets, which is publicly available in UCI in [17].

Credit approval data set concerns credit card applications, consisting of nine nominal-valued and six numeric-valued features. The 690 samples are classified into two classes with 307 and 483 respectively as good and bad. It contains 67 missing values on both numeric and categorical features, which are ignored in distance calculation and neuron update. We normalized the Dataset as preprocessing step for this work.

4.2 Performance Metrics

In order to evaluate a binary decision task we first define a contingency matrix representing the possible outcomes of the classification, as shown in Table 1. Error rate ($\frac{b+c}{a+b+c+d}$) measure defined based on this contingency table.

Table 1. Contingency table for binary classification.

	Class Positive	Class Negative
Assigned Positive	a (True Positives)	b (False Positives)
Assigned Negative	c (False Negatives)	d (True Negatives)

4.3 Experimental Setup

We used Matlab to run SOM on data set with SOM-map [8×8] (see Figure 4) for 30 times to make the results statistically significant.

– We used 80% randomly chosen samples in the train set and 20% randomly chosen on test set and Labeled each cluster with voted method based on majority class. The results showed that Error rate was between 0.10 to 0.25 with mean 0.17 and standard deviation 0.037. The confusion matrix is shown in Table 2.

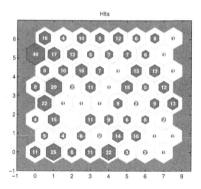

Fig. 4. Result of SOM mapping corresponding to one iteration

Table 2. Confusion matrix for Voted Method(Mean for 30 times).

	Class Positive	Class Negative
Assigned Positive	64.96 (True Positives)	11.51 (False Positives)
Assigned Negative	11.58 (False Negatives)	47.58 (True Negatives)
81% (Accuracy)	84% (Sensitivity)	80% (Specificity)

– We used 80% randomly chosen samples in the train set and 20% randomly chosen on test set and Labeled each cluster with feed forward Neural Network. The results showed that Error rate was between 0.11 to 0.18 with mean 0.15 and standard deviation 0.02. The confusion matrix is shown in Table 3.

Table 3. Confusion matrix for our proposed Method(Mean for 30 times).

	Class Positive	Class Negative
Assigned Positive	67.20 (True Positives)	12.80 (False Positives)
Assigned Negative	10.5 (False Negatives)	47.5 (True Negatives)
83% (Accuracy)	86% (Sensitivity)	78% (Specificity)

It can be observed that the least Error and high Accuracy is achieved with our approach.

5   Conclusion

SOMs have been widely used in data clustering as valuable tools due to the unique properties on data summarization and visualization. Normally, standard SOMs label each cluster based on majority class in it called voted method.

In this paper, we presented an approach to handle labeling for each cluster with a feed forward Neural Network, which uses the previous learning step of the found topological grid clusters. Thus, the NN inputs are the outputs of the Self-organising Map learned in the previous stage. The performance of the proposed algorithm showed less error rate in average than the baseline voted method. Moreover it presented low standard deviation which demonstrates more accuracy, stability and robustness.

In future work, the transfer of learning task should be deepen by considering approaches other than inductive learning. In addition, SOM parameters, in particular distance metrics, as well as neural network parameters should be further optimized.

Acknowledgments. Sirvan Khalighi is gratefully acknowledged for the helpful comments.

References

1. Kohonen, T.: Self-Organized Formation of Topologically Correct Feature Maps. Biological Cybernetics **43**, 59–69 (1982)
2. Greene, W.H.: A statistical model for credit scoring. NYU Working Paper No. EC-92-29 (1992)
3. Davidian, D.: Feed forward neural network. Google Patents, US Patent 5, 438, 646 (1995)
4. Dietterich, T.G., Bakiri, G.: Solving multiclass learning problems via error-correcting output codes. Journal of Artificial Intelligence Research, 263–286 (1995)
5. Vapnik, V.N.: Statistical Learning Theory. John Wiley & Sons, Inc. (1998)
6. Thomas, L.C.: A survey of credit and behavioural scoring: forecasting financial risk of lending to consumers. International Journal of Forecasting **16**(2), 149–172 (2000)
7. Japkowicz, N.: Supervised learning with unsupervised output separation. In: International Conference on Artificial Intelligence and Soft Computing, vol. 3, pp. 321–325 (2002)
8. Merkevičius, E., Garšva, G., Simutis, R.: Forecasting of credit classes with the self-organizing maps. Informaciens Technologijos **4**(33), Valdymas (2004)
9. Yu, L., Wang, S., Lai, K.K.: Credit risk assessment with a multistage neural network ensemble learning approach. Expert Systems with Applications **34**(2), 1434–1444 (2008)
10. Khashman, A.: Neural networks for credit risk evaluation: Investigation of different neural models and learning schemes. Expert Systems with Applications **37**(9), 6233–6239 (2010)
11. Yap, B.W., Ong, S.H., Husain, N.H.M.: Using data mining to improve assessment of credit worthiness via credit scoring models. Expert Systems with Applications **38**(10), 13274–13283 (2011)
12. Chen, N., Ribeiro, B., Vieira, A.S., Duarte, J., Neves, J.C.: A genetic algorithm-based approach to cost-sensitive bankruptcy prediction. Expert Systems with Applications **38**(10), 12939–12945 (2011)

13. Karem, F., Dhibi, M., Martin, A.: Combination of supervised and unsupervised classification using the theory of belief functions. In: Denoeux, T., Masson, M.-H. (eds.) Belief Functions: Theory and Applications. AISC, vol. 164, pp. 85–92. Springer, Heidelberg (2012)
14. Chen, N., Ribeiro, B.: A consensus approach for combining multiple classifiers in cost-sensitive bankruptcy prediction. In: Tomassini, M., Antonioni, A., Daolio, F., Buesser, P. (eds.) ICANNGA 2013. LNCS, vol. 7824, pp. 266–276. Springer, Heidelberg (2013)
15. Merkevičius, E., Garšva, G., Simutis, R.: Forecasting of credit classes with the Self-Organizing Maps. Information Technology and Control **33**(4) (2015)
16. SOM Figure from. http://www.pitt.edu/~is2470pb/Spring05/FinalProjects/Group1a/tutorial/som.html
17. UCI Machine Learning Repository. archive.ics.uci.edu/ml/datasets/Statlog+%28Australian+Credit+Approval%29

Echo State Networks

Using Echo State Networks to Classify Unscripted, Real-World Punctual Activity

Doug P. Hunt[(✉)] and Dave Parry

Auckland University of Technology, Auckland, New Zealand
{dphunt,dparry}@aut.ac.nz

Abstract. This paper employs an Echo State Network (ESN) to classify unscripted, real-world, punctual activity using inertial sensor data collected from horse riders. ESN has been shown to be an effective black-box classifier for spatio-temporal data and so we suggest that ESN could be useful as a classifier for punctual human activities and as a result a potential tool for wearable technologies. The aim of this study is to provide an example classifier, illustrating the applicability of ESN as a punctual activity classifier for the chosen problem domain. This is part of a wider set of work to build a wearable coach for equestrian sport.

Keywords: Echo State Network · Punctual activity classification · Spatio-temporal · Equestrian sport · Wearable technology · Wearable coach

1 Introduction

Most activity classifiers that have been proposed that use spatio-temporal input data from on-body inertial sensors have classified durative/cyclic activities such as running, rowing, standing, walking while much less has been attempted for punctual, non-cyclic activities such as picking up or placing an item.

Recent work done on classifying punctual activities from on-body inertial sensors, including [5,11,13] has used a common, multi-staged approach as described in [8, p.3] where the input data is segmented into windows, then each window has statistical or frequency domain features extracted from it such as mean, variance, energy and entropy. These features are then used as input into the classification engine, as opposed to the inertial readings themselves.

This work proposes Reservoir Computing (RC) methods as a possible classification engine for non-windowed, unprocessed spatio-temporal data and follows on from earlier work, namely [3,9,10] where we demonstrated that RC methods are conceptually useful and where we successfully demonstrated RC methods could successfully classify scripted activities from two participants.

This latest work demonstrates that an Echo State Network (ESN) classifier is capable of classifying unscripted, real-world activity from equestrian sport based on non-windowed data captured from a number of different participants across a period of several months at differing geographic locations.

L. Iliadis and C. Jayne (Eds.): EANN 2015, CCIS 517, pp. 369–378, 2015.
DOI: 10.1007/978-3-319-23983-5_34

Section 2 of this paper describes the application domain including our goal so that other practitioners can understand our motivation and assess the relevancy of this work. We also define the activity to be classified, with an illustration and identify the entities involved with our goal scenario. Section 3 describes the experimental set up including the approach, equipment used, participants, activity setting/scenario, data obtained, main characteristics of the data, data preprocessing and a description of the classification engine. Section 4 presents the results obtained and section 5 discusses our conclusions and future work.

In reporting this work we have followed the recommendations in "*How to do Good Research in Activity Recognition*" [7].

2   Application Domain

The overall goal of our recent research is to construct a Punctual Activity Classification System (ACS) to classify "Mounting", a single, boundary activity that is a common part of equestrian sport.

Identifying the precise start and end of any reasonably complex activity can be non-trivial [7] and without an agreed definition two observers may have somewhat differing ideas of what an activity is and when it starts and ends. Figure 1 shows a typical stirrup mount sequence and this could often be described as: *place left foot in stirrup; while holding the reins in the left hand, place left hand on saddle pommel (or similar position); place right hand on saddle cantle (or similar); lift the body and right leg off the ground (some hops or bounces may be involved); swing the right leg over the saddle cantle while leaning forward and moving the right hand forward; sit into the saddle; place right foot in stirrup.*

With this in mind and as we do not know of any existing agreed definition of "Mounting" for activity classification purposes, our definition of mounting starts after any hops or bounces, when the rider lifts their right leg (and body) off the ground, continues while they swing their leg over the cantle and ends when the rider puts their weight into the saddle or both stirrups.

We developed this definition by working backwards from the point where most observers would agree that a rider is mounted on a horse, *when the rider first puts their weight into the saddle or both stirrups*, to a point where there was reasonable consistency across riders. We deemed this to be the last time when they lift their right leg off the ground. Mounting a horse via a stirrup mount can be a strenuous endeavour and for a shorter rider or a taller horse, the rider may need to bounce a number of times before they accumulate enough impulsion to lift their own weight and body up and over the horse. By starting our definition after these bounces, we encapsulate that part of the mounting sequence that is most consistent while also encompassing the essence of the mount.

We are interested in classifying mounting within equestrian sport as our long term goal is to build a wearable coaching system [10] that provides coaching feedback to riders as they ride. By classifying mounting, we can define a start boundary for riding. Other boundary activities will also need to be classified but mounting is a logical place to start.

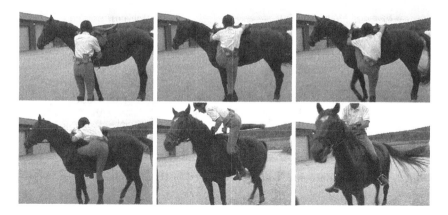

Fig. 1. A series of photos showing a participant stirrup mounting a horse, the left-most photo on the second line is the start of our definition of mounting.

When using a continuously active coaching system, there is no point in providing riding feedback before the rider mounts. Of course, it is possible for a rider to be mounted on a horse but not riding, the rider may simply be resting or doing other activities that are not riding. However it is not possible to be riding without first being mounted on a horse. This means that Mounting is a relevant and significant activity within the realm of horse riding.

The entities involved in this research are six participants (JC, HH, EE, HE, MZ, LL and AW) and their horses. All participants were adult females who volunteered and were recruited using Snowball techniques. None of the participants were known to the researchers prior to the work.

3   Technical Details

3.1   Approach

Most of the prior research into classifying human activities using inertial sensor data has concentrated on durative or cyclic activities such as walking, running, standing, lying down, cycling. Much of the earliest work using inertial sensors for activity classification such as [1,2,12] came from researchers with a health background who had an interest in how active a person (patient) was and this set the scene and happened to coincide with durative/cyclic activities.

More recently, researchers with an auditing intent (did the worker fit the screw on the car?) have become interested in activity classification and we are seeing work such as [5,11,13] that includes both durative and punctual activity. However, the success of prior activity classification researchers who primarily used sliding windowed techniques to segment their input data and who then calculated (mainly) statistical features from the data segments and used those for classification has influenced current research and so many contemporary

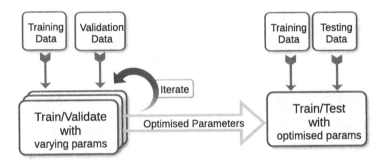

Fig. 2. Train, Validate and Test datasets used in the optimise and test cycle.

researchers continue to use the techniques that were developed for durative activities for the recent work with punctual activities. What has been missing from the activity classification research dialogue has been a clear differentiation between durative and punctual activities and an acknowledgement that techniques that are successful in one area may not be ideal in the other area.

The recent success of RC machine learning models with other forms of unsegmented spatio-temporal data such as speech inspired us to try these methods with inertial based, unsegmented, spatio-temporal punctual activity data. RC methods short–term memory facility may well be useful in this area. Our earlier successes with scripted activities has encouraged us to further test our hypothesis using data captured from real-world, unscripted actions.

Our approach is to use an ESN based machine learning tool as a black-box punctual activity classifier. The ESN undergoes supervised training when we feed it with labelled, unsegmented training data that includes instances of the punctual activity of interest and we then test the ability of the ESN to correctly classify the punctual activity by running previously unseen inertial data through the model and noting where it classifies the punctual activity. The ESN requires a number of parameters to be set to optimise the model to match the data characteristics and so our data is divided into three parts; learning, validation and testing. The learning and validation data is utilised to optimise the ESN parameters through an iterative adjustment phase and then the previously unseen test data is run through the optimised model. This is illustrated in Figure 2.

3.2 Data

The data used for this research includes part of one, scripted laboratory session and ten unscripted, real life riding sessions from seven participants (and horses) and was captured using a wrist mounted, commercially available, six degrees of freedom inertial sensor from SparkFun worn on the participant's right wrist.

During the ten real life sessions, the participants' activities were completely unscripted apart from a small synchronisation activity at the start or each session and were, as close as observed sessions could be, typical, everyday riding sessions with their own horse in their own environment. Each session was videoed so that activities could be manually classified. The data was collected in differing geographic locations in Sweden over a three month period. Data collection was done as part of a Masters project and was captured under ethical approval processes specified by our university.

Table 1. Train, Validate and Test datasets showing participants and sessions dates

Participant	Train	Validate	Test
JC	Lab End	Lab Start	0719
HH	0719		0812
EE		0912	
HE	0912	0913	
MZ			0829 and 0902
LL			0830
AW			0902

The sensor outputs three axis of magnetic, accelerometer and gyroscope data based on a 10 bit analogue to digital converter; a 16 bit sample number, start and stop characters; giving a total of 12 data fields per sample. Sensor readings were sampled at 10Hz and captured to an on-body receiver for later analysis. For this part of our research, only the Gyroscope data was used.

Sensor readings are linearly translated from a range of $[0 : 1023]$ to $[-1 : +1]$. The ESN is sensitive to input scale and so for consistency across data we prefer the input to be scaled within an expected range. We do not, however, translate the sensor readings into angular velocity, nor do we remove outliers.

During the data capture sessions the riders wore the sensor on their right wrist using a Velcro bandage. Other than ensuring consistent orientation, we deliberately took no care to place the sensor in a consistent position on a participant's wrist. Participants were free to adjust the sensor for comfort and most did adjust the sensor in some way after initial fitting. Our decision not to be exact in our placement of the sensor was driven by our desire to replicate a real-world situation as closely as possible, where participants choosing to wear a sensor of this type would inevitably fit it is slightly differing positions. In all real-world sessions participants used their own horse and tack. The real-world sessions do not include any examples of unusually tall or short horses or riders.

We decided to use some scripted mounts during the ESN model training and validation so as to preserve the real-world sessions for testing. To ensure that the ESN model was not biased towards the much larger number of scripted mounts from the laboratory session we split that dataset into two parts and only selected two mounts from the start for the validation data and two mounts from the end for the training data. No scripted mounts were used during testing. This is summarised in Table 1.

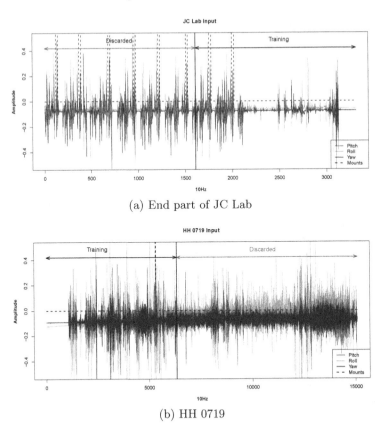

(a) End part of JC Lab

(b) HH 0719

Fig. 3. Examples of Gyroscope Data and Mounts used for Training.

Figure 3 depicts two parts of the three part training data. The figures show Gyroscope readings with the mount points depicted by the dotted lines. All three parts of the training data are concatenated into a single file when used to train the ESN. Sections of each part (file) are discarded to partially reduce the bias caused by the overwhelming majority of samples having been recorded during non-mount activity. The discarded sections are considered to be adequately represented by the non-discarded data. The validation dataset is constructed in a similar manner, using different data, as outlined in Table 1. The testing dataset is a concatenation of data not previously used during either training or validation and each dataset is used in full with no sections of the testing data discarded.

3.3 ESN Model Description

ESN were first introduced in [4]. We outline the concept of the method here and refer to the review on ESN and related RC methods in [6] and our prior work [3], for additional details. We use the same nomenclature as the review article.

The ESN learns a functional mapping from an input time series $\mathbf{u}(t) \in \mathbb{R}^{N_u}$ to a target $\mathbf{y}_{\text{target}}(t) \in \mathbb{R}^{N_y}$ based on a training data set $\{\mathbf{u}(t), \mathbf{y}_{\text{target}}(t)\}$ with $t = 1, \ldots, T$ where T is the size of the training set. The ESN has N_u input neurons, N_x reservoir (hidden) neurons and N_y output neurons. The reservoir neurons are either fully interconnected with connection weights specified by a weight matrix $\mathbf{W} \in \mathbb{R}^{N_x \times N_x}$. Matrix $\mathbf{W}$ is initialized with random (uniform) weights and then scaled by the spectral radius $\rho(\mathbf{W})$ which is the largest eigen value of $\mathbf{W}$. All N_u input neurons are connected to all reservoir neurons via connection weights defined by input matrix $\mathbf{W}_{\text{in}} \in \mathbb{R}^{N_x \times N_u}$. The weights of the input matrix are initialized as either -1 or 1 and then scaled by a scaling factor.

At time step t, the input $\mathbf{u}(t)$ is fed into a neural network. The output of all reservoir neurons in the network is computed:

$$\mathbf{x}'(t) = f(\mathbf{W}_{\text{in}}\mathbf{u}(t) + \mathbf{W}\mathbf{x}(t-1)) \tag{1}$$
$$\mathbf{x}(t) = (1-a)\mathbf{x}(t-1) + a\mathbf{x}'(t), \tag{2}$$

function f being a neuron activation function usually defined as the (element-wise) hyperbolic tangent $\tanh(\cdot)$, and factor $a \in \mathbb{R}$ being a leaking rate that controls the contribution of the previous neural output to its current state.

The output vector at time step t is linearly transformed into the final output of the network using an output weight vector connecting all reservoir and, optionally, all input neurons with N_y output neurons:

$$\mathbf{y}(t) = f_{\text{out}}(\mathbf{W}_{\text{out}}[\mathbf{x}(t)]) \tag{3}$$

f_{out} is the activation function of the output neurons which is usually chosen as the identity. The learning task is defined as an optimization problem in which the difference between $\mathbf{y}(t)$ and $\mathbf{y}_{\text{target}}(t)$) is minimised using linear or ridge regression to compute $\mathbf{W}_{\text{out}}$.

$$\mathbf{W}_{\text{out}} = \mathbf{Y}_{\text{target}}\mathbf{X}^T(\mathbf{X}\mathbf{X}^T + \alpha^2\mathbf{I})^{-1} \tag{4}$$

$\mathbf{I} \in \mathbb{R}^{N_x}$ is the identity matrix and $\alpha \in \mathbb{R}$ is a regularization factor that is tuned for optimal results. Matrix $\mathbf{Y}_{\text{target}} \in \mathbb{R}^{N_y \times T}$ contains all vectors $\mathbf{y}_{\text{target}}$ concatenated into a matrix. The connection weights $\mathbf{W}$ are not modified by the learning rule and only the output weights $\mathbf{W}_{\text{out}}$ are updated during training. Using a linear learning rule makes the training process more efficient.

The reservoir has the effect of transforming the input signal into a high-dimensional intermediate feature space represented by the output at any time t. Although linear methods are used to transform the feature vector into a desired target output, the mapping of the input $\mathbf{u}(t)$ to output $\mathbf{y}(t)$ remains non-linear.

4   Results

We used an evolutionary algorithm, i.e. a Particle Swarm Optimiser (PSO) to do a search of the parameter space for the ESN, to find a suitable configuration.

Table 2. Optimised Run Time Parameters for ESN

Input values	3 vectors of data
Number of neurons	336
Input scaling	0.925157
Neuron leak rate	0.137871
Regression Regularisation	0.000001
Spectral Radius	0.945490

Table 3. Precision and Recall at 0.6 Cut-Off

	Base Class	Mount Class
Precision	0.9996	0.0737
Recall	0.9981	0.2920

The parameter ranges that were used by the PSO are: Number of neurons $[N_x]$ **100 : 360**; Regularisation $[\alpha]$ **0.00001 : 5**; Input scaling $[s\,(W_{in})]$ **0.0001 : 1**; Leaking rate $[a]$ **0.0001 : 1** and Spectral radius $[\rho(W)]$ **0.6 : 1**. The ESN parameters from this process with the lowest fitness score are listed in table 2.

For testing, the model was trained using the parameters from Table 2 and with the training data specified. The trained model was then tested using the concatenated test file (see Table 1). The output from the test run is shown in figure 4. A Mount is deemed classified if the ESN output is 0.6 or greater for at least one sample during the Mount segment.

Five of the seven mount sequences are associated with an ESN response of 0.6 or greater, giving a classification rate of 0.714 at this cut-off level. This represents a satisfactory result based on the area under the ROC curve of 0.94184. However, high True Positive rates (0.7 and above, see Figure 5a) are only achieved at low response levels from the ESN model at around 0.2. This is much lower than our selected 0.6 cut-off level. As the cut-off level is raised from 0.2 towards 0.6, the Precision rate rises slowly while the Recall rate drops quickly (Figure 5b). Looking at True versus False Positives, we would have to raise the cut-off level well above 1.0 simply to get the total number of False Positives below 100 with this particular ESN model. Most of the False Positives are associated with two participant sessions, 0830-3 LL and 0902-1 AW and the False Positives mostly occur prior to mounting. Both sessions included extensive riding preparation activities including brushing the horse.

Table 4. Confusion Matrix at 0.6 Cut-off

Actual	Predicted	
	Base Class	Mount Class
Base Class	217295	415
Mount Class	80	33

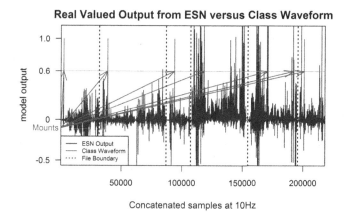

Fig. 4. ESN Output showing mounts via red arrows

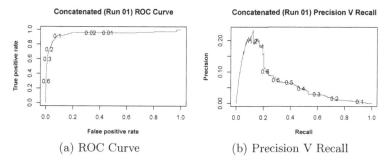

(a) ROC Curve   (b) Precision V Recall

Fig. 5. Classification Performance Graphs

5 Conclusions and Future Directions

We have shown an ESN based, continuous classification method to detect spatio-temporal patterns in real-world gyroscope data from unscripted actions within Equestrian sports. Despite having a relatively small population of training data and the rarity of the class data, the method reported useful but unspectacular results. This is our first attempt to use real–life inertial data using RC methods and we intend to pursue this method to see if we can improve these results.

In addition, future studies will compare RC classification methods with existing kernel based methods such as Support Vector Machines (SVM) and other established classification methods to see how RC methods compare. We also intend to compare unsegmented data against more traditional segmented data to test our hypothesis that punctual activities may require different classification methods than durative activities. We also plan to extend our data collection so that we have more examples in this area.

References

1. Bussmann, J., Veltink, P., Martens, W., Stam, H.: Activity monitoring with accelerometers. In: Dynamic Analysis Using Body Fixed Sensors. Congress book, pp. 13–17. McRoberts BV Den Haag (1994)
2. Foerster, F., Smeja, M., Fahrenberg, J.: Detection of posture and motion by accelerometry: a validation study in ambulatory monitoring. Computers in Human Behavior **15**(5), 571–583 (1999)
3. Hunt, D.P.L., Parry, D., Schliebs, S.: Exploring the applicability of reservoir methods for classifying punctual sports activities using on-body sensors. In: ACSC 2014. Australian Computer Society, Auckland, January 2014
4. Jaeger, H.: The "echo state" approach to analysing and training recurrent neural networks. Tech. rep., Fraunhofer Institute for Autonomous Intelligent Syst., Bremen, Germany (2001)
5. Junker, H., Amft, O., Lukowicz, P., Troster, G.: Gesture spotting with body-worn inertial sensors to detect user activities. Pattern Recognition **41**(6), 2010–2024 (2008)
6. Lukosevicius, M., Jaeger, H.: Reservoir computing approaches to recurrent neural network training. Computer Science Review **3**(3), 127–149 (2009)
7. Plötz, T.: How to do good research in activity recognition. In: How to do Good Research in Activity Recognition; Workshop in Conjunction with Pervasive 2010, Heksinki, Finland, p. 4 (2010)
8. Preece, S.J., Goulermas, J.Y., Kenney, L.P.J., Howard, D., Meijer, K., Crompton, R.: Activity identification using body-mounted sensors-a review of classification techniques. Physiological Measurement **30**(4), R1 (2009). 0119
9. Schliebs, S., Hunt, D.: Continuous classification of spatio-temporal data streams using liquid state machines. In: Huang, T., Zeng, Z., Li, C., Leung, C.S. (eds.) ICONIP 2012, Part IV. LNCS, vol. 7666, pp. 626–633. Springer, Heidelberg (2012)
10. Schliebs, S., Kasabov, N., Parry, D., Hunt, D.: Towards a wearable coach: classifying sports activities with reservoir computing. In: Iliadis, L., Papadopoulos, H., Jayne, C. (eds.) EANN 2013, Part I. CCIS, vol. 383, pp. 233–242. Springer, Heidelberg (2013)
11. Stiefmeier, T., Ogris, G., Junker, H., Lukowicz, P., Troster, G.: Combining motion sensors and ultrasonic hands tracking for continuous activity recognition in a maintenance scenario. In: 2006 10th IEEE International Symposium on Wearable Computers, pp. 97–104. IEEE (2006)
12. Veltink, P.H., Bussmann, H., de Vries, W., Martens, W., Van Lummel, R.C.: Detection of static and dynamic activities using uniaxial accelerometers. IEEE Transactions on Rehabilitation Engineering **4**(4), 375–385 (1996). 0312
13. Ward, J., Lukowicz, P., Troster, G., Starner, T.: Activity recognition of assembly tasks using body-worn microphones and accelerometers. IEEE Transactions on Pattern Analysis and Machine Intelligence **28**(10), 1553–1567 (2006)

An Echo State Network-Based Soft Sensor of Downhole Pressure for a Gas-Lift Oil Well

Eric Aislan Antonelo$^{(\boxtimes)}$ and Eduardo Camponogara

Department of Automation and Systems Engineering,
Federal University of Santa Catarina, Florianpolis, Brazil
erantone@elis.ugent.be

Abstract. Soft sensor technology has been increasingly used in industry. Its importance is magnified when the process variable to be estimated is key to control and monitoring processes and the respective sensor either has a high probability of failure or is unreliable due to harsh environment conditions. This is the case for permanent downhole gauge (PDG) sensors in the oil and gas industry, which measure pressure and temperature in deepwater oil wells. In this paper, historical data obtained from an actual offshore oil well is used to build a black box model that estimates the PDG downhole pressure from platform variables, using Echo State Networks (ESNs), which are a class of recurrent networks with powerful modeling capabilities. These networks, differently from other neural networks models used by most soft sensors in literature, can model the nonlinear dynamical properties present in the noisy real-world data by using a two-layer structure with efficient training: a recurrent nonlinear layer with fixed randomly generated weights and a linear adaptive readout output layer. Experimental results show that ESNs are a promising technique to model soft sensors in an industrial setting.

Keywords: Echo state network · Soft sensor · Gas-lift oil well · Reservoir computing

1 Introduction

With the advancement of powerful machine learning methods in the last decades, soft sensor technology has been made possible for a broad range of industries. Soft sensors aim to provide an additional source of information for process variables which are not reliable enough or whose expensive sensors can fail permanently, for instance, in harzadous environments. These soft sensors are predictive models built with methods which can infer an output $y(t)$ based on a number of input measurements $\mathbf{u}(t)$ [1]. The most common way of building soft sensors is through system identification using historical time series which show the (likely nonlinear) relationship between process variables. The resulting soft sensors are called data-driven since they are empirically obtained. Grey-box models and black-box models can be used to fit the empirical data. However, black-box models are more

© Springer International Publishing Switzerland 2015
L. Iliadis and C. Jayne (Eds.): EANN 2015, CCIS 517, pp. 379–389, 2015.
DOI: 10.1007/978-3-319-23983-5_35

often used for soft sensor technology than grey-box models [2], which have in artificial neural networks (ANNs) their most important and frequently used method [3]. This is because ANNs can efficiently model nonlinear relationships in process variables, which is usually the case for real-world processes.

Deepwater or low pressure oil wells usually require gas-lift technology in order to extract the oil from the well. The artificially injected gas diminishes the density of the well fluid, which, in turn, makes possible its extraction with the created difference in pressure. The downhole pressure is essential in assessing the dynamics of the oil well and must be monitored for controlling the productivity and stability of the well through the gas-lift flow rate as well as the production choke. The problem comes from the fact that permanent downhole gauge (PDG) sensors have a prohibitive cost for maintainance or replacement [4], and also that their premature failure is not uncommmon. Additionally, pertubations and noise can affect the PDG sensor measurements, making it not always a reliable information source.

Therefore, ANN-based soft sensors represent an important and alternative way of monitoring these variables. As the objective is to model an unknown dinamical system from real-world data, it is necessary that the used model maintains an internal state as a dynamical system does. This can be directly achieved by using a recurrent neural network (RNN), considered an universal approximation method for dynamical systems. An alternative is to use a tapped delay line with feedforward networks, which provides a finite window of past inputs, but provides no internal state for the network as the RNN does. Most models in literature use this last approach [5] or alternatively NARMAX models [6], since training an RNN with backpropagation-through-time is not trivial due to slow convergence properties and existence of bifurcations during training.

This paper aims to build a soft sensor of downhole pressure for gas-lift oil wells using a particular model of RNNs which exhibit fast training without local optima. Analog recurrent networks of these type are called Echo State Networks (ESNs) [7]. Their distinct feature is to separate the network in two main layers: one randomly generated pool of recurrent nonlinear neurons (called the reservoir [1]), and a linear adaptative readout layer (see Fig. 1). As the recurrent reservoir has its weights randomly generated, only the readout output layer needs to be trained, usually using linear regression methods such as the Least Squares method which has global convergence properties. Thus, the complexity of training recurrent weights is nonexistent. It is also possible to see the reservoir as a nonlinear dynamical kernel which projects the input into a high-dimensional nonlinear space, facilitating learning complex models. Methods which share this type of feature are called in literature Reservoir Computing (RC) methods [8].

In [9], a simulated vertical riser model was identified using ESNs as a black-box model. The ESN, by using feedback from the output to the reservoir layer, was able to sustain oscillations as well as steady states with only a single ESN. The input consisted of a single variable, the opening of the production choke, and the output to be estimated was the downhole pressure. In the current

---

[1] The term reservoir is not related to reservoirs in oil and gas industry.

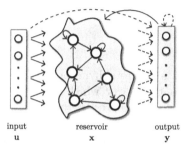

Fig. 1. Reservoir Computing (RC) network. The reservoir is a non-linear dynamical system usually composed of recurrent sigmoid units. Solid lines represent fixed, randomly generated connections, while dashed lines represent trainable or adaptive weights.

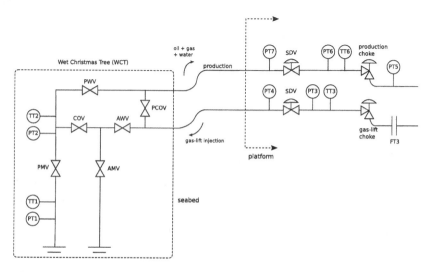

Fig. 2. Oil well scheme showing the location of sensors and chokes. FT3 is a flow rate sensor; PT# and TT# are pressure and temperature sensors. SDV stands for ShutDown Valve.

paper, instead of using simulation data, the ESN-based model is built from real-world data obtained from a particular deepwater oil well. The PDG downhole pressure is estimated based on input from 8 sensor measurements plus 2 choke openings (gas-lift and production chokes). All these input measurements come from the platform variables since seabed variables are not always available in offshore oil wells (see Fig. 2 for a scheme with position of sensors and chokes). The main contribution of this paper lies in the analysis of the powerful RC approach in modeling real-world noisy data from the gas and oil industry.

In the following section, the ESN model and training is presented. Section 3 presents the procedure to build the ESN-based soft sensor. Next, experimental results are shown in Section 4 and the conclusion drawn in Section 5.

2   Reservoir Computing

2.1   ESN Model

An ESN is composed of a discrete hyperbolic-tangent RNN, the reservoir, and of a linear readout output layer which maps the reservoir states to the actual output. Let n_i, n_r and n_o represent the number of input, reservoir and output units, respectively, $\mathbf{u}[n]$ the n_i-dimensional external input, $\mathbf{x}[n]$ the n_r-dimensional reservoir activation state, $\mathbf{y}[n]$ the n_o-dimensional output vector, at discrete time n. Then the discrete time dynamics of the ESN is given by the state update equation

$$\begin{aligned}
\mathbf{x}[n+1] =& (1-\alpha)\mathbf{x}[n] + \alpha f(\mathbf{W}_r^r \mathbf{x}[n] + \mathbf{W}_i^r \mathbf{u}[n] + \\
& \mathbf{W}_o^r \mathbf{y}[n] + \mathbf{W}_b^r),
\end{aligned} \quad (1)$$

and by the output computed as:

$$\mathbf{y}[n+1] = g\left(\mathbf{W}_r^o \mathbf{x}[n+1] + \mathbf{W}_i^o \mathbf{u}[n] + \mathbf{W}_o^o \mathbf{y}[n] + \mathbf{W}_b^o\right) \quad (2)$$

$$= g\left(\mathbf{W}^{\text{out}}\left(\mathbf{x}[n+1], \mathbf{u}[n], \mathbf{y}[n], 1\right)\right) \quad (3)$$

$$= g\left(\mathbf{W}^{\text{out}}\mathbf{z}[n+1]\right), \quad (4)$$

where: α is the leak rate [10,11]; $f(\cdot) = \tanh(\cdot)$ is the hyperbolic tangent activation function, commonly used for ESNs; g is a post-processing activation function (in this paper, g is the identity function); $\mathbf{W}^{\text{out}}$ is the column-wise concatenation of $\mathbf{W}_r^o$, $\mathbf{W}_i^o$, $\mathbf{W}_o^o$ and $\mathbf{W}_b^o$; and $\mathbf{z}[n+1] = (\mathbf{x}[n+1], \mathbf{u}[n], \mathbf{y}[n], 1)$ is the extended reservoir state, i.e., the concatenation of the state, the previous input and output vectors and a bias term, respectively.

The matrices $\mathbf{W}^{\text{to}}_{\text{from}}$ represent the connection weights between the nodes of the complete network, where r, i, o, b denotes *reservoir*, *input*, *output*, and *bias*, respectively. All weight matrices representing the connections to the reservoir, denoted as $\mathbf{W}^r$, are initialized randomly (represented by solid arrows in Figure 1), whereas all connections to the output layer, denoted as $\mathbf{W}^o$, are trained (represented by dashed arrows in Figure 1). For the experiments in this paper, output feedback and bias to reservoir are not used ($\mathbf{W}_o^r$ and $\mathbf{W}_b^r$ are not present). Additionaly, $\mathbf{W}_o^o$, $\mathbf{W}_b^o$ and $\mathbf{W}_i^o$ are also absent. Thus, equations (1) and (2) become:

$$\mathbf{x}[n+1] = (1-\alpha)\mathbf{x}[n] + \alpha f\left(\mathbf{W}_r^r \mathbf{x}[n] + \mathbf{W}_i^r \mathbf{u}[n]\right) \quad (5)$$

$$\mathbf{y}[n+1] = g\left(\mathbf{W}^{\text{out}}\mathbf{x}[n+1]\right). \quad (6)$$

The non-trainable connection matrices $\mathbf{W}_r^r, \mathbf{W}_i^r$ are usually generated from a Gaussian distribution $N(0,1)$ or a uniform discrete set $\{-1,0,1\}$. During this random initialization, sparsity can be obtained by using a parameter called connection fraction $c^{\text{to}}_{\text{from}}$ which determines the percentage of nonzero weights in the respective connection matrix $\mathbf{W}^{\text{to}}_{\text{from}}$. Another parameter in this procedure is the input scaling $v^{\text{to}}_{\text{from}}$ which is a scalar multiplied by the respective matrix $\mathbf{W}^{\text{to}}_{\text{from}}$.

This *scaling* of matrices is important because it influences greatly the reservoir dynamics [8] and, in this way, must be tuned for optimal performance in most cases.

The weights from the reservoir connection matrix $\mathbf{W}_r^r$ are obtained randomly through a Normal distribution $(N(0,1))$ and then rescaled such that the resulting system is stable but still exhibits rich dynamics. A general rule to create good reservoirs is to set the reservoir weights such that the reservoir has the *Echo State Property* (ESP) [12], i.e., a reservoir with fading memory. A common method used in literature is to rescale $\mathbf{W}_r^r$ such that its spectral radius $\rho(\mathbf{W}_r^r) < 1$ [12]. Although it does not guarantee the ESP, in practice it has been empirically observed that this criterium works well and often produces analog sigmoid ESNs with ESP for any input.

It is important to note that spectral radius closer to unity as well as larger input scaling makes the reservoir more non-linear, which has a deterioration impact on the memory capacity as side-effect [13]. The scaling of these non-trainable weights is a parameter which should be chosen according to the task at hand empirically, analyzing the behavior of the reservoir state over time, or by grid searching over parameter ranges.

Most temporal learning tasks require that the timescale present in the reservoir match the timescales present in the input signal as well as in the task space. This matching can be done by the use of a leak rate $(\alpha \in (0,1])$ and/or by resampling the input signal. For instance, low leak rates yield reservoirs with more memory which can *hold* the previous stimuli for longer time spans.

2.2   Training

Training the RC network means finding $\mathbf{W}^{\text{out}}$ in (2) or (6), that is, the weights for readout output layer from Fig. 1. For that, the reservoir is driven by an input sequence $\mathbf{u}(1), \ldots, \mathbf{u}(n_s)$ which yields a sequence of extended reservoir states $\mathbf{z}(1), \ldots, \mathbf{z}(n_s)$ using (1) (the initial state is $\mathbf{x}(0) = \mathbf{0}$). The desired target outputs $\hat{\mathbf{y}}[n]$ are collected row-wise into a matrix $\hat{\mathbf{Y}}$. The generated extended states are collected row-wise into a matrix $\mathbf{X}$ of size $n_s \times (n_r + n_i + n_o + 1)$ if using (1) or $n_s \times n_r$ if using (5).

Then, the training of the output layer is done by using the **Ridge Regression** method [14], also called *Regularized Linear Least Squares* or *Tikhonov regularization* [15]:

$$\tilde{\mathbf{W}}^{\text{out}} = (\mathbf{X}^\top \mathbf{X} + \lambda \mathbf{I})^{-1} \mathbf{X}^\top \hat{\mathbf{Y}} \tag{7}$$

where $\tilde{\mathbf{W}}^{\text{out}}$ is the column-wise concatenation of $\mathbf{W}_r^o$, and the optional matrices $\mathbf{W}_i^o$, $\mathbf{W}_o^o$ and n_s denotes the total number of training samples.

In the generation of $\mathbf{X}$, a process called **warm-up drop** is used to disregard a possible undesired initial transient in the reservoir starting at $\mathbf{x}(0) = \mathbf{0}$. This is achieved by dropping the first n_{wd} samples so that only the samples $\mathbf{z}[n]$, $n = n_{wd}, n_{wd} + 1, ..., n_s$ are collected into the matrix $\mathbf{X}$.

The learning of the RC network is a fast process without local minima. Once trained, the resulting RC-based system can be used for real-time operation on moderate hardware since the computations are very fast (only matrix multiplications of small matrices).

Error Measure. For regression tasks, the Normalized Root Mean Square Error (NRMSE) is used as a performance measure and is defined as:

$$\text{NRMSE} = \sqrt{\frac{\langle (\hat{y}[n] - y[n])^2 \rangle}{\sigma_{\hat{y}[n]}^2}}, \tag{8}$$

where the numerator is the mean squared error of the output $y[n]$ and the denominator is the variance of desired output $\hat{y}[n]$.

3   Soft Sensor Through ESNs

The task is to infer the downhole pressure at the PDG sensor, located in the seabed, from sensor measurements obtained at the more easily accessible platform location (see Fig. 2). The sets of input and output variables are given in Table 1. In this work, the ESN or RC network is used to learn a dynamical mapping between the input variables $\mathbf{u}(t)$ (which, in principle, consists of 10 inputs normalized to the interval $[0, 1]$, corresponding to the 8 platform variables from Table 1 plus the openings of the gas-lift choke and production choke) and the output variable $y(t)$ (PDG pressure sensor). The available dataset contains 5 months of measurement data: 08/2010, 01/2011, 07/2011, 11/2011 and 12/2011. The measured downhole pressure for the complete period can seen in Fig. 3.

The soft sensor is built according to one of the following approaches: using all the available sampled data (5 months); and using only a limited number of samples such as from only one month. Considering the former case, i.e., when the whole dataset is used, the exhibited dynamical nonlinear behavior range is very wide. This is because the measurements can also include cases when the

Table 1. Process variables

Tag	Process variable	Location	Variables Set
PT1	Downhole pressure	Seabed	Output
TT1	Downhole temperature	Seabed	—
PT2	WCT pressure	Seabed	—
TT2	WCT temperature	Seabed	—
PT3	Pressure before SDV	Platform	Input
TT3	Temperature before SDV	Platform	Input
FT3	Instantaneous gas-lift flow rate	Platform	Input
PT4	Pressure after SDV	Platform	Input
PT5	Pressure after production choke	Platform	Input
PT6	Pressure before production choke	Platform	Input
TT6	Temperature before production choke	Platform	Input
PT7	Pressure before SDV	Platform	Input

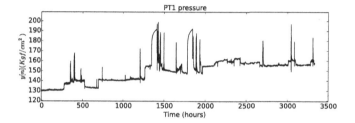

Fig. 3. Downhole Pressure (PT1) measured over the complete period of 5 months.

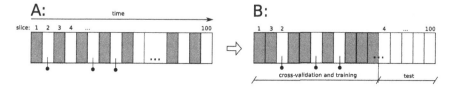

Fig. 4. Scheme showing how the dataset is divided for training and test. Grey-shaded rectangles plus randomly chosen white rectangles (indicated by sticks) are selected for training, while white ones are used for test. See text for more information.

well is closed, causing abnormal behavior, and also that a prolonged period will include different or evolving relationships between variables for samples very distant from each other in time. Taking this into account, we propose a method to select a representative subset of the dataset for training and the rest for test. The method is graphically represented in Fig. 4 and described in the following. We decide that 70% of the dataset is selected for training, while the rest 30% is used for test. For that, first, the dataset is divided into two groups, gray and white rectangles, which are alternatingly chosen in the time axis, as *individual sample slices* each containing 2,000 measurements (the sampling rate is one measurement per minute). Now, we have two sets, each having 50% of the dataset (situation A in the figure). To form the training dataset, we add more 20% of randomly chosen white rectangles to the the training dataset, indicated in Fig. 4 by little sticks. The resulting operation leads to the re-arranged sample slices used for training and test (Situation B in the plot), where now the temporal order is only valid between the samples contained in the slices.

4   Experimental Results

4.1   Whole Dataset

Using the method described in the previous section and shown in Fig. 4, 70% of the dataset was selected for training and validation whereas the rest 30% of the data was reserved for testing. With the selected 70%, a grid-search experiment was accomplished to find the best performing parameters: input scaling v_i^r, leak rate α, and spectral radius $\rho(\mathbf{W}_r^r)$. Fixed parameters were as follows: size of reservoir $n_r = 200$ and all connection fractions were $c = 1$. Each run of the grid search was made through a 5-fold random cross-validation procedure, considering 3 randomly generated reservoirs, where the ridge regression parameter λ was optimized for each generated reservoir. The values found were as follows: $v_i^r = 0.2$, $\alpha = 0.5$, and $\rho(\mathbf{W}_r^r) = 0.5$.

Next, the RC network was trained using above parameters on the whole training dataset, and the generalization results were obtained evaluating the trained network on the test dataset (30% of the samples). Fig. 5 shows these results comparing the predicted network output (given by a light blue line) to the target output (or the measured downhole pressure PT1, given by a black line). Each vertical dashed line marks the frontier between one sample slice and the next one (see Section 3), each one containing 2,000 consecutive samples (measured each minute). Fig. 5 (a) shows the prediction over the complete test set, revealing that most of the operating points of the downhole pressure were correctly estimated. The dynamical properties of this signal varied considerably taking into account all the 5 months. Some nonlinear dynamical behaviors present in the dataset were feasible to be learned while other periods showed inconsistencies (Fig. 5 (b)), probably due to abnormal and less frequent behaviors, which may not be inferred from the input variables. The NRMSE on test data was 5.03 for a reservoir of 200 units and 4.81 for a reservoir of 400 units.

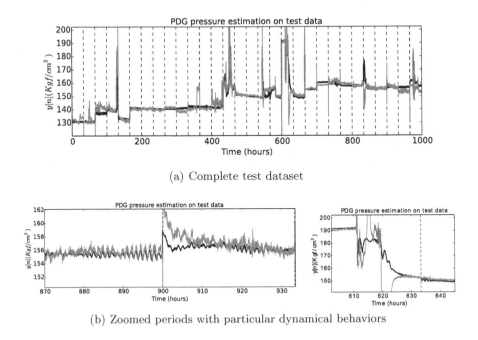

(a) Complete test dataset

(b) Zoomed periods with particular dynamical behaviors

Fig. 5. Results considering 5 months of well sample data. The estimated and real PDG pressures are given by blue and black solid lines, respectively. Each vertical dashed line delimits slices containing 2,000 minutes of samples each.

Furthermore, to verify which variables were most important for the estimation task, a procedure called backward variable removal was executed. It consists of starting with all 10 variables as inputs, and removing the one that results in the least error at that iteration. At the next iteration, the same logic is executed. Fig. 6 (a) shows the results of this procedure. The minimal error is reached when variables FT3, PT6, PT3, and PT4 are removed. The most important variables are TT6 (which was not removed during the complete process), P.C (production choke), G.C (gas-lift choke).

4.2   One Month Dataset

Considering a smaller section of the dataset, corresponding to the samples obtained on December 2011, an RC network of 200 reservoir units was trained with the same parameters from the previous section: $v_i^r = 0.2$, $\alpha = 0.5$, and $\rho(\mathbf{W}_r^r) = 0.5$. The regularization parameter λ was chosen using a randomly selected validation set. In Fig. 6 (b), the predicted downhole pressure for the last 30% samples of the month can be seen using an RC network trained on the first 70% samples. The blue line shows that the estimation is considerably good when compared to the target PT1 downhole pressure. The test NRMSE is 5.84 (note that the NRMSE has a denominator equal to the variance of the target signal, which means that a smaller variance will increase the NRMSE, which is the case here compared to previous section's results).

5   Conclusion

In this work, ESNs (RC networks) have been used to construct a black-box model of a soft sensor in deepwater oil wells. The task was to infer the PDG downhole pressure in an actual oil well based on readings from the platform sensor variables. A single ESN was used for such task, considering a dataset of either 5 months or 1 month of samples. The task reached reasonable results for the first experiment, although using 5 months was more difficult since there were undetected

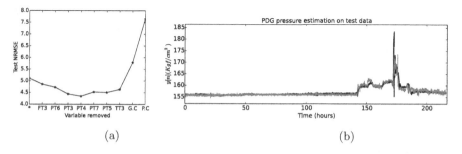

(a)   (b)

Fig. 6. (a) Backward variable removal results showing the test NRMSE as variables are gradually removed. (b) Results considering the last month (December 2011) for both training and test. The lines represent the same concepts as in Fig. 5.

outliers and inconsistencies which were not removed in the current work. The second experiment, modeling the sensor with data from only a month, showed very good results using the same parameters (input scaling, spectral radius, leak rate and reservoir size) for the ESN as in the first experiment. This is because the total data used to train the RC model is very close in the temporal dimension to the test data. This indicates that, if data is always available, the RC model can benefit from a self-update mechanism (e.g. online learning) for achieving superior performance.

Future work will tackle several points: removal of outliers and inconsistent sampling periods for training dataset selection; architectural network changes to accommodate different timescales present in the process variables (i.e., for improving modeling small oscillations in apparently steady states conditions); and building of an ESN-based observation model which can be used to correct the process model output.

References

1. Fortuna, L., Graziani, S., Rizzo, A., Xibilia, M.G.: Soft sensors for monitoring and control of industrial processes. Springer Science & Business Media (2007)
2. Sbarbaro, D., Ascencio, P., Espinoza, P., Mujica, F., Cortes, G.: Adaptive soft-sensors for on-line particle size estimation in wet grinding circuits. Control Engineering Practice 16(2), 171–178 (2008)
3. Fujiwara, K., Kano, M., Hasebe, S.: Development of correlation-based pattern recognition algorithm and adaptive soft-sensor design. Control Engineering Practice 20(4), 371–378 (2012)
4. Eck, J., et al.: Downhole monitoring: the story so far. Oilfield Review 11(3), 18–29 (1999)
5. Teixeira, B.O., Castro, W.S., Teixeira, A.F., Aguirre, L.A.: Data-driven soft sensor of downhole pressure for a gas-lift oil well. Control Eng. Practice 22, 34–43 (2014)
6. Billings, S.A.: Nonlinear system identification: NARMAX methods in the time, frequency, and spatio-temporal domains. Wiley (2013)
7. Jaeger, H., Haas, H.: Harnessing nonlinearity: predicting chaotic systems and saving energy in wireless telecommunication. Science 304(5667), 78–80 (2004)
8. Verstraeten, D., Schrauwen, B., D'Haene, M., Stroobandt, D.: An experimental unification of reservoir computing methods. Neural Networks 20(3), 391–403 (2007)
9. Antonelo, E.A., Camponogara, E., Plucenio, A.: System identification of a vertical riser model with echo state networks. In: 2nd IFAC Workshop on Automatic Control in Offshore Oil and Gas Production (2015)
10. Jaeger, H., Lukosevicius, M., Popovici, D.: Optimization and applications of echo state networks with leaky integrator neurons. Neur. Netw. 20(3), 335–352 (2007)
11. Schrauwen, B., Defour, J., Verstraeten, D., Van Campenhout, J.: The introduction of time-scales in reservoir computing, applied to isolated digits recognition. In: de Sá, J.M., Alexandre, L.A., Duch, W., Mandic, D.P. (eds.) ICANN 2007. LNCS, vol. 4668, pp. 471–479. Springer, Heidelberg (2007)

12. Jaeger, H.: The echo state approach to analysing and training recurrent neural networks. Technical Report GMD Report 148, German National Research Center for Information Technology (2001)
13. Verstraeten, D., Dambre, J., Dutoit, X., Schrauwen, B.: Memory versus non-linearity in reservoirs. In: Proc. of the IEEE IJCNN, pp. 1–8, July 2010
14. Bishop, C.M.: Pattern Recognition and Machine Learning (Information Science and Statistics). Springer, August 2006
15. Tychonoff, A., Arsenin, V.Y.: Solutions of Ill-Posed Problems. Winston & Sons, Washington (1977)

Robust Bone Marrow Cell Discrimination by Rotation-Invariant Training of Multi-class Echo State Networks

Philipp Kainz[1,2]([✉]), Harald Burgsteiner[3], Martin Asslaber[4],
and Helmut Ahammer[1]

[1] Institute of Biophysics, Medical University of Graz, Graz, Austria
philipp.kainz@medunigraz.at
[2] Institute of Neuroinformatics, University of Zurich and ETH Zurich,
Zurich, Switzerland
[3] Institute for eHealth, Graz University of Applied Sciences, Graz, Austria
[4] Institute of Pathology, Medical University of Graz, Graz, Austria

Abstract. Classification of cell types in context of the architecture in tissue specimen is the basis of diagnostic pathology and decisions for comprehensive investigations rely on a valid interpretation of tissue morphology. Especially visual examination of bone marrow cells takes a considerable amount of time and inter-observer variability can be remarkable. In this work, we propose a novel rotation-invariant learning scheme for multi-class Echo State Networks (ESNs), which achieves very high performance in automated bone marrow cell classification. Based on representing static images as temporal sequence of rotations, we show how ESNs robustly recognize cells of arbitrary rotations by taking advantage of their short-term memory capacity.

Keywords: Computer-assisted pathology · Histopathological image analysis · Bone marrow cell classification · Echo state networks

1 Introduction

Assessment of cellularity of different cell types in context of architecture in tissue specimen is the initial step of histopathological diagnosis. Especially in bone marrow specimen several thousand cells of multiple classes have to be interpreted by the hematopathologist and the distribution of cell classes has to be reported. This diagnostic step is typically performed in Hematoxylin-Eosin stained histopathological sections, based on cell morphology and spatial cell distribution. These qualitative and semiquantitative classification approaches are heavily dependent on the observer's experience. Moreover, in hematopoietic cell lineages the disparities between development stages are frequently indistinct and even equal maturation

The authors would like to thank Michael Pfeiffer from the Institute of Neuroinformatics, University of Zurich and ETH Zurich, for fruitful discussions.

L. Iliadis and C. Jayne (Eds.): EANN 2015, CCIS 517, pp. 390–400, 2015.
DOI: 10.1007/978-3-319-23983-5_36

levels of different cell lineages share morphological characteristics. Consequently, both inter- and intra-observer variability can be considerable, affecting the accurate diagnosis of reactive or even premalignant, and early malignant changes. Thus, automated recognition systems exhibiting low variance and predictable high classification accuracy are highly desirable. Since virtual slide microscopy using whole slide images scanned at high magnification is emerging to a standard in pathology departments [1], computer-aided pathology with such systems can easily be implemented in the routine diagnostic process.

Over the recent years, a lot of research has been conducted on cell classification in histopathological images [2]. Several approaches use statistical pattern recognition and classification techniques like support vector machines or Bayesian classifiers to learn object features [3–6]. Handcrafted features mostly are tied to specific applications, require prior knowledge and experience, and are not easily transferable to other problems. Geometrical and morphological features require accurate segmentation, which perhaps is even harder than recognition from raw cell images. On the other hand, working directly on raw image data provides a rich object representation and is only limited by physical constraints of the imaging modality. Cell classification and counting has been performed with feed-forward neural networks (FF-NN) in [7]. Recently, convolutional neural networks were applied to leukocyte counting [8]. Mature blood cell recognition from peripheral blood smears has been done in [6], and others worked on bone marrow material classifying pathological megakaryocytes [9], leukocytes [10], or myelocytes from myelogenous leukemia [3]. However, the quantification of blood cell maturation in the bone marrow has not been sufficiently studied yet.

Under the umbrella-term of *reservoir computing*, two related training methods for recurrent neural networks (RNN) have been developed independently: Echo State Networks (ESN, [11]) and Liquid State Machines (LSM, [12]). Both approaches use a randomly and recurrently connected pool of hidden units, and learn to classify the observed temporal activities by adapting the readout weights only. While LSM is considered a biologically more realistic model, applying ESN is usually easier due to less hyper-parameters. Image classification using a combination of ESN and FF-NN has been presented in [13].

Generally, cells can appear under varying in-plane rotations. In a conventional rotation-invariant learning setting, one could train exhaustively on samples representing independent rotations of the cell without taking the coherence between consecutive rotations into account. Motivated by how RNN can capture appearance information in temporal features, we propose a different rotation-invariant learning scheme for cell classification using ESN, see Fig. 1. We show that it is possible to train an ESN with standard ridge regression directly on raw image data in a way that its classification accuracy is completely independent from the rotation of the cell. Hence, this approach does not rely on handcrafted features.

2   Methods

We propose an ESN approach for rotation-invariant blood cell maturity recognition. Our supervised learning system is trained with labeled image patches,

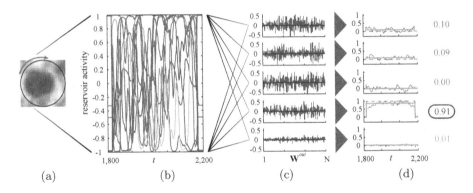

Fig. 1. Rotation-invariant multi-class ESN training scheme. Rotation of a cell patch (a) causes reservoir activity over time (b). (c) Learned readout weights for each class. (d) The readout neuron with the highest mean output (green curve) finally determines the class. The blue dashed curve is the binary target function, which is set to one for the correct class and zero everywhere else. The red curve is the actual network output.

which are classified as one of multiple foreground classes, or background. In this work, we omit cell detection and focus on classifying cropped image patches.

2.1 Multi-Class Echo State Networks

Echo State Networks (ESN) are a way to train RNN for temporal prediction tasks [11]. Many different ESN architectures have been proposed [14], but in this work we focus on the classical architecture [11]. The reservoir, a randomly connected RNN composed of N units, models short-term memory and non-linear input expansion. To be an universal function approximator, the reservoir weights $\mathbf{W} \in \mathbb{R}^{N \times N}$ must be scaled, such that the spectral radius $\rho(\mathbf{W}) < 1$ [11]. In practice, the spectral radius defines how fast the reservoir activity vanishes [15].

Fig. 2 shows a multi-class ESN architecture. The L-dimensional input at time t, given by $\mathbf{u}(t) = [u_1(t), u_2(t), \ldots, u_L(t)]^{\mathrm{T}}$, and a bias unit are connected to the reservoir via the input weights $\mathbf{W}^{in} \in \mathbb{R}^{N \times (1+L)}$ and cause observable activity (temporal features). $\mathbf{W}^{in}$ and $\mathbf{W}$ may be sparse and remain fixed after random initialization and meeting task-specific scaling criteria [16].

A binary classification task can already be performed by a single readout unit, which is trained towards recognizing one of two classes. This simple scheme can easily be extended to solve multi-class problems: each class $\mathcal{C}_c, c \in \{1, \ldots, K\}$ is represented by a readout unit, which is trained in a *one-versus-all* scheme.

The reservoir state update at time step $t + 1$ is computed as

$$\mathbf{x}(t + 1) = (1 - \alpha)\mathbf{x}(t) + \alpha \tanh(\mathbf{W}^{in}[1; \mathbf{u}(t)] + \mathbf{W}\mathbf{x}(t)), \qquad (1)$$

where $\mathbf{x}(t)$ denotes the state vector at the previous time step t. The leaking rate α defines the short-term memory capacity, i.e. how strong the reservoir activity at time step t influences activity at $t + 1$. The ESN can be set to a generative mode, where the input $\mathbf{u}(t)$ is switched off, such that

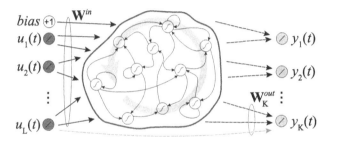

Fig. 2. Multi-class ESN architecture. At each time step t, linear input units (green) feed input $\mathbf{u}(t)$ into the reservoir via input weights $\mathbf{W}^{in}$. Each linear readout unit (blue) corresponds to a specific class. The reservoir consists of N units with sigmoid (hyperbolic tangent) activation function. Readout-to-reservoir feedback connections are omitted. The input layer is fully connected to the each readout unit (grey dashed arrow at the bottom) and provides contextual information on the original input in parallel to the temporal features. After learning readout weights $\mathbf{W}_c^{out}$, output $y_c(t)$ can be determined at each readout unit.

$$\mathbf{x}(t+1) = (1-\alpha)\mathbf{x}(t) + \alpha \tanh(\mathbf{W}\mathbf{x}(t)). \tag{2}$$

During a recording time Ψ, input and state vectors are concatenated in a state matrix $\mathbf{X} \in \mathbb{R}^{(1+L+N)\times\Psi}$. Readout weights $\mathbf{W}_c^{out} \in \mathbb{R}^{(1+L+N)}$ for $\mathcal{C}_c$ are learned via ridge regression with Tikhonov regularization ($\beta = 10^{-2}$)

$$\mathbf{W}_c^{out} = \mathbf{Y}_c^{target}\mathbf{X}^{\mathrm{T}}(\mathbf{X}^{\mathrm{T}}\mathbf{X} + \beta\mathbf{I})^{-1}, \tag{3}$$

where $\mathbf{Y}_c^{target} \in \mathbb{R}^{\Psi}$ denotes the desired target function of class $\mathcal{C}_c$, $\mathbf{X}^{\mathrm{T}}$ the transpose of the state matrix, and $\mathbf{I}$ the identity matrix.

A piece-wise constant target function is regressed for each class $\mathcal{C}_c$ at a recorded time step t, which is given by

$$y_c^{target}(t) = \begin{cases} 1 & \text{if } \mathbf{u}(t) \in \mathcal{C}_c \\ 0 & \text{otherwise} \end{cases}. \tag{4}$$

Each readout unit produces an output $\mathbf{Y}_c = \mathbf{W}_c^{out}\mathbf{X}$, and a score over a predefined inference period Ω, computed as the mean output $\bar{y}_c = \frac{1}{|\Omega|}\sum_{t=-\Omega/2}^{\Omega/2} y_c(t)$. The unit with the highest score then determines the winner class

$$c^* = \arg\max_c\{\bar{y}_c\}. \tag{5}$$

2.2 Rotation-Invariant Cell Classification

Given an image patch $I(\mathbf{x}) \in \mathbb{R}^2$, $\mathbf{x} = (x, y)$ containing a single cell at its center, we need to transform it into time-dependent input for the ESN classifier. In order to generate temporal input from $I(\mathbf{x})$, we take advantage of the fact that cells can occur in arbitrary rotations within tissue. Fig. 3 illustrates the generation of

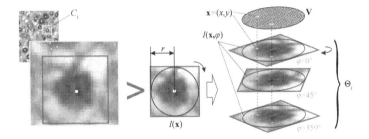

Fig. 3. Illustration of the input stream generation for the rotation-invariant learning scheme. A patch of cell C_i is extracted from a virtual slide. It is then normalized to a predefined size $2r \times 2r$ to fit a receptive field **V**. A static image patch $I(\mathbf{x})$ is transformed into a stream Θ_i by concatenating subsequent rotations $I(\mathbf{x}, \varphi)$. For each rotation, **V** forwards the pixel intensity within the incircle of $I(\mathbf{x}, \varphi)$ into the reservoir.

a temporal input stream Θ_i for a cell C_i by concatenating subsequent rotations of the patch $I(\mathbf{x}, \varphi)$. While rotating by an angle φ, we ignore the patch corners and just consider the pixels within the incircle radius r. For this purpose, a receptive field **V** of radius r is defined as the input layer and forwards the raw pixel intensity into the reservoir. All patches are normalized to a fixed size of $2r \times 2r$ beforehand.

Rotation-invariance is achieved by letting the reservoir generate features for each $I(\mathbf{x}, \varphi)$, $\varphi = 0, \ldots, 359$, starting at an arbitrary angle φ_0 that relates to a cell's arbitrary orientation in a slide. These reservoir states are harvested by evaluating Eq. (1) and the target function is approximated at each recorded time step. After the network saw all $I(\mathbf{x}, \varphi)$, the final class is determined with Eq. (5).

The generative mode of ESNs also facilitates skipping $\Delta\varphi$ rotations after receiving external input $I(\mathbf{x}, \varphi)$. Due to the memory and slowly decaying reservoir activity we still can obtain discriminative features using Eq. (2), even without external input. The reservoir activity usually approaches a resting state without external input and thus a properly selected $\rho(\mathbf{W})$ ensures that there is enough activity left before the next input $I(\mathbf{x}, \varphi + \Delta\varphi)$ is presented.

Smoothing of temporal features is controlled by the leaking rate α. If $\alpha \ll 1$, features of previous rotations may significantly overlap with subsequent ones and may cause over-smoothed states. On the other hand, if $\alpha = 1$, the memory will be turned off and features for each $I(\mathbf{x}, \varphi)$ are completely independent from any previous input. Thus, α must be chosen having this trade-off in mind. In section 3.2 the classification behaviour of these approaches is examined.

3   Experimental Results

3.1   Dataset and Implementation Details

We challenge our approach on a non-neoplastic human bone marrow cell dataset composed of three consecutive maturation stages in granulopoiesis as well as one

Table 1. The non-neoplastic bone marrow dataset used for method evaluation consists of a total of $n = 944$ patches. Myelocytes, metamyelocytes and band cells are consecutive maturation stages of white blood cells and exhibit a small inter-class distance.

	Group	Name	n_c	Sample Patches
$\mathcal{C}_1$	Granulopoiesis	Band Cell	200	
$\mathcal{C}_2$	Granulopoiesis	Metamyelocyte	144	
$\mathcal{C}_3$	Granulopoiesis	Myelocyte	200	
$\mathcal{C}_4$	Erythropoiesis	Orthochromatic Normoblast	200	
$\mathcal{C}_{bg}$	–	Background	200	

class from erythropoiesis, see Table 1. Biological samples were taken from the human iliac crest, embedded in acrylate, cut into slices of ≈ 2 m, and stained with Hematoxylin-Eosin. Cell patches were extracted from virtual slides of two patients (digitized at 40× magnification) and labeled by an expert pathologist. All cells were at the same object scale. Usually in cell classification, the object scale within a class does not vary by more than a factor of approximately ± 0.2, which can still be compensated by our approach. The total number of original patches was extended by a factor of six using non-linear warping transformations (circular distortion by ± 35) as well as horizontal flipping, resulting in 744 foreground patches of four classes. However, we ensured that both transformed images and originals were unique and that rotation of the extended dataset did not introduce any duplicates. Additionally, we used 200 random background patches from the same virtual slides as control class. All patches were converted to grey scale and normalized to 20×20 pixels. The input layer was connected to all incircle ($r = 10$) pixels of that patch, resulting in $L = 332$ network inputs. In total, the dataset comprises $n = 944$ patches.

The dataset was randomly shuffled at the beginning of the experiments. In order to avoid learning the sequence of cell classes rather than the appearance of each cell, we introduced zero-input periods of random length between two subsequent image patches. They are furthermore required to let the reservoir 'forget' about the previous image and learn each instance separately. Please note that this is a different concept than setting the network to generative mode after presenting an image input, because we do not present another stimulus while it is not completely at resting state.

The ESN hyper-parameters were successively optimized for the cell recognition task in 10-fold cross-validation (CV) experiments according to their influence to achieve higher performance [15]. We used dense connectivity in $\mathbf{W}^{in}$, while $\mathbf{W}$ remained sparse (70%). The normalized pixel intensity was bounded within $[0, 1]$, so we shifted and scaled it to $[-1, 1]$ to avoid the linear part of the sigmoid activation function of the reservoir units. For a presented $I(\mathbf{x})$, all readout units showed rather high activity in the first and last recording step.

Since this did not contribute to the actual classification, we bounded the inference window Ω to start 5% after the first and to end 5% before the last sample.

Overall performance is reported as accuracy weighted by the class distribution ($ACC = \frac{1}{n}\sum_c n_c TP_c$) and the corresponding standard deviation (SD). Class-wise performance is reported as precision ($PRC_c = TP_c/(TP_c+\sum_c FP_c)$), recall ($REC_c = TP_c/(TP_c + \sum_c FN_c)$), specificity ($SPC_c = TN_c/(TN_c + \sum_c FP_c)$), and F1-score, with TP_c as true positives, FP_c false positives, TN_c true negatives, and FN_c false negatives.

3.2 Rotation-Invariant Model Evaluation

The proposed rotation-invariant learning approach was evaluated in five independent 10-fold CV experiments on (i) all rotation angles and (ii) different times of running in generative mode. In order to fix a suitable amount of short-term memory, prevent over-smoothing the temporal state space, and account for reservoir decay, we employed the following parameter tuples ($\Delta\varphi, \rho(\mathbf{W}), \alpha$) for these experiments: $(0, 0.6, 0.85)$, $(5, 0.8, 0.85)$, and $(10, 0.95, 0.85)$. We collected 360 states for each sample, starting at rotation $\varphi_0 = 0$.

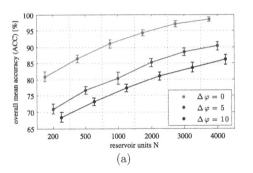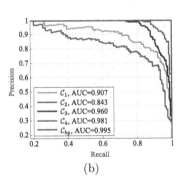

(a)   (b)

Fig. 4. Performance evaluation. (a) ACC over five independent 10-fold CV experiments, error bars refer to 3 standard deviations. Larger values of $\Delta\varphi$ require larger reservoirs to generate proper temporal features. (b) Precision-recall curve for one CV experiment with $N = 1000$ and $\Delta\varphi = 0$. Here, C_1 and C_2 show inferior performance, which is observed to increase with larger reservoir sizes. Performance depends on multiple parameters, thus we cannot evaluate w.r.t. a single one. Lower values of α, however, may cause more stationary features (due to less fading) and too high values of α result in more uncorrelated states.

Fig. 4(a) and Table 2 summarize CV accuracy of these distinct configurations. Compared to $\Delta\varphi = 0$, it requires larger reservoirs to get similar performance for $\Delta\varphi > 0$. Furthermore, we observe that the challenging task of discriminating subsequent maturation stages (i.e. C_1-C_3) requires larger reservoirs to capture the subtle differences in the cells' appearance. In general, metamyelocytes (C_2) seem to be harder to learn, expressed by lower precision for $N < 2000$, while for instance C_4 and C_{bg} already show high precision and recall, see Fig. 4(b).

Table 2. Model evaluation accuracy, reported in percent as ACC (SD), cf. Fig. 4(a).

$\Delta\varphi$	$N = 200$	$N = 500$	$N = 1000$	$N = 2000$	$N = 3000$	$N = 4000$
0	80.92 (0.55)	86.38 (0.40)	90.98 (0.43)	94.43 (0.28)	97.15 (0.32)	98.53 (0.21)
5	70.96 (0.50)	76.72 (0.36)	80.36 (0.62)	85.21 (0.35)	88.64 (0.39)	90.32 (0.38)
10	68.43 (0.53)	73.30 (0.64)	77.45 (0.38)	80.96 (0.45)	83.66 (0.54)	86.13 (0.51)

3.3 Robustness for Random φ_0

In the CV experiments it was confirmed that increasing the reservoir size increases the classification performance on all classes. Using $\varphi_0 = 0$ and $\Delta\varphi = 5$ as reference, we examined whether the very same test patches could be recognized equally well when the rotation started at a random φ_0. We split the shuffled dataset into training (66%) and test set (34%) and evaluated for $N = 4000$.

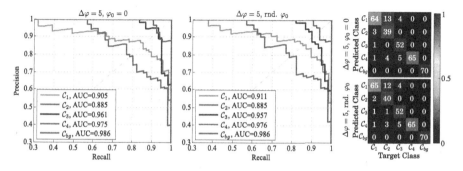

Fig. 5. Classifier performance on the test set ($n = 321$, $N = 4000$, $\Delta\varphi = 5$). Results are of equal overall accuracy for a patch rotation starting at $\varphi_0 = 0$ (90.34%, left) and random φ_0 (90.97%, middle). Confusion matrices for $\varphi_0 = 0$ (right, top) and random φ_0 (right, bottom). Only four cells (1.25%) were classified differently by these schemes, indicating that recognition accuracy does not depend on φ_0. The color bar encodes the class-wise precision (rows), colors towards red correspond to less FP_c.

In Fig. 5 we report precision-recall curves and the confusion matrices on the test set. The recognition performance of individual classes does not depend on the initial rotation φ_0 of the cell, as performance measures in Table 3 show. More importantly, these results indicate that the ESN is able to recognize the very same cell in 98.75% of all cases with similarly high accuracy, even if φ_0 randomly falls within $\Delta\varphi$, where the network is in generative mode.

4 Discussion and Conclusions

While previous work [17] focused on binary classification of similar cells, this paper showed that this approach robustly generalizes to multi-class problems as

Table 3. Class-wise performance on the fixed split ($n = 321$, $\Delta\varphi = 5$, $N = 4000$). The last row contains the weighted mean (w.m.) according to the class distribution in the test set. C_4 and C_{bg} are recalled perfectly, and C_{bg} also shows perfect precision.

	PRC_c		REC_c		SPC_c		F1-score	
	$\varphi_0 = 0$	rnd. φ_0	$\varphi_0 = 0$	rnd. φ_0	$\varphi_0 = 0$	rnd. φ_0	$\varphi_0 = 0$	rnd. φ_0
C_1	0.790	0.803	0.928	0.942	0.936	0.937	0.853	0.867
C_2	0.927	0.952	0.696	0.714	0.996	0.993	0.796	0.816
C_3	0.981	0.963	0.853	0.853	0.996	0.992	0.912	0.904
C_4	0.867	0.878	1.000	1.000	0.961	0.965	0.927	0.935
C_{bg}	1.000	1.000	1.000	1.000	1.000	1.000	1.000	1.000
w.m.	0.912	0.918	0.903	0.902	0.975	0.980	0.902	0.908

well. The model evaluation revealed that learning the temporal features actually works better for $\Delta\varphi = 0$ than for $\Delta\varphi > 0$ when using the same ESN complexity. Yet ESNs can be run in generative mode with a fraction of image rotations $(1/|\Delta\varphi|)$, but larger reservoirs are required to increase performance. Future work could focus on evaluating the maximum period in generative mode before the test error significantly increases and performance equals random guessing.

We see advantages in using ESN for cell recognition: multiple classes can be learned in a global optimum fashion from a randomly connected RNN, which is driven by raw image data. Compared to other gradient descent-based training methods [18,19], training via ridge regression is simple. Besides the proposed inference scheme, more sophisticated (non-linear) schemes may lead to superior performance. Nevertheless, obtaining good hyper-parameters is highly task-specific and remains a tedious duty. Deep learning models, especially deep convolutional neural networks [20], are considered as promising candidate for both segmentation and classification in future research. While they also operate on raw images, they may be more robust in capturing the high intra-class variance while coping with small inter-class distance of blood cell maturation.

The classification performance of the ESN on the given bone marrow dataset must be interpreted carefully, though. Firstly, due to the minimal inter-class distance caused by continuous maturation (i.e. C_1-C_3) cell appearance may be very similar and exacerbates finding a good (linearly) separating hyper-plane. Secondly, even after several years of experience, it is a non-trivial task for expert pathologists to make a hard distinction between consecutive maturation stages. However, our experimental results showed that non-consecutive maturation stages are recognized almost perfectly even by less complex models.

The proposed method is attractive for applications in biomedical diagnostics due to its high accuracy and low variance. An important measure in most medical application settings besides recall is a high specificity, as it is expressed by our recognition system. Cell segmentation and manual feature design is not required in this approach and it can more easily be transferred to other problems than rigid standard image processing approaches. Moreover, integration into an automated, image-based pathology workflow seems feasible.

References

1. Al-Janabi, S., Huisman, A., Van Diest, P.J.: Digital pathology: current status and future perspectives. Histopathology **61**(1), 1–9 (2011)
2. Gurcan, M.N., Boucheron, L.E., Can, A., Madabhushi, A., Rajpoot, N.M., Yener, B.: Histopathological image analysis: a review. IEEE Rev. Biomed. Eng. **2**, 147–171 (2009)
3. Staroszczyk, T., Osowski, S., Markiewicz, T.: Comparative analysis of feature selection methods for blood cell recognition in leukemia. In: Perner, P. (ed.) MLDM 2012. LNCS, vol. 7376, pp. 467–481. Springer, Heidelberg (2012)
4. Markiewicz, T., Osowski, S., Marianska, B., Moszczynski, L.: Automatic recognition of the blood cells of myelogenous leukemia using SVM. In: IJCNN, pp. 2496–2501 (2005)
5. Theera-Umpon, N., Dhompongsa, S.: Morphological granulometric features of nucleus in automatic bone marrow white blood cell classification. IEEE Trans. Inf. Technol. Biomed. **11**(3), 353–359 (2007)
6. Sabino, D.M.U., Costa, L.F., Rizzatti, E.G., Zago, M.A.: Toward leukocyte recognition using morphometry, texture and color. In: ISBI, pp. 121–124 (2004)
7. Sjöström, P.J., Frydel, B.R., Wahlberg, L.U.: Artificial neural network-aided image analysis system for cell counting. Cytometry **36**(1), 18–26 (1999)
8. Habibzadeh, M., Krzyżak, A., Fevens, T.: White blood cell differential counts using convolutional neural networks for low resolution images. In: Rutkowski, L., Korytkowski, M., Scherer, R., Tadeusiewicz, R., Zadeh, L.A., Zurada, J.M. (eds.) ICAISC 2013, Part II. LNCS, vol. 7895, pp. 263–274. Springer, Heidelberg (2013)
9. Ballarò, B., Florena, A.M., Franco, V., Tegolo, D., Tripodo, C., Valenti, C.: An automated image analysis methodology for classifying megakaryocytes in chronic myeloproliferative disorders. Med. Image Anal. **12**(6), 703–712 (2008)
10. Nilsson, B., Heyden, A.: Segmentation of complex cell clusters in microscopic images: Application to bone marrow samples. Cytometry A **66A**(1), 24–31 (2005)
11. Jaeger, H.: The"echo state"approach to analysing and training recurrent neural networks - with an erratum note. GMD Report 148 (2001)
12. Maass, W., Natschläger, T., Markram, H.: Real-time computing without stable states: A new framework for neural computation based on perturbations. Neural Comput. **14**(11), 2531–2560 (2002)
13. Woodward, A., Ikegami, T.: A reservoir computing approach to image classification using coupled echo state and back-propagation neural networks. In: ICIVC, pp. 543–458 (2011)
14. Lukoševičius, M., Jaeger, H.: Reservoir computing approaches to recurrent neural network training. Comp. Sci. Review **3**(3), 127–149 (2009)
15. Lukoševičius, M.: A practical guide to applying echo state networks. In: Montavon, G., Orr, G.B., Müller, K.-R. (eds.) Neural Networks: Tricks of the Trade, 2nd edn. LNCS, vol. 7700, pp. 659–686. Springer, Heidelberg (2012)
16. Verstraeten, D., Dambre, J., Dutoit, X., Schrauwen, B.: Memory versus non-linearity in reservoirs. In: IJCNN, pp. 1–8 (2010)
17. Kainz, P., Mayrhofer-Reinhartshuber, M., Burgsteiner, H., Asslaber, M., Ahammer, H.: Echo state networks for granulopoietic cell recognition in histopathological images of human bone marrow. Biomedizinische Technik **59**(S1), S492–S495 (2014)

18. Steil, J.J.: Online reservoir adaptation by intrinsic plasticity for backpropagation-decorrelation and echo state learning. Neural Networks **20**(3), 353–364 (2007)
19. Werbos, P.J.: Backpropagation through time: what it does and how to do it. Proc. IEEE **78**(10), 1550–1560 (1990)
20. LeCun, Y., Kavukcuoglu, K., Farabet, C.: Convolutional networks and applications in vision. In: ISCAS, pp. 253–256 (2010)

Author Index